96.00

Second Messengers in Plant Growth and Development

Plant Biology

Titles in the Series

581.3 $Se24c$

Second Messengers in Plant Growth and Development

Editors

Wendy F. Boss
Department of Botany
North Carolina State University
Raleigh, North Carolina

D. James Morré
Department of Medicinal Chemistry
and Pharmacognosy
Purdue University
West Lafayette, Indiana

ALAN R. LISS, INC., New York

Address all Inquiries to the Publisher
Alan R. Liss, Inc., 41 East 11th Street, New York, NY 10003

Library of Congress Cataloging-in-Publication Data

Second messengers in plant growth and development.

(Plant biology ; v. 6)
Includes index.
1. Growth (plants).   2. Plants—Development.
I. Boss, Wendy F.   II. Morré, D. James, 1935–
III. Series.
QK731.S44   1988   582.3'1   88-22966
ISBN 0-8451-1805-6

A0000048137444

Contents

Contributors

James Michael Anderson, Agricultural Research Service, United States Department of Agriculture, and Departments of Crop Science and Botany, North Carolina State University, Raleigh, NC 27695-7631 [181]

D.P. Blowers, Department of Botany, University of Edinburgh, EH9 3JH, Edinburgh, Scotland [1]

Thomas Boller, Abteilung Pflanzenphysiologie, Botanisches Institut der Universität Basel, CH-4056 Basel, Switzerland [227]

Wendy F. Boss, Department of Botany, North Carolina State University, Raleigh, NC 27695 [xi, 29]

Alain M. Boudet, Centre de Physiologie Végétale, Université Paul Sabatier, Unité Associée au C.N.R.S. No. 241, F-31062 Toulouse, France [213]

Frederick L. Crane, Department of Biological Sciences, Purdue University, West Lafayette, IN 47907 [115]

Hubert Felle, Botanisches Institut I der Justus Liebig-Universität, D-6300 Giessen, Federal Republic of Germany [145]

Tom J. Guilfoyle, Department of Biochemistry, University of Missouri, Columbia, MO 65211 [315]

R. Gupta, Department of Botany, University of Delhi, Delhi, India [257]

E. Hartmann, Department of Botany, Johannes Gutenberg University Mainz, 6500 Mainz, Federal Republic of Germany [257]

Benjamin A. Horwitz, Department of Biology, Technion-Israel Institute of Technology, Haifa 32000, Israel [289]

Dieter Marmé, Institute of Biology III, University of Freiburg, 7800 Freiburg, Federal Republic of Germany [57]

D. James Morré, Department of Medicinal Chemistry and Pharmacognosy, Purdue University, West Lafayette, IN 47907 [xi, 81]

Raoul Ranjeva, Centre de Physiologie Végétale, Université Paul Sabatier, Unité Associée au C.N.R.S. No. 241, F-31062 Toulouse, France [213]

Günther F.E. Scherer, Botanisches Institut, Universität Bonn, D-5300 Bonn 1, Federal Republic of Germany [167]

A.J. Trewavas, Department of Botany, University of Edinburgh, EH9 3JH, Edinburgh, Scotland [1]

The numbers in brackets are the opening page numbers of the contributors' articles.

Foreword

This book, Volume 6 in the Plant Biology series, is the first that is not based on a symposium or conference proceedings. Rather, it was planned and authors were selected to cover what the editors and I deemed to be an important and exciting area of research: the effect(s) of secondary messengers on plant cells. This topic is particularly timely because there is a great deal of new and interesting information about secondary messengers and their effects on plant cells.

It is my hope that this book will serve as a fine introduction to students and investigators who want to enter this area of study and that it will summarize and organize current knowledge about secondary messengers for researchers already working in this field of study. The studies reported here will no doubt have an impact on research into plant biochemistry, molecular and cell biology, biophysiology, and plant physiology.

It has been a pleasure to work with the editors of this volume, Dr. Wendy F. Boss and Dr. D. James Morré, whose enthusiasm for the topic and for science in general inspired everyone involved with the book.

Nina Strömgren Allen
Winston-Salem, NC
July, 1988

Preface

You all have learned reliance
On the sacred teachings of science,
So I hope, through life, you never will decline
In spite of philistine
Defiance
To do what all good scientists do.
Experiment!
Make it your motto day and night.
Experiment,
And it will lead you to the light.

—From "Experiment" by Cole Porter

While plants are very sensitive to their environment and respond to external stimuli, the mechanisms of signal transduction in plants are for the most part ill defined. Scientists often assume that, because most plants are sessile, they do not have mechanisms for rapidly responding to environmental stimuli or to hormonal stimuli transmitted from one organ to another. This concept was reinforced when evidence for cAMP as a second messenger in plants was not forthcoming. However, as technology progressed so that rapid measurements of small changes in cellular metabolites, pH, calcium, and membrane potential could be measured within minutes after a stimulus was applied, it became evident that plants do indeed show rapid metabolic changes in response to stimuli. Therefore, the question again can be asked: What are the mechanisms in plants for transmitting stimuli from outside the membrane into the cells' interior and for transducing this signal to a form that can be perceived and amplified by components of the cytosol?

The focus of this book is on the identity and roles of potential second messengers in plants. The chapters not only discuss established second messenger systems such as calcium but also look ahead to the possibility that plants will be found to possess second messengers and response mechanisms that are unique. These messengers and receptors may provide valuable entrées to the discovery of molecules potentially amenable to genetic modulation.

The purpose of the book is not only to catalog the available information but also to look to the future. The authors were encouraged to present new ideas at the forefront of their areas of study, to pose questions that need to be addressed, and to set criteria for future research. The success of the book is to be judged more according to what it stimulates with regard to future research rather than to what it presents of the present state of knowledge of the field.

The book begins with a discussion of protein phosphorylation in second messenger responses and continues with discussions of the biochemistry of changes in phosphoinositides and calcium-calmodulin regulated responses. Stimuli, including auxin, light, and pathogens, and potential signaling systems such as redox reactions, pH, and calcium are discussed. In addition, consideration is given to acetylcholine, jasmonic acid, and platelet activating factor-like lipids as having possible roles in cellular signaling and to the vacuole as a potential source of second messengers. Lastly, the need to bridge the gap between the agonist response and the regulation of chromosomal activity is considered. It is hoped that this book will stimulate new thinking and inspire readers to investigate the role of second messengers in plant growth and development—that is, to experiment.

Wendy F. Boss
D. James Morré

Second Messengers in Plant Growth
and Development, pages 1–28
© 1989 Alan R. Liss, Inc.

1. Second Messengers: Their Existence and Relationship to Protein Kinases

D.P. Blowers and A.J. Trewavas

*Department of Botany, University of Edinburgh, EH9 3JH Edinburgh,
Scotland*

INTRODUCTION

Plants live in an environment that fluctuates continually. In the wild, the constituents of the environment, light intensity and quality, CO_2, O_2, minerals (about 17 are required and are unevenly distributed in the soil), temperature, water availability, humidity and wind, soil structure, other plants and predatory animals, insects, and microorganisms vary from minute to minute and appear frequently in unbalanced mixtures for optimal growth. Plant movements and growth ensure continual realignment of cells with the gravity vector. All of these environmental variables are known to influence growth and development and thus are sensed in some way or another by plant cells. In addition to these external variables, plant cells are bathed in a complex mixture of sugars, oligosaccharides, amino acids, minerals, water, vitamins, and growth regulators in a solution of variable pH. So far as we are aware, the composition of this mixture varies both qualitatively and quantitatively and fluctuates rapidly in its rate of movement. Plant cells, therefore, require metabolic control systems capable of dealing with these fluctuations and of yet maintaining growth and development in a form of homeostatic resistance to the disorganising effects of the environment.

Cellular Sensing and Response to External Information

There are considerable physiological data concerning the response of plants to external environmental variables. These responses may be classi-

Address reprint requests to Dr. A.J. Trewavas, Department of Botany, University of Edinburgh, Mayfield Road, Edinburgh EH9 3JH, Scotland.

fied as changes in structure/morphology resulting from developmental plasticity or institution/breakage of dormancy (Trewavas, 1986). The adjustments can be finely tuned. For example, the numbers of fruits and leaves on trees can be adjusted, by abscission, to the levels of photosynthate or available water. The number of root meristems can be adjusted to nitrate, phosphate, or water. The shape, size, and thickness of stems, flowers, and leaves can vary plastically, although within certain boundaries. Developmental plasticity seems a novel characteristic confined to sessile organisms.

For these environmental variables to have an effect, they must be sensed by the metabolic system of the cell and a controlled response issued to the perturbation. Information has to cross the outer barrier of the cell, the *plasma membrane,* from the primary source. This information is called *primary information,* or *primary messages,* to distinguish it from secondary messengers and later effects induced inside the responding cell. This terminology is commonly used and is derived from animal models. It clearly envisages a linear train of information from outside to inside. The analogy may be weak, since living plants enmesh with the environment to a much greater extent than animals.

Little is known of how external signals are sensed by plant cells. Photomorphogenetically active light is sensed by phytochrome, blue light receptors, or other unknown pigment proteins; growth regulators and sugars may have specific receptor proteins. Nitrate may be sensed by nitrate reductase (Trewavas, 1983a). But nothing is known of the others. Is there a temperature or water receptor; might O_2 be sensed through oxidation reactions, CO_2 through carboxylation reactions or pH changes, and so on? Considering the specific effects of these environmental variables on development, the level of ignorance about even these simple facts is remarkable.

In this chapter we examine in more detail the primary/secondary messenger concept as it was first enunciated, update it, estimate the usefulness of applying it to plants, and then consider protein kinase activity as an integrating element in such transduction processes. The chapter will finish by indicating how protein kinase may serve as a metabolic switch in development.

THE FIRST/SECOND MESSENGER CONCEPT

Sutherland and co-workers (1968) seem to have been the first to use the term *second messenger.* This followed from Sutherland's brilliant unravelling of the biological function of cyclic AMP in adrenergic responses. First messages were considered to be action potentials, neurotransmitters, and hormones, and these initiated their biological effects in sensitive tissues by the intracellular production of a second messenger, cyclic AMP. Similar parallels have been drawn between animal hormones and plant growth substances (so-called hormones), and it has often been thought that plant hormones too should have an equivalent, all-embracing, cell-activating second messenger such as cyclic AMP. In contrast with this simple plant/animal parallel, we will suggest instead that growth substances are cellular second messengers that mediate environmental influences and thus, perhaps, an effective plant equivalent of cyclic AMP.

Are Hormones First Messages?

The notion that hormones are chemical messengers seems deeply rooted. A survey of endocrinologists in the 1930s (Huxley, 1935) showed that the majority supported this definition, although the term *hormone* itself means simply *to arouse*. Superficially, the hormone system structure seems to agree with an information transfer function. Adrenalin is released into the blood stream after nervous stimulation of the adrenal gland, and there, in turn, it affects some 8–12 remote tissues where metabolic changes are instituted. Thus the information transmission structure, source–channel–receivers, is apparently exemplified (Trewavas and Allan, 1987).

It is pertinent, however, to ask whether this structure is really essential to the biological function of hormones, and the answer is no. It is not actually thought that the adrenal gland communicates with other tissues via adrenalin. If anything, the channel of information transfer is between the nervous system and adrenalin-sensitive tissues. The element critical to hormonal function is that whereas all tissues are exposed to adrenalin, only a limited number respond, that is, are sensitive to adrenalin. Coordinately, these produce a defined behavioural response. It is almost certainly easier to make tissues selectively sensitive to hormones than to draw up complicated nervous connections, but there seems no biological reason why adrenalin should not be released in each individual sensitive tissue. There is considerable evolutionary advantage to be gained, since the mix of tissues sensitive to a particular hormone can be quickly changed (unlike the laying down of nervous pathways), thus altering the whole organism response. The real essence of hormone function, then, is to coordinate; the origin of the hormone is not, in most instances, really relevant. It is a convenience to localise synthesis in one tissue, since it simplifies control of release. The fact that some hormones actually work in their tissue of origin and occur in some single-celled organisms (Collier et al., 1987; Csaba, 1980; Leopold and Nooden, 1984) confirms this view.

A further important consideration in hormone systems to the messenger definition is the characteristic of the channel of communication. In mammals this is the bloodstream, the composition and temperature of which are carefully controlled. Any information can then be transferred with the minimum of destruction by *noise,* that is, by extraneous environmental variation. The effect of noise can be minimised by a monitoring of the responsive tissue by the source. Indeed, in the nerve/hormone complex in mammals this seems to occur frequently via feedback loops. Anachronistically, the greater the noise input, the greater the need for monitoring to counteract this. It is here that the greatest difficulty with the concept of plant hormones originates.

Animal hormone models have been transferred into plants (e.g., Jacobs, 1979) without recognising the fundamental constructional and functional differences between these groups of organisms. In contrast with the unitary structure of mammals, higher plants are modular organisms in which the repetitious growing elements compete with each other for resources (Trewavas, 1983b, 1986; Trewavas and Allan, 1987). Negative controls, that is, competition, seem to be the prime means of tissue communication. More important, it is very difficult to see how any simple information transmission system in plants could withstand the rigours of environmental fluctuation and size and tissue number plasticity, which are the reality for plant growth in the wild.

Monitoring of the response by the source would be essential in such a situation, but there is no evidence that direct monitoring occurs; phenotypic plasticity suggests it does not. Furthermore, there are now numerous examples where plant hormones/growth substances have their effects in the tissue of origin. Again, the essential element in the growth substance system is the selectivity of response by individual tissues, rather than the source of growth substance. The function of growth substances is likely to be in deciding which tissues survive the increased competition for internal resources that result from poor growth conditions (Trewavas, 1986, 1987a). In other words, it is a growth factor function.

The notion of hormones as primary messengers misdirects attention to the level of the hormone as the key element, rather than to the sensitivity of the tissue and to the coordination as chief functions. However, in mammals the majority of hormones arrive outside the cell, and in some way, that *primary* information has to be transferred inside.

The Earliest Second Messenger System

After arrival of adrenalin at the surface of the sensitive cell, the set of events believed to happen is summarised in Figure 1. The combination of the

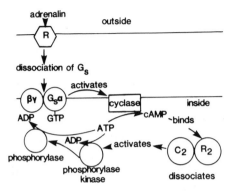

Fig. 1 Outline of reaction sequence of effects of adrenalin on catalytic activity of phosphorylase. Adrenalin combines with a specific receptor. The complex dissociates a G protein. The α-subunit binds GTP, and this activates adenyl cyclase. Intracellular production of cyclic AMP activates a cyclic AMP-sensitive protein kinase, which phosphorylates and activates phosphorylase kinase. This latter enzyme phosphorylates and activates phosphorylase b converting it to phosphorylase a.

hormone with a receptor changes the conformation of a transduction protein (G protein). Combination with GTP stabilises the altered and now active state of the G protein and this, in turn, activates adenyl cyclase, causing the intracellular synthesis of cyclic AMP. Cyclic AMP-dependent protein kinase is activated, thus phosphorylating phosphorylase kinase. The latter enzyme converts enzymatically inactive phosphorylase b to active phosphorylase a, initiating the breakdown of glycogen and the subsequent synthesis of glucose. Thus, the original signal on the outside is transduced across the membrane to the inside of the cell, and information is conveyed further via a new molecule, cyclic AMP. Magnitude amplification (Koshland et al., 1982) of the initial signal, adrenalin, can occur at four steps in the sequence.

1. One molecule of adrenalin/receptor complex can activate many G-proteins.

2. One molecule of adenyl cyclase can produce many molecules of cyclic AMP.

3. One molecule of active cyclic AMP-dependent protein kinase can phosphorylate and activate many molecules of inactive phosphorylase kinase.

4. One molecule of active phosphorylase kinase can phosphorylate and activate many molecules of inactive phosphorylase.

Thus the original weak signal, adrenalin, which may reach a concentration no higher than nanomolar can be amplified to produce many more active enzyme molecules.

What Requirements Necessitate the Presence
of a Second Messenger System?

This question can be directly asked of the adrenalin system. The answer will help us understand whether or not plant cells will need second messengers.

With adrenalin, the speed of the response seems critical. Adrenalin is a hydrophilic molecule. This ensures that on release it is only diluted into a defined blood volume and is not partitioned (at a much slower pace) into the greater cellular volume. Hydrophobic hormones such as steroids partition into cells and have intracellular receptors with strong affinity for the hormone, reflecting their greater degree of dilution. However, the required response time to steroids is very much slower. An advantage to confining adrenalin in the blood is the speed with which it can be excreted, thus removing the stimulus.

Even so, the dilution of adrenalin that accompanies release poses problems for sensing that can be solved either by having receptors of exceptional affinity or by amplification of a weak signal. The former is true for steroid hormones but the disadvantage for rapidly acting hormones is that high-affinity receptors would greatly prolong the activated state of the cell after decline of hormone levels. Thus a weak signal is amplified, and amplification can be easily reversed. For the purposes of rapid amplification and transduction, the plasma membrane seems ideal. The cellular volume occupied by the plasma membrane is near 0.1%, but the lipid:protein ratio is approximately 1:1. The plasma membrane is a very concentrated soup of proteins, and the movement of these proteins is restricted to the membrane phase. Thus rapid transduction of a signal, which requires the production of novel interactions of a protein receptor with any of a number of different transduction G-Proteins, could occur with great speed. According to this concept, the cytosol can be regarded either as a weak solution of proteins, in which instance novel interactions between receptor/transduction proteins would be slow, or as consisting of a protein scaffold on which enzymes and other proteins are immobilised, thus limiting the possibilities for interaction.

Interactions between the bimolecular lipid layer and membrane proteins constrain the spatial configuration (sidedness) of membrane proteins. The plasma membrane as the effective outer skin of the cell is probably the sole cellular membrane in which the spatial direction of the protein constituents is uniquely specified with respect to the cellular axis. Sensing by the constituents of the plasma membrane is likely to be involved when cells find it necessary to assess the vectorial direction of an incoming signal.

Environmental and Extracellular Signals May Necessitate Second Messengers in Plants

The requirements for amplification of weak light and gravity signals would seem to necessitate second messengers. It has been suggested that P_{FR} (active phytochrome) is transduced by Ca^{2+} ions as a second message (Roux et al., 1986). The phenomenon of plagiogeotropism suggests that gravity sensors must likewise be structurally fixed, that is, membrane located. The blue light receptor for phototropic responses must also be membrane located. Some evidence associates transduction of gravity signals with Ca^{2+} as a second messenger (Lee et al., 1983). Mild mechanical stress to plant tissues may lead to an enhanced synthesis of ethylene, which thus becomes a possible second messenger. The enzymes involved in ethylene synthesis also seem to be membrane located; although the membranes involved may be tonoplast membranes (Guy and Kende, 1984). Torsion effects on cells could be transmitted to other organelles through the cytoskeleton.

Plant development can be very sensitive to water availability and humidity and it can be argued that a turgor sensor is needed. However, the reduction in cell volume consonant with loss of water would reduce the pressure between the plasma membrane and the cell wall, as well as altering cytoskeletal arrangements. Studies of hearing show how exquisitely sensitive the plasma membrane can be to slight pressure changes. Enzymatic changes certainly accompany reductions in available water, so transduction of information would seem to be essential (Hanson, 1982). Abscisic acid might act as a second messenger in these circumstances.

Circulating carbohydrates and oligosaccharides can be expected to have plasma membrane receptors. Membrane-bound invertase might act as a specific receptor for sucrose levels. Removal of the cell wall leads to its regeneration, and damage to the cell wall by invading organisms leads to cytoplasmic enzymatic changes, such as chitinase and phytoalexin synthesis, and wound related phenomena induced in cells distant from the wound (Boller et al., 1983; MacNicol, 1976). Some of these enzymatic changes involve lipid mobilisation (Theologis and Laties, 1980). Thus information concerning cell wall structure and composition seems to be available to the cytoplasm, implying the presence of receptors on the cell surface; perhaps these receptors directly interact with carbohydrates. Phytoalexin inhibition of calcium/calmodulin-dependent protein kinase activity has also been reported (Amin et al., 1987). Transmission of primary information and its transduction to second messengers seem to be essential for the occurrence of these phenomena.

Hydrophobic or amphipathic molecules, which can penetrate the plasma membrane with ease, seem unlikely to have second messenger requirements unless the subsequent metabolic effects, of for example, an auxin or gibberellin are considered to represent second messengers. Temperature might also directly affect the metabolic system inside the cell. The developmental effects of oxygen might be sensed by enzymes such as those producing the redox potential or peroxidase in the plasma membrane (Barr et al., 1985; Sticher et al., 1981) or cell wall, as well as internally. The effects of CO_2 might operate through cytoplasmic or organelle H^+ ion shifts through increased carboxylation reactions; again, this simply is not known. The effects of nitrate might be mediated through conversion to ammonium ions. Ammonium ions are known to have developmental effects in tissue cultures (Tran Thanh Van, 1981).

Broadening the Definition: Second and Third Messengers

It should be seen from this brief discussion that secondary messengers seem generally to be the way in which information is conveyed from a membrane, usually the plasma membrane, to the cytoplasm. Anything that conveys information about the state of the plasma membrane to the cytoplasm seems likely to fall into the second messenger category. Calcium ions are regarded by many as second messengers. This example shows that it is not necessary to chemically modify a molecule on its transit through the plasma membrane for the molecule to be classed as a second messenger; it is merely necessary to change its flux. The rate of entry of calcium can be modified by a variety of membrane changes, for example, membrane potential or channel phosphorylation; thus information about the general plasma membrane state is conveyed. The sensitivity of calcium-responsive reactions can be altered in a variety of ways, thus complicating the messenger view (Rasmussen, 1983).

A number of minerals requiring specific transport systems in the plasma membrane to enter the cell can provide information as to plasma membrane states, since their rate of uptake is dependent on the membrane state. The further requirement is that these minerals must be sensed and have metabolic effects if they are to be classed as second messengers. At present, boron seems to be a good possibility (Clarkson and Hanson, 1980). The distinction between second messengers and other aspects of metabolism is fine. It may be possible to categorise external stimuli by whether or not the cellular effects require an intermediate stage of amplification, but at present there is insufficient evidence to draw valid conclusions. We simply do not know.

A further complication emerges when the highly compartmented nature of plant cells is considered. Counting membranes as compartments, there are at least 24 (Trewavas, 1986); coordinate action by a number of these would require a molecule equally at home in membrane and soluble compartments. Second messengers such as calcium are likely to have their effects limited to the soluble compartment; conveyance through further cell membranes may require tertiary messengers. Release of calcium from other intracellular stores seems to require that an inositol phosphate type of second messenger be released from the plasma membrane, no doubt reflecting the relative immobility of calcium in the cytoplasmic phase (Thomas, 1982).

Amphipathic molecules, such as growth substances, would seem likely to be equally at home in membrane and soluble compartments. In this respect they might act as general intracellular coordinators, moving freely from the subcellular compartment in which they are synthesised. It has already been suggested that growth substances act to help the cellular response to various imbalances in the major internal primary parameters: sugar, nitrogen, water, oxygen, and wounding (Trewavas, 1986). If growth substance synthesis is limited to one cell compartment, then growth substances might act as general second messengers in plant cells (rather like cyclic AMP in animal cells), indicating the change in primary information that crosses the plasma membrane. If so, this would represent an entirely novel biological situation and it would also help to explain why applied growth substances can often have dramatic metabolic effects. The fact that there are at least five growth substances (second messengers) would reflect the much more profound impact the environment has on individual plant cells.

Specification of Primary and Secondary Messenger Notions in Plants Is an Artificial Isolation From a Complex Network of Interactions

The plant world can be divided into three compartments; these compartments differ in the strength or number of controls that operate on them.

The first compartment is the atmosphere or soil space between individual plants. The various environmental parameters are atmospheric composition, temperature, humidity, wind speed, light, and soil composition/structure. We tend to think of these as being purely environmental (i.e., totally uncontrolled). However, plants respond to the environment they perceive and through the processes of photosynthesis, transpiration, overgrowing, structural change, shading, and reflection modify each of these parameters in turn. The extent of control may be quite weak but is definitely detectable. Lovelock's *Gaia* (1979) reveals that global atmospheric controls operate on simple feedback principles, with plants the major living input. Information comes into the plant but it also comes out again in a different form.

The second compartment represents the internal but extracellular space. The boundary to the outside is the epidermis and cuticle—on the inside the plasma membrane. This compartment contains numerous enzymes; it is likely that its composition will be more carefully controlled than outside. Again, information must cross the plasma membrane, but, equally, it must return to the extracellular space in a modified form. Does this returning information comprise tertiary messengers?

The third compartment is the plasma membrane-bounded compartment, which, in turn, is very subdivided by other membranes. Many controlling elements will operate here, and information from these will return outside to the two other compartments.

In this view the plant is not an isolated entity bombarded with a one-way linear transmission of information from outside to inside but part of a complex set of interactions in which a continual and altering two-way interchange of information leads to compartments whose composition is subject to different levels of both information and control. The first/second message concept should be seen in this context.

Does the Primary/Secondary Messenger Sequence in Adrenergic Responses Involve Amplification, Linear Trains of Events, and All-or-None Control?

Certain simple features concerning the sequence of events in Figure 1 seem to be generally accepted:

A. There is considerable amplification between the signal and final glucose production.

B. A linear sequence of events follows the appearance of adrenalin at the plasma membrane.

C. The system is an all-or-none control. Concentrations of hormones and cyclic AMP are considered dominant regulators, and artificial increases of cyclic AMP should trigger completely or partially the same cellular event as the first message.

A more modern appraisal now suggests that none of these accepted features are thought to be correct as first conceived. What has changed this view is the sheer complexity of interactions and controls now known to operate.

The extent of amplification is constrained by guanosine triphosphatases (GTPases) in the plasma membrane, hydrolysis of cyclic AMP by phosphodiesterase, hydrolysis of various phosphorylated proteins by protein phosphatases, and enzymes converting glucose-1-phosphate (G-1-P) back to glycogen. Compared with its potential, estimated amplification is at present quite moderate (Koshland et al., 1982).

The notions of a linear train of events and an all-or-none switch have been disproved, respectively, by the extreme reticulation in the network of interactions and the complexity of controls shown to operate. The immediate enzymes, including those that synthesise glycogen, and small molecules immediately involved in the network probably number over 50 (Chock et al., 1980), and there are connections via these enzymes and molecules to the whole metabolic network. There is allosteric control by numerous metabolites such as UDPG, G-1-P, glucose-6-phosphate (G-6-P), GTP, AMP, glucose, glycogen, and various metal ions. There are multiple phosphorylations by the protein kinases that phosphorylate more than one substrate at different sites. At least seven enzymes in the network are phosphorylated/dephosphorylated. Specific-site phosphorylation of enzymes can change the affinity for their substrates, although this depends on the level of other molecules. There are four regulatory proteins involved in protein phosphatase activity and one of these proteins is subject to phosphorylation. Regulatory proteins affect protein kinase activity. The rate of protein phosphatase activity is determined by competition amongst the available phosphoproteins and ligand effects that stabilise phosphorylated enzyme substrates.

The role of cyclic AMP is now less certain, since some hormones may increase cyclic AMP levels without an apparent effect on phosphorylase kinase (Krebs and Beavo, 1979). Calmodulin is bound to phosphorylase kinase, and calcium ions are known to interact with this calmodulin. Phosphorylation of this enzyme changes the sensitivity to calcium ions. The entry of calcium through channels to the cytoplasm is affected by membrane potential, phosphorylation, and G-proteins (Houslay, 1987), and expulsion is calmodulin dependent. Calmodulin is linked to the system via calmodulin-dependent phosphodiesterase, calcium-activated protein kinase and phosphatase, and through calmodulin availability to many other metabolic events. A calcium-activated protease can activate phosphorylase kinase. Even protein kinase C is involved, since this enzyme can phosphorylate phosphorylase kinase and adenyl cyclase (Yoshimasa et al., 1987), and many stimuli result in diglyceride activation of protein kinase C (Kikkawa and Nishizuka, 1986).

The whole metabolic structure is thus a highly reticulate network and it is now thought that such networks do not operate in an all-or-none fashion but, instead, finely control the rate of glucose release (or their final output). Because the network interacts with so many other molecules and thus, effectively, the whole metabolic network, it serves primarily as a metabolic integration system, which assesses the whole metabolic network condition. The network is described as a cascade (one enzyme affects another) and there are multisite interactions with allosteric effectors, which sense fluctuations in the concentrations of multitudes of metabolites. Through these interactions a change in the concentration of *any one or more* allosteric effectors leads to a smooth adjustment of enzyme levels and glucose output. The effect of any one molecule on network flux, in turn, is dependent on the state of the whole. These cascades serve as rate amplifiers with signal amplification.

Is cyclic AMP a dominant control? Changing the concentration of any interacting molecule will alter the flux of material through the network. Altering calcium, calmodulin, or diglyceride levels, which in turn modify protein kinase C sensitivity to calcium or a host of allosteric molecules will modify flux rate. The effects of a change in cyclic AMP levels then will depend on the levels of all other contributing components. Thus the notion that second messengers provide some overall autocratic control is misleading; cyclic AMP is just another contributing influence. We do not know the flexibility of the whole network to manipulation by individual metabolites, but, experimentally, adding enough of any chemical may greatly decrease/increase the flux through the network. Experiments in which cyclic AMP (or hormones) are added in excess of endogenous levels may be greatly misleading as to the true state of control.

The means to assess the strength of control by cyclic AMP are available. These methods (Kacser and Burns, 1973) require tiny variations in each of the contributing substances and assessment of the effect on the whole flux rate. When such assessments have been made, control is generally found to be distributed throughout the network.

In conclusion, covalent modification of enzymes by phosphorylation (and other post-translational means) is the basis of a complex network of interlocking cascades that can respond to common biological signals and thereby permit coordinated and synchronous control of many biochemical functions.

A major function of a cascade, therefore, seems to be to provide numerous points of interaction with other molecules, thereby increasing integration with other metabolic events.

SECOND MESSENGER PROTEIN KINASE SYSTEMS IN PLANTS AND ANIMALS

The roles of protein kinases in signal transduction are twofold. First, they have an obvious function in amplifying weak signals; second they provide an integrating function. That is, by lengthening the number of steps in the signal transduction process they can integrate metabolic information to a much greater degree and provide a much better and smoother control to the metabolic system. This latter property would still be valuable, even if the initial signal is strong.

On this basis, we think there may be cellular protein kinases or protein phosphatases that are responsive to Ca^{2+}, K^+, H^+, redox potential, lipids, temperature, light, pressure, gravity sensing, oligosaccharides, sucrose, oxygen, nitrate, sulphate, and possibly CO_2, membrane potential, and polyamines. Protein kinase or protein phosphorylation sensitive to Ca^{2+} (Hetherington and Trewavas, 1982), H^+ (Tognoli and Basso, 1987), redox potential (Bennet, 1983), lipid (Schafer et al., 1985), light (Wong et al., 1986), and polyamines (Veluthambi and Poovaiah, 1984b) are already known. Figures 2 and 3 show the effect of temperature, pH, and auxin on membrane protein phosphorylation in the pea. Clearly there is opportunity in this area for potentially significant studies to understand metabolic integration. In the following section we examine the one well-characterised second messenger in plants, calcium, in relation to protein kinases. Special consideration is given to the role of tyrosine phosphorylation and autophosphorylation in signal transduction.

Calcium-Dependent Protein Kinases in Plants

Calcium-dependent protein kinases have been found in the soluble fractions of wheat embryo (Polya and Davies, 1982; Polya and Micucci, 1984), silver beet leaves (Polya et al., 1985a), soybean (Putnam-Evans et al., 1986; Harmon et al., 1987), and carrot (Refeno et al., 1982). Calcium-activated protein kinases associated with plant membranes have been found in the pea (Hetherington and Trewavas, 1982), zucchini (Salimath and Marmé, 1983), ryegrass (Polya et al., 1984), corn (Veluthambi and Poovaiah, 1984a), tobacco pollen (Polya et al., 1985b), and sycamore (Macherel et al., 1986;

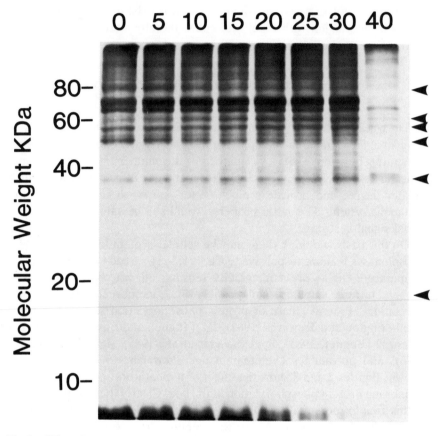

Fig. 2. Effect of temperature on pea bud membrane in vitro protein phosphorylation. Washed pea bud microsomal membranes were incubated in [$\gamma - ^{32}P$] ATP and 100 μM free Ca^{+2} ions at 0°, 5°, 10°, 15°, 20°, 25°, 30° and 40°C, as indicated above. Incubation times were set so that in each case the reaction was terminated at maximum incorporation. Phosphorylated proteins were separated by sodium dodecylsulphate– polyacrylamide gel electrophoresis (SDS PAGE) and autoradiographed. Arrows mark labelled bands showing significant variation with temperature up to 30°C. Note considerable difference in phosphorylation patterns between 30°C and 40°C. Hetherington and Trewavas, unpublished.

Teulieres et al., 1985). However, only for the wheat embryo, soybean, and pea protein kinases has the activity been associated with a protein of known molecular weight. Protein kinase activities in the plasma membrane and tonoplast have been demonstrated.

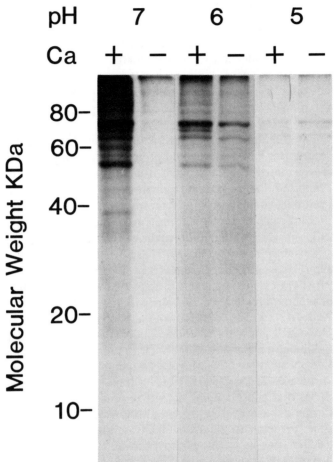

Fig. 3.   Effects of pH variation on pea bud membrane in vitro protein phosphorylation. Washed pea bud microsomal membranes were incubated in $[\gamma - {}^{32}P]$ ATP in the presence and absence of 100 μM free Ca^{2+} ions at pH 7, pH 6, and pH 5. Changing pH from 7 to 6 leads to virtual complete abolition of calcium sensitivity. Addition of 1 mM indoleacetic acid to all of the above reactions was without effect. Hetherington and Trewavas , unpublished.

For some of the above enzymes the involvement of calmodulin in Ca^{2+} activation has been invoked; in others the activation would appear to be in a novel calcium-binding protein or a direct interaction of the protein kinase with Ca^{2+}.

Substrates for the Calcium-Activated Protein Kinases of Plants

Identification of a protein kinase activity dependent on Ca^{2+} is only the beginning of a second messenger system. For the primary stimulus and second messenger production to be meaningful, they must be directly related to a cellular response. Numerous calcium-dependent protein phosphorylations can be observed (Fig. 2, 3). Clearly, the identification of these substrates is the key to unravelling the cellular response to protein kinase activation.

To date only three positive and several other tentative identifications of native substrates for calcium-activated protein kinases in plants have been made, and even these are identified only from in vitro labelling. 1) Refeno et al. (1982) reported that the enzyme quinate:NAD^{+1} oxidoreductase, catalyzing the reversible oxidation of quinate, was regulated by reversible phosphorylation by a calcium-dependent protein kinase. However, the calcium requirement is in excess of that found in the cytosol. 2) the calcium-dependent protein kinase of Blowers et al. (1985) uses itself as a substrate; that is, it autophosphorylates and thus regulates its own activity (Blowers and Trewavas, 1987). 3) The calcium-activated protein kinase of Harmon et al. (1987) also autophosphorylates in a similar fashion to that of Blowers and Trewavas (1987). The K^+/H^+ ATPase of corn root has been reported to be regulated by reversible Ca^{2+} and calmodulin-dependent phosphorylation (Zocchi, 1985). In addition, it is possible that a stage in the activation of the enzyme isofluoroside phosphate synthase involves calcium-dependent phosphorylation of a protease (Kauss, 1983). Equally tentative is the suggestion by Moreau (1986) that phospholipase activity in potato leaves is regulated by Ca^{2+} and calmodulin-activated protein phosphorylation.

What other proteins are likely candidates? With the apparent involvement of calcium ions in secretion and polarity, the proteins of the cytoskeleton are obvious choices.

Calcium channels, Ca^{2+} ATPase and initiation factors for protein synthesis represent additional good future possibilities that are currently being examined. Other phosphorylated plant enzymes and proteins that have been previously detected (Ranjeva and Boudet, 1987) now need to be checked to see if calcium-regulated protein kinase is responsible for their phosphorylation. Good, obvious candidates here are histone H1, pyruvate dehydrogenase, and PEP carboxylase.

Polyamine-Stimulated Protein Kinase

Polyamine-stimulated protein kinases have been identified in soluble fractions from wheat germ (Yan and Tao, 1982), corn coleoptile (Veluthambi and Poovaiah, 1984b), and isolated pea nuclei (Datta et al., 1986). Additionally, ornithine decarboxylase is thought to be regulated by a reversible phosphorylation by a polyamine-dependent protein kinase (Kuehn and Atmar, 1982).

Polyamines are clearly present in plant tissues and appear to be involved in a host of physiological responses (Galston, 1983; Slocum et al., 1984). However, whether or not they really function as second messengers is a point of some debate.

Tyrosine Protein Kinase in Signal Transduction

During the late 1970s a novel type of protein kinase activity, tyrosine kinase, was discovered. This new type of protein kinase phosphorylated tyrosine residues in proteins. Prior to this time, protein kinases were considered to phosphorylate only protein-bound serine and threonine, although there were limited reports on the so-called acid-labile histone phosphates thought to be phosphohistidine and phospholysine (Chen et al., 1974).

Polyoma tumor antigen extracts were found to contain a protein kinase activity capable of phosphorylating tyrosine and the report of Eckhart et al. (1979) seems to be the first of its kind. In the following 3 years an RNA virus (retrovirus) and the receptors for three animal polypeptide hormones or growth factors—epidermal growth factor (EGF) (Ushiro and Cohen, 1980), platelet-derived growth factor (PDGF) (Ek et al., 1982), and insulin (Kasuga et al., 1982)—were found to be membrane-associated tyrosine-specific protein kinases. Many other retroviral products (oncogenes) have now been characterised, and a very substantial number are protein tyrosine kinases (Nishimura and Sekiya, 1987). Neoplastic cell growth appears in part to be the result of an overexpression of these tyrosine kinases, which are thought to function normally as part of the cellular control of cell division.

One substrate for a tyrosine kinase, vinculin, has been identified; it is a protein concerned with the attachment of the cytoskeleton to membranes. The other major substrates seem to be the tyrosine kinases themselves, since they all autophosphorylate. The possible relationship of these enzymes with cellular calcium levels has been outlined already (Hetherington and Trewavas, 1984).

Are tyrosine kinases present in plant cells? Only one report of tyrosine phosphate (Elliott and Geytenbeek, 1985) and one report of tyrosine phosphorylation (Toruella et al., 1986) in plants have appeared in the literature. However, no tyrosine kinase activity could be detected by us in microsomal membranes isolated from pea root, pea shoot, zucchini hypocotyl, or crown gall callus (Blowers and Trewavas, 1986). Interestingly, the work of Toruella et al. (1986) found a major portion of the tyrosine kinase activity to be located in the nucleus rather than in other cellular fractions. The level of tyrosine phosphate, even in animals, is small and this will complicate its identification.

The possibility of a role for tyrosine phosphorylation in the control of plant cell growth remains open.

Autophosphorylating Protein Kinases

Since the protein kinases are protein in nature, they are potential substrates for their own activity. Such self-phosphorylation is generally termed autophosphorylation and is a relatively common property of the protein kinases (Krebs, 1983). In animal systems, phosphorylase kinase (Wang et al., 1976), type II cyclic AMP-dependent protein kinase (Rosen and Erlichman, 1975), myosin light-chain kinase (Wolf and Hofmann, 1980), cyclic GMP-dependent protein kinase (De Jonge and Rosen, 1977), casein kinase II (Geahlen et al., 1986), glycogen synthase kinase (Yamauchi and Fujisawa, 1985), type II calcium/calmodulin-dependent protein kinase (Kuret and Schulman, 1985), AP50 of the clathrin coat complex (Keen et al., 1987), protein kinase C (Newton and Koshland, 1987), EGF receptor (Ushiro and Cohen, 1980), the platelet-derived growth factor receptor (Ek et al., 1982), and the insulin receptor (Kasuga et al., 1982) have all been shown to autophosphorylate. In plants only four examples have been reported. Wong et al. (1986) found that purified preparations of phytochrome contain protein kinase activity leading to phosphorylation of phytochrome. Whether phytochrome is the protein kinase is the crucial but difficult-to-answer question. The redox-state-dependent thylakoid protein kinase of Bennet (1983) and Coughlan and Hind (1986), the plasma membrane calcium- and calmodulin-dependent protein kinase of Blowers et al. (1985), and the soluble calcium-dependent protein kinase of Harmon et al. (1987) also have been shown to autophosphorylate.

The autophosphorylation may be of two kinds: 1) that occurring between individual enzyme molecules or complexes—termed *intermolecular autophosphorylation;* 2) that occurring within the same enzyme molecule (or very tightly bound complex)—termed *intramolecular autophosphorylation.* Autophosphorylation, in some instances, is capable of altering the kinase activity and the importance of this will be discussed later. If no change in kinase activity or specificity can be found after autophosphorylation, then the reaction becomes futile. It is possible that many of the examples of autophosphorylation from animal systems where no change in kinase properties can be found are simply the result of an in vitro incubation of the protein kinase in the absence of other native substrates.

Coughlan and Hind (1986) present no data concerning the nature of their autophosphorylation in thylakoid protein kinase and could find no effect on subsequent enzyme activity directed towards both native and exogenous substrates. Likewise, Harmon et al. (1987) present no data concerning these properties in their enzyme. The kinase of Blowers et al. (1985) exhibits intramolecular autophosphorylation, resulting in subsequent inhibition of the phosphorylation of an exogenous substrate (Blowers and Trewavas, 1987). Other effects of the autophosphorylation on the affinity for endogenous substrates has yet to be determined. Figure 4 shows dilution kinetics for this protein kinase and the effect of intramolecular autophosphorylation on catalytic activity towards histone H1.

Autophosphorylation as a Control and Memory System

Cellular responses to cytoplasmic calcium might: 1) depend on the presence of a continued elevated calcium concentration, or 2) be prolonged after an elevated concentration has declined. It is known that calcium-regulated protein kinase can increase the rate of autophosphorylation by use of calcium ions. Three effects of autophosphorylation have been noted. 1) Autophosphorylation can modify catalytic activity to make the protein kinase independent of associated calcium ions. 2) Autophosphorylation can increase the sensitivity of the protein kinase to available calcium (sensitivity modulation, Rasmussen, 1983), and it can modify substrate specificity. 3) Autophosphorylation may greatly accelerate catalytic (and autophosphorylative) activity in a form of positive feedback.

Autophosphorylation is a means of prolonging the effects of a transient calcium stimulus and will be important when cytotoxic effects of elevated calcium might cause complications.

An example of the effects of autophosphorylation is the type II calcium- and calmodulin-dependent protein kinase isolated from brain tissue. Activation of the protein kinase by calcium and calmodulin leads to the production of an autonomous enzyme (Kuret and Schulman, 1985), which has altered substrate specificity (Bronstein et al., 1986). Further examples are to be found in the tyrosine kinase activities of hormone receptors. For instance, combining insulin with its receptor greatly increases tyrosine kinase activity (Rosen et al., 1983). The resulting autophosphorylation stabilises this "active" condition so that it is no longer dependent on insulin.

Another somewhat related property of protein kinases is their ability to translate between the cytosol and membranes. Saitoh and Schwartz (1985)

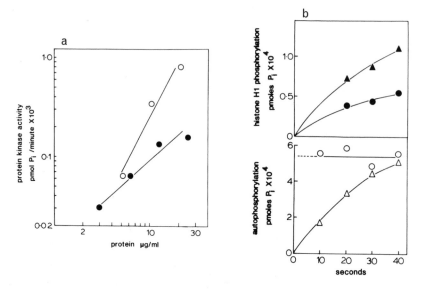

Fig. 4   Van't Hoff plot illustrating intramolecular autophosphorylation of a calcium-dependent protein kinase. The protein kinase preparation was diluted to give protein concentrations as indicated and assayed for protein kinase activity in the presence of a constant amount of $[\gamma - {}^{32}P]$ATP, approximately 100 μM free calcium ions and 50 μg ml^{-1} bovine calmodulin. For histone phosphorylation, the histone H1 was added at a starting concentration of 0.8 mg ml^{-1} and co-diluted with the protein kinase. Labelling for 10 seconds at 0°C was taken as an estimate of the initial rate and preparations were separated by sodium dodecylsulphate polyacrylamide gel electrophoresis (SDS-PAGE). The appropriate bands were excised and incorporated phosphate estimated by Cerenkov counting. Histone H1 phosphorylation (o) has a gradient of approximately 2, typical for an unattached substrate. However, autophosphorylation (●) has a gradient of approximately 1, indicative of intramolecular autophosphoryation. b: Modification of protein kinase activity by prior autophosphorylation. Reactions, without dilution, were performed as described above. A comparison of histone H1 phosphorylation by untreated protein kinase (▲) and protein kinase subjected to prior autophosphorylation for 1 min (●) clearly reveals the induced inhibition. Prior autophosphorylation with $[\gamma - {}^{32}P]$ ATP was followed by addition of histone H1 and further $[\gamma - {}^{32}P]$ATP. In the absence of prior autophosphorylation, twice the usual amount of $[\gamma - {}^{32}P]$ATP was added after 1 min. The lower graph of protein kinase autophosphorylation upon addition of $[\gamma - {}^{32}P]$ATP (Δ) and with more $[\gamma - {}^{32}P]$ ATP added to the enzyme subjected to prior autophosphorylation (o) indicates the status of the enzyme during the histone phosphorylation assays.

give evidence for the autophosphorylation of membrane-associated protein kinase in *Aplysia* and release of the subsequent autonomous enzyme to the cytosol. Clearly, the range of substrates available to the free enzyme is much enhanced. Similarly, Willmund et al. (1986) demonstrated that type II calcium- and calmodulin-dependent protein kinase in *Drosophila* showed a phosphorylation-dependent translocation from the cytoskeleton to the cytosol. Correlation with behaviour patterns of the flies showed that this process was possibly involved in a short-term memory system.

One of the great interests in autophosphorylation is the extent to which such coupled systems might act as metabolic switches. Figure 5 suggests a possible scheme of intracellular calcium regulation, which involves a calcium-dependent autophosphorylating protein kinase. It is proposed that a plasma membrane calcium-channel protein complex permits calcium entry into the cell. The subsequent elevated cytosolic free calcium ion concentration activates a protein kinase, possibly via calmodulin. Phosphorylation of the calcium channel inhibits further calcium entry whilst phosphorylation of the calcium pump (Ca^{2+}-ATPase) stimulates the active removal of calcium from the cell. The intracellular-free calcium ion concentration is thus returned to its initial status. However, calcium-activated autophosphorylation of the protein kinase inhibits its activity towards other substrates, for example, the calcium channel and Ca^{2+}- ATPase. Such a system, with combined competing phosphatases, should initiate calcium entry oscillations, which have been observed (Woods et al., 1986; Felle, Chapter 6).

Lisman (1985) has proposed a molecular mechanism for long-term binary memory storage involving an autophosphorylating protein kinase. This system, as applicable to neurons, has recently been reviewed by Schwartz and Greenberg (1987). However, such a system is entirely possible in plant cells.

The mechanism describes a molecular switch that can store information indefinitely despite the complete turnover of the molecules that make up the switch. Central to the mechanism is a kinase that is activated by autophosphorylation and is capable of intermolecular autophosphorylation.

The switch may operate as follows:

In response to a stimulus, possibly an increase in cytosolic free calcium, some molecules of the protein kinase autophosphorylate, increasing catalytic activity and thus they become capable of phosphorylating still more protein kinase molecules. Autophosphorylating protein kinase will, however, be counterbalanced by protein phosphatase. With a low stimulus the numbers of "activated" protein kinase molecules will be insufficient to counterbalance the "deactivating" influence of the protein phosphatase. The net effect is that the enzyme remains in a "deactivated" state. However, if the stimulus is sufficiently large, a threshold is passed. The numbers of phosphorylated protein kinase molecules outweigh the effects of protein phosphatase; all of the protein kinase molecules are phosphorylated rapidly.

OUTSIDE

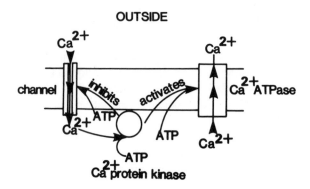

INSIDE

Fig. 5. Hypothetical scheme for control of calcium channel function and Ca^{2+}-ATPase by a calcium-dependent plasma membrane bound protein kinase. The scheme suggests that increased entry of calcium ions to the cytoplasm activates the protein kinase, which phosphorylates and inhibits channel activity and phosphorylates and activates the Ca^{2+}ATPase. Subsequent calcium activated autophosphorylation of the protein kinase inhibits further catalytic function. Such a system could give rise to oscillations in calcium entry, depending on calcium requirements for activation and autophosphorylation.

This stable, phosphorylated state continues, provided ATP is readily available and protein phosphatase is constant. A decline in ATP or an increase in protein phosphatase will lead to the crossing of another threshold in which the protein kinase population is now dephosphorylated. There are thus at least two stable states of the system, and the protein kinase is referred to as bistable. Thus the bistable protein kinase is capable of "remembering" a transient activating stimulus. For the activated state to be "forgotten," ATP synthesis should decline and/or protein phosphatase increase.

There are a number of possible variations of this model. Autophosphorylation might change substrate specificity, for example, or simply result in a protein kinase incapable of phosphorylating other proteins or even lose sensitivity to calcium ions.

The most intriguing aspect is the suggestion some 9 years earlier by Lewis et al. (1977) of a very similar bistable switch for development that operates on the same principles. However, this paper was unable to include the molecular involvement of protein kinase autophosphorylation. Lewis et al. (1977) were concerned about explaining the switching of cells during development and this gives great significance to the Lisman (1985) model.

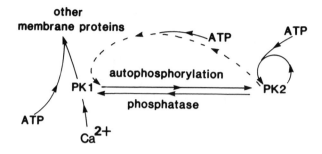

GERMINATION   DORMANCY

Fig. 6. A hypothetical scheme suggesting how a bistable calcium-activated protein kinase could control seed dormancy. The scheme envisages a protein kinase which in the auto-phosphorylated state preferentially phosphorylates itself but in the unphosphorylated state is calcium sensitive and can phosphorylate other proteins. The requirements for such bistability have been outlined by Lisman (1985). The calcium activated plasma membrane bound protein kinase of pea buds has some of the characteristics required by the model (Blowers and Trewavas, 1987; Trewavas, 1987b).

Evidence directly supporting the Lisman model has appeared in the literature (Lou et al., 1986). However, most protein kinase auto-phosphorylation appears to be intramolecular rather than intermolecular, although in most instances a very low level of intermolecular auto-phosphorylation would probably pass unnoticed.

One of us has suggested a variant Lisman model to explain aspects of seed dormancy breakage (Trewavas, 1987b). This is shown in Figure 6. It is suggested that the autophosphorylated protein kinase is insensitive to activation by calcium ions, is capable of only phosphorylating itself, and is to be found in this configuration in dormant seeds. Seed dormancy can be broken by inhibitors of ATP synthesis. The effect of dephosphorylating the protein kinase, which would follow such a treatment, is to render the enzyme sensitive to calcium ions and capable of phosphorylating other proteins. Many other stages of plant development might involve an equivalent mechanism.

CONCLUSION

This brief discussion should illustrate the potential of protein kinase investigation to increase the understanding of many features of plant growth and development. However, this chapter has also highlighted our ignorance about so much. The real value of describing second messengers in plants is the very large lacuna in knowledge to which it points; inevitably, those who have worked on animal systems have provided much of the framework we use to increase our understanding of the subject. Calcium, protein kinases, and second messengers are central to the goal of trying to understand how plant cells respond to the environment. The potential is there; it waits to be fulfilled.

REFERENCES

Amin M, F Kurosaki, A Nishi (1987): Carrot phytoalexin inhibits Ca^{2+}, calmodulin-dependent protein phosphorylation in carrot cells. Phytochemistry 21:51–53.

Barr R, TA Craig, FL Crane (1985): Transmembrane ferricyanide reduction in carrot. Biochim Biophys Acta 812:49–54.

Bennet J (1983): Regulation of photosynthesis by chloroplast protein phosphorylation. Philos Trans R Soc Lond [Biol] 302:113–125.

Blowers DP, AM Hetherington, AJ Trewavas (1985): Isolation of plasma membrane bound calcium/calmodulin regulated protein kinase from pea using Western blotting. Planta 166:208–215.

Blowers DP, AJ Trewavas (1986): Tyrosine specific protein kinases in plant tissues. In Trewavas AJ (ed): "Molecular and Cellular Aspects of Calcium in Plant Development." New York: Plenum, pp 339–340.

Blowers DP, AJ Trewavas (1987): Autophosphorylation of plasma membrane-bound calcium calmodulin-dependent protein kinase from pea seedlings and modification of catalytic activity by autophosphorylation. Biochem Biophys Res Commun 143:691–696.

Boller T, A Gelvi, T Mauch, U Vogelli (1983): Chitinase in bean leaves: Induction by ethylene purification, properties and possible function. Planta 157:22–31.

Bronstein JM, DB Farber, CG Wasterlain (1986): Autophosphorylation of calmodulin kinase II: Functional aspects. Fed Eur Biochem Soc Lett 196:135–138.

Chen C-C, DL Smith, BB Bruegger, RM Halpern, RA Smith (1974): Occurrence and distribution of acid labile histone phosphates in regenerating rat liver. Biochemistry 13:3785–3789.

Chock PB, SG Rhee, ER Stadtman (1980): Interconvertible enzyme cascades in cellular regulation. Annu Rev Biochem 49:813–843.

Clarkson DF, JB Hanson (1980): The mineral nutrition of higher plants. Annu Rev Plant Physiol 31:239–298.

Collier E, A Watkinson, CF Cleland, J Roth (1987): Partial purification and characterization of an insulin-like material from spinach and Lemna gibba G3. J Biol Chem 262:6238–6247.

Coughlan SJ, G Hind (1986): Purification and characterization of a membrane bound protein kinase from spinach thylakoids. J Biol Chem 261:11378–11385.

Csaba G (1980): Phylogeny and ontogeny of hormone receptors: The selection theory of receptor formation and hormonal imprinting. Biol Rev 55:47–63.

Datta N, LK Hardison, SJ Roux (1986): Polyamine stimulation of protein phosphorylation in isolated pea nuclei. Plant Physiol 82:681–684.

De Jonge HR, OM Rosen (1977): Self phosphorylation of cyclic 3',5'-monophosphate dependent protein kinase from bovine lung. J Biol Chem 252:2780–2783.

Eckhart W, MA Hutchinson, T Hunter (1979): An activity of phosphorylating tyrosine in polyoma T antigen immunoprecipitates. Cell 18:925–933.

Ek B, B Westermark, A Wasteson, C-H Heldin (1982): Stimulation of tyrosine specific phosphorylation by PDGF. Nature 295:419–420.

Elliott DC, M Geytenbeek (1985): Identification of products phosphorylation in T37-transformed cells and comparison with normal cells. Biochem Biophys Res Commun 845:317–323.

Galston AW (1983): Polyamines as modulators of plant development. Biosci 33:382–388.

Geahlen RF, M Anostario Jr, PS Low, ML Harrison (1986): Detection of protein kinase activity in SDS polyacrylamide gels. Anal Biochem 153:151–158.

Guy M, H Kende (1984): Conversion of l-aminocyclopropane-1-carboxylic acid to ethylene by isolated vacuoles of *Pisum sativum* L. Planta 160:281–287.

Hanson AD (1982): Metabolic responses of mesophytes to plant water deficits. Annu Rev Plant Physiol 33:163–203.

Harmon AC, C Putnan-Evans, MJ Cormier (1987): A calcium-dependent but calmodulin-independent protein kinase from soybean. Plant Physiol 83:830–837.

Hetherington A, AJ Trewavas (1982): Calcium dependent protein kinase in pea shoot membranes. Fed Eur Biochem Soc Lett 145:67–71.

Hetherington A, AJ Trewavas (1984): The regulation of membrane bound kinases by phospholipid and calcium. In Boudet AM (ed): "Membranes and Compartmentation in the Regulation of Plant Functions: Annual Proceedings of the Phytochemical Society of Europe." Oxford Clarendon, 24:181–195.

Houslay MD (1987): Ion channels controlled by guanine nucleotide regulatory proteins. Trends Biochem Sci 12:167–168.

Huxley JS (1935): Chemical regulation and the hormone concept. Biol Rev 10:427–441.

Jacobs WP (1979): Plant Hormones and Plant Development. Cambridge: Cambridge University Press.

Kacser H, JA Burns (1973): The control of flux. In Davies DD (ed): "Rate Control of Biological Processes: Society of Experimental Biology Symposium." Cambridge: Cambridge University Press, 27:65–105.

Kasuga M, Y Zick, DL Blithe, M Crettaz, CR Kahn (1982): Insulin stimulates tyrosine phosphorylation of the insulin receptor in a cell free system. Nature 298:667–669.

Kauss H (1983): Volume regulation in *Poteriochromas*. Plant Physiol 71:169–172.

Keen JH, MH Chestnut, KA Beck (1987): The clathrin coat assembly polypeptide complex: Autophosphorylation and assemply activities. J Biol Chem 262:3864–3871.

Kikkawa U, Nishizuka Y (1986): The role of protein kinase C in transmembrane signalling. Annu Rev Cell Biol 2:149–178.

Koshland DE, A Goldbeter, JB Stock (1982): Amplification and adaptation in regulatory and sensory systems. Science 217:220–225.

Krebs EG (1983): Historical perspectives in protein phosphorylation and a classification system for protein kinases. Philos Trans R Soc Lond [Biol] 302:3–11.

Krebs EG, JA Beavo (1979): Phosphorylation and dephosphorylation of enzymes. Annu Rev Biochem 48:923–959.

Kuehn GD, VJ Atmar (1982): Post-translational control of ornithine decarboxylase by polyamine-dependent protein kinase. Fed Proc 41:3078–3083.

Kuret J, H Schulman (1985): Mechanism of autophosphorylation of the multifunctional calcium/calmodulin dependent protein kinase. J Biol Chem 260:6427–6433.

Lee JS, TJ Mulkey, ML Evans (1983): Gravity induced polar transport of calcium across root tips of maize. Plant Physiol 73:874–876.

Leopold AC, LG Nooden (1984): Hormonal regulatory systems in plants. In Scott TK (ed): "Hormonal Regulation of Development: Encyclopedia of Plant Physiology," New Series. Berlin: Springer-Verlag, 10:4–23.

Lewis J, JMW Black, L Wolpert (1977): Thresholds in development. J Theor Biol 65:579–590.

Lisman JE (1985): A mechanism for memory storage insensitive to molecular turnover: A bistable autophosphorylating kinase. Proc Natl Acad Sci USA 82:3055–3057.

Lou LL, SJ Lloyd, M Schulman (1986): Activation of multifunctional Ca^{2+}/calmodulin-dependent protein kinase by autophosphorylation: ATP modulates production of an autonumous enzyme. Proc Natl Acad Sci USA 83:9497–9501.

Lovelock JE (1979): Gaia: "A New Look at Life on Earth." Oxford: Oxford University Press.

Macherel D, A Viale, T Akazawa (1986): Protein phosphorylation in amyloplasts isolated from suspension cultured cells of sycamore *(Acer pseudoplatanus L.)*. Plant Physiol 80:1041–1044.

MacNicol PK (1976): Rapid metabolic changes in the wounding response of leaf phase following excision. Plant Physiol 57:80–84.

Moreau RA (1986): Regulation of phospholipase activity in potato leaves by calmodulin and protein phosphorylation–dephosphorylation. Plant Sci 47:1–9.

Newton AC, DE Koshland Jr (1987): Protein kinase C autophosphorylates by an intrapeptide reaction J Bid Chem 262:10185–10188.

Nishimura S, T Sekiya (1987): Human cancer and cellular oncogenes. Biochem J 243:313–327.

Polya GM, JR Davies (1982): Resolution of a calcium activated protein kinase from wheat germ. Fed Eur Biochem Soc Lett 150:167–171.

Polya GM, V Micucci (1984): Partial purification and characterization of a second calmodulin dependent protein kinase from wheat germ. Biochim Biophys Acta 785:68–74.

Polya GM, V Micucci, A Gantinas, S Basiliadis (1985a): Plant leaf calcium dependent protein kinase, cGMP promoted protein phosphorylation and cGMP inhibited phospho-hydrolase. Proc Aust Biochem Soc 17:88.

Polya GM, V Micucci, AL Rae, PJ Harris, AE Clarke (1985b): Calcium dependent protein phosphorylation in germinated pollen of *Nicotiana alata*. Proc Aust Biochem Soc 17:74.

Polya GM, A Schibeci, V Micucci (1984): Phosphorylation of membrane proteins from *Lolium multiflorum* (ryegrass) endosperm cells. Plant Sci Lett 36:51–57.

Putnam-Evans CL, AC Harmon, MJ Cormier (1986): Calcium-dependent protein phosphorylation in cultured soybean cells. In Trewavas AJ (ed): "Molecular and Cellular Aspects of Calcium in Plant Development." New York: Plenum, pp 99–106.

Ranjeva R, AM Boudet (1987): Phosphorylation of proteins in plants: Regulatory effects and potential involvement in stimulus/response coupling. Annu Rev Plant Physiol 38:73–93.

Rasmussen H (1983): Pathways of amplitude and sensitivity modulation in the calcium messenger system. In Cheung WY (ed): "Calcium and Cell Function," Vol IV. New York: Academic Press, pp 1–61.

Refeno G, R Ranjeva, AM Boudet (1982): Modulation of QORase activity through reversible phosphorylation in carrot cell suspensions. Planta 154:193–198.

Rosen OM, J Erlichman (1975): Reversible autophosphorylation of cyclic 3'-5'-AMP dependent protein kinase from bovine cardiac muscle. J Biol Chem 250:7788–7794.

Rosen OM, R Herrera, Y Olowe, LM Petruzzilli, MH Cobb (1983): Phosphorylation activates the insulin receptor tyrosine protein kinase. Proc Natl Acad Sci USA 80:3237–3240.

Roux SJ, RD Wayne, N Datta (1986): Role of calcium ions in phytochrome responses: An update. Plant Physiol 66:344–348.

Saitoh T, JH Schwartz (1985): Phosphorylation dependent subcellular translocation of a calcium/calmodulin dependent protein kinase produces autonomous enzyme in *Aplysia* neurons. J Cell Biol 100:835–842.

Salimath BP, D Marme (1983): Protein phosphorylation and its regulation by calcium and calmodulin in membrane fractions from zucchini hypocotyls. Plant 158:560–568.

Schafer A, F Bygrave, S Matzemauer, D Marme (1985): Identification of a calcium and phospholipid dependent protein kinase in plant tissue. Fed Eur Biochem Soc Lett 187:25–28.

Schwartz JH, SM Greenberg (1987): Molecular mechanisms for memory: Second messenger induced modifications of protein kinases in nerve cells. Annu Rev Neurosci 10:459–476.

Slocum RD, R Kaur-Sawhney, AW Galston (1984): The physiology of biochemistry of polyamines in plants. Arch Biochem Biophys 235:283–303.

Sticher L, C Penel, H Greppin (1981): Calcium requirement for the secretion of peroxidases by plant cell suspensions. J Cell Sci 48:345–350.

Sutherland EW, GA Robinson, RW Butcher (1968): Some aspects of the biological role of adenosine 3'-5'-monophosphate (cAMP). Circulation 37:279–306.

Teulieres C, G Alibert, R Ranjeva (1985): Reversible phosphorylation of tonoplast proteins involves tonoplast-bound calcium-calmodulin-dependent protein kinase(s) and protein phosphastase(s). Plant Cell Rep 4:199-201.

Theologis A, GG Laties (1980): Membrane lipid breakdown in relation to wound induced and cyanide resistant respiration in tissue slices. Plant Physiol 66:890–896.

Thomas WVN (1982): "Techniques in Calcium Research." London: Academic.

Tognoli L, B Basso (1987): The fusicoccin-stimulated phosphorylation of a 33KDa polypeptide in cells of *Acer pseudoplatanus* as influenced by extracellular and intracellular pH. Plant Cell Env 10:233–239.

Torruella M, LM Casano, RH Vallejos (1986): Evidence of activity of tyrosine kinase(s) and of the presence of phosphotyrosine in pea plantlets. J Biol Chem 261:6651–6653.

Tran Thanh Van KM (1981): Control of morphogenesis in in vitro cultures. Annu Rev Plant Physiol 32:291–311.

Trewavas AJ (1983a): Nitrate as a plant hormone. In Jackson MB (ed):. "British Plant Growth Regulator Group," Monograph 9. British Plant Growth Regulator Group, Wantage, Oxon. pp 97–110.

Trewavas AJ (1983b): Plant growth substances, metabolic flywheels for plant development. Cell Biol Int Rep 7:569–575.

Trewavas AJ (1986): Resource allocation under poor growth conditions: A major role for growth substance in developmental plasticity. In Jennings DH, AJ Trewavas (eds): "Plasticity in Plants." Cambridge: Company of Biologists Ltd. pp 31–76.

Trewavas AJ (1987a): Sensitivity and sensory adaptation in growth substance responses. In Hoad G, Lenton JR, Jackson M, Atkin RK, (eds): "Hormone Action in Plant Development: A Critical Appraisal." London: Butterworths, pp 19–39.

Trewavas AJ (1987b): Timing and memory processes in seed embryo dormancy: A conceptual paradigm for plant development questions. Bioessays 6:87–92.

Trewavas AJ, E Allan (1987): An appraisal of the contribution of growth substances to plant development. In Wisiol K (ed): "Plant Growth Modelling for Resource Management." Boca Raton, FL: CRC Vol 2 pp 25–49.

Ushiro H, SJ Cohen (1980): Identification of phosphotyrosine as a product of EGF activated protein kinase in A431 cell membranes. J Biol Chem 255:8363–8365.

Veluthambi K, BW Poovaiah (1984a): Calcium and calmodulin-regulated phosphorylation of soluble and membrane proteins from corn coleoptiles. Plant Physiol 76:359–365.

Veluthambi K, BW Poovaiah (1984b): Polyamine stimulated phosphorylation of proteins from corn *(Zea mays)* coleoptiles. Biochem Biophys Res Commun 122:1374–1380.

Wang JM, JT Stull, T-S Huang, EG Krebs (1976): A study on the autoactivation of rabbit muscle phosphorylase kinase. J Biol Chem 251:4521-4527.

Willmund R, M Mitschulat, K Scheider (1986): Long term modulation of Ca^{2+} stimulated autophosphorylation and subcellular distribution of the Ca^{2+}/calmodulin dependent protein kinase in the brain of *Drosophila*. Proc Natl Acad Sci USA 83;9789–9793.

Wolf H, F Hofmann (1980): Purification of myosin light chain kinase from bovine cardiac muscle. Proc Natl Acad Sci USA 77:5852–5855.

Wong Y-S, H-C Cheng, DA Walsh, JC Lagarias (1986): Phosphorylation of *Avena* phytochrome in vitro as a probe of light induced conformational changes. J Biol Chem 261:12089–12097.

Woods NM, KSR Cuthbertson, PH Cobbold (1986): Repetitive rises in cytoplasmic free calcium in hormone stimulated hepatocytes. Nature 319:600–602.

Yamauchi T, H Fujisawa (1985): Self-regulation of calmodulin-dependent protein kinase II and glycogen synthase by autophosphorylation. Biochem Biophys Res Commun 129: 213–219.

Yan T-F, M Tao (1982): Purification and characterizaition of a wheat germ protein kinase. J Biol Chem 257:7037–7043.

Yoshimasa T, DR Sibley, M Bouvier, RJ Letkowitz, MG Caron (1987): Cross-talk between cellular signalling pathways suggested by phorbol ester induced adenylate cyclase phosphorylation. Nature 327:67–68.

Zocchi G (1985): Phosphorylation-dephosphorylation of membrane proteins controls the microsomal H + ATPase activity of corn roots. Plant Sci 40:153–159.

Second Messengers in Plant Growth
and Development, pages 29–56
© 1989 Alan R. Liss, Inc.

2. Phosphoinositide Metabolism: Its Relation to Signal Transduction in Plants

Wendy F. Boss

Department of Botany, North Carolina State University, Raleigh, North Carolina 27695

INTRODUCTION

The term *phosphoinositide* is used to describe a group of phospholipids, which include phosphatidylinositol (PI), phosphatidylinositol-4-mono-phosphate (PIP), phosphatidylinositol-4,5-bisphosphate (PIP$_2$), and their respective lysolipids, lysoPI (LPI), lysoPIP (LPIP), and lysoPIP$_2$ (LPIP$_2$) (Fig. 1). Current interest in the metabolism of phosphoinositides in plants has arisen from studies of animal cells, which have demonstrated that these lipids are a source of second messengers (Michell, 1975; Berridge and Irvine, 1984; Hokin, 1985; Majerus et al., 1986). For a review of individual animal response mechanisms involving the phosphoinositides, see Putney (1986). While the regulation of the metabolism of the phosphoinositides may prove to be different in plant and animal cells, the animal pathways provide a framework for investigating a role for phosphoinositides in signal transduction in plants.

The paradigm for animal cells begins with an external stimulus, which is perceived at the plasma membrane by a receptor (Fig. 2). The receptor activitates a PIP$_2$-specific phospholipase C or phosphodiesterase. In animal cells, activation of the phospholipase C is enhanced by GTP (Litosch et al., 1985; Smith et al., 1986). It has been proposed that a G protein may increase the binding of soluble lipases to the membrane, thereby enhancing PIP$_2$ hydrolysis (Majerus et al., 1986). The PIP$_2$ present on the cytosolic side of

Address reprint requests to Dr. Wendy F. Boss, Box 7612, Botany Department, North Carolina State University, Raleigh, NC 27695.

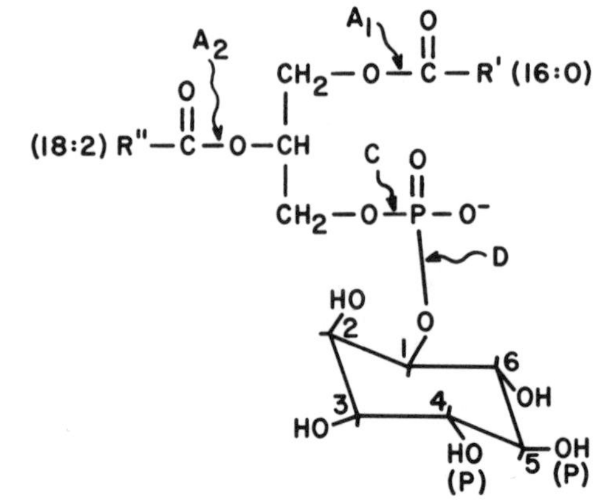

Fig. 1. The structures of carrot phosphoinositides. Phosphatidylinositol (PI) is shown. Phosphatidylinositol-4-monophosphate (PIP) is phosphorylated at the number 4 hydroxyl group. Phosphatidylinositol-4,5-bisphosphate (PIP$_2$) is phosphorylated at the number 4 and 5 hydroxyls. The letters A$_1$, A$_2$, C, and D indicate sites of action of the respective phospholipases. Lysolipids are produced as a result of phospholipase A$_1$ and A$_2$ activity.

the plasma membrane is hydrolyzed to inositol-1,4,5-trisphosphate (IP$_3$) and diacylglycerol (DAG). Each of these metabolites is a potential second messenger, initiating a cascade of metabolic events. For example, IP$_3$ mobilizes calcium from intracellular stores, specifically from the endoplasmic reticulum (ER) (Streb et al., 1983; Prentki et al., 1984; Oron et al., 1985), causing an increase in cytosolic calcium. The increase in cytosolic calcium can activate calcium and calcium/calmodulin-dependent enzymes such as kinases, lipases, and ATPases. The DAG formed as a result of the phospholipase C activates protein kinase C (Nishizuka, 1984). DAG enhances the affinity of the protein kinase for calcium, so that resting levels of calcium are sufficient to activate the kinase (Kaibuchi et al., 1981). Thus, although calcium-dependent ATPases rapidly lower cytosolic calcium to resting levels, there are long-term effects on the physiology of the cell.

At one time it was hypothesized that cAMP was a second messenger in plants; however, neither adenylate cyclase (Yunghaus and Morré, 1977) nor cAMP-dependent protein kinase, the physiological receptor for cAMP, have been found in plants (Brown and Newton, 1981). Additionally, cAMP has never been found to be required for any physiological response (Hepler and Wayne, 1985). Calcium plays an important role in regulating plant growth and development (for reviews see Hanson, 1984; Hepler and Wayne, 1985; Marmé and Dieter, 1983; Roux and Slocum, 1982; Kauss, 1987; Poovaiah and Reddy, 1987; Trewavas, 1986). Many physiological responses

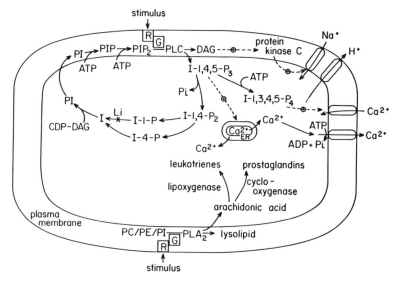

Fig. 2.   A model of phosphoinositide metabolism in animal cells.
R, receptor; G, GTP-regulated protein; PLC, phospholipase C; PLA$_2$, phospholipase A$_2$; PI, phosphatidylinositol; PIP, phosphatidylinositol-4-monophosphate; PIP$_2$, phosphatidylinositol-4,5-bisphosphate; DAG, diacylglycerol; I-1,4,5-P$_3$, inositol-1,4,5-trisphosphate; I-1,4-P$_2$, inositol-1,4-bisphosphate; I-1-P, inositol-1-phosphate; I-4-P, inositol-4-phosphate; I-1,3,4,5-P$_4$, inositol-1,3,4,5-tetrakisphosphate; PC, phosphatidylcholine; PE, phosphatidylethanolamine.

such as cell division (Wolniak et al., 1983), tropic responses (Pickard, 1985; Evans 1985), phytochrome-mediated responses (Roux et al., 1986), secretion (Jones and Carbonell, 1984; Akazawa and Mitsui, 1985), and protoplast fusion (Grimes and Boss, 1985) appear to involve an increase in cytoplasmic calcium, which precedes the response.

While an influx of external calcium may contribute to the stimulus-response coupling, the phosphoinositide paradigm offers additional opportunities for specificity and signal amplification. 1) It would provide plants with a mechanism for responding to external stimuli. This mechanism would depend on the presence of the phosphoinositides, rather than on external calcium. 2) It would introduce a potential for multiple responses. These responses would vary not only with the stimulus but also with the intracellular calcium reserves and thereby would be sensitive to the physiological state of the cell. 3) Continued response to the stimulus would depend on the ability of the cells to regenerate the phosphoinositides, which would provide an additional mechanism of feedback control.

This chapter will investigate the evidence for the involvement of phospho-inositides in signal transduction in plant cells. First, the occurrence and location of the phosphoinositides and inositol phosphates in plants will be discussed. Second, the pathways for the biosynthesis and metabolism of the phosphoinositides will be described. Third, data will be presented indicating that the mechanism for the involvement of phosphoinositides in physiological responses in plant cells is as yet undefined.

MYO-INOSITOL METABOLISM IN PLANTS

When considering phosphoinositide metabolism in plants, one cannot avoid an appreciation of the multifunctional role of myo-inositol in plant metabolism. This has been reviewed by Loewus and Loewus (1983). There are several aspects of myo-inositol metabolism that are of particular importance to stimulus—response studies. These are discussed in the following paragraphs.

Myo-inositol Synthesis and Degradation

Myo-inositol-1-phosphate is synthesized from glucose-6-phosphate by the enzyme myo-inositol-1-phosphate synthase and is dephosphorylated to myo-inositol. For simplicity, myo-inositol will be referred to as inositol for the remainder of this chapter. Of interest is the insensitivity of inositol-1-phosphate phosphatase of plant cells to lithium. In pollen, for example, 50 mM of lithium is required for 50% inhibition of the phosphatase (Gumber et al., 1984), while in many animal cells, inositol-1-phosphate phosphatase is inhibited to the same extent by 0.8 mM lithium (Hallcher and Sherman, 1980). There are several advantages to being able to decrease the rate of metabolism of IP. For example, the decrease in degradation of IP causes a decrease in degradation of IP_2 and IP_3; therefore, adding lithium to animal cells during stimulation enhances the recovery of all of the inositol phosphates. In addition, by inhibiting the degradation of IP, inositol is no longer available for the synthesis of phosphatidylinositol. In a system such as the blowfly salivary gland, where there is little free inositol, adding lithium during stimulation inhibits the resynthesis of PIP_2 and thereby eliminates continued stimulation by 5-hydroxytryptamine (Fain and Berridge, 1979). Adding inositol to the lithium-treated salivary gland results in the recovery of the 5-hydroxytryptamine-stimulated secretory response. There is no evidence, however, that lithium enhances the recovery of inositol phosphates from plants (Morse et al., 1987a), and inositol is not rate-limiting in the biosynthesis of plant phosphatidylinositol.

In addition to synthesizing inositol, plant cells can synthesize inositol hexaphosphate (IP_6 or phytic acid). IP_6 is found in dormant tissue and seeds and serves as a storage form for phosphate. Similarly, plants can readily metabolize IP_6 to other inositol phosphates. The prevalence of the kinases and phosphatases leads to a potential for many forms of inositol phosphate. Curiously, none of the intermediates in the biosynthesis of IP_6 from IP have ever been isolated (Loewus and Loewus, 1983). Presumably, IP and the intermediates in the pathway remain bound to an enzyme complex until their phosphorylation to IP_6 is complete.

Inositol Metabolites

Analysis of inositol phosphates in plant cells is further complicated by the fact that inositol is metabolized to glucuronic acid which is used in the biosynthesis of noncellulosic wall polysaccharides, glycoproteins, gums, and mucilages (Loewus and Loewus, 1983). This means that when labeling plant cells with [^{3}H]inositol, a significant portion of the inositol will be incorporated into nonlipid components. Analysis of carrot suspension culture cells after 6 hr of labeling with [^{3}H]inositol confirmed the incorporation of [^{3}H] into arabinose, xylose, and glucuronic acid (Verma and Dougall, 1979). In germinating wheat kernels, it was found also that a major portion of the [^{3}H]inositol was incorporated into polysaccharides (Maiti and Loewus, 1978). The [^{3}H] components of the cell wall have been found to co-elute with the inositol phosphates using anion exchange chromatography (Coté et al., 1987). The problem is exemplified in Figure 3 (Rincon, unpublished results), where the cell wall was isolated from [^{3}H]inositol-labeled carrot cells and analyzed by anion exchange chromatography according to Berridge et al. (1983). Note the recovery of [^{3}H] compounds under conditions that would elute IP_2 and IP_3. These data emphasize the need for precise identification of the inositol phosphates.

Inositol also is used in the biosynthesis of analogues important for the storage and transport of the growth regulator auxin (Cohen and Bandurski, 1982). Finally, while inositol can be incorporated into phosphatidylinositol via the enzyme CDP-DAG:inositol transferase, free inositol can be exchanged with the inositol of PI via an inositol exchange reaction described by Sexton and Moore (1981) and Sandelius and Morré (1987). The diversity of pathways for the metabolism of inositol in plants lends added complexity to the interpretation of data with regard to their role in phosphoinositide metabolism and signal transduction.

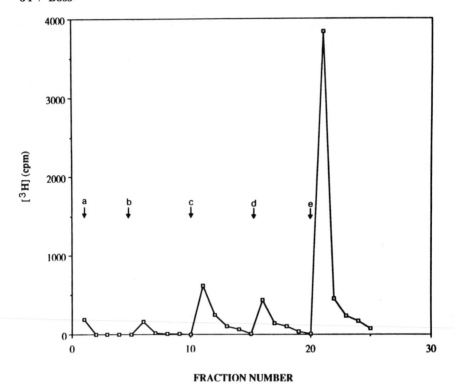

FRACTION NUMBER

Fig. 3.   Anion exchange chromatography of [³H]-labeled components of the cell wall fraction isolated from wild carrot cells. The cells were labeled overnight with myo-[2-³H]inositol and the cell wall isolated according to Asamizu and Nishi (1979). The aqueous extract from the trichloroacetic acid-treated cell wall was applied onto the Dowex-1 column and analyzed according to Berridge et al. (1983). Arrows indicate the sequential additions of the elutants: **a:** deonized water; **b:** 5 mM Na⁺ tetraborate-60 mM Na⁺ formate; **c:** 0.2 M NH₄ formate-0.1 M formic acid, **d:** 0.4 M NH₄⁺ formate 0.1 M formic acid; **e:** 1.0 M NH₄⁺ formate-0.1 M formic acid. Fractions of 2 ml were collected. IP, IP₂, and IP₃ standards would elute with solvents **c, d,** and **e,** respectively.

ISOLATION AND IDENTIFICATION OF PLANT PHOSPHOINOSITIDES

Polyphosphoinositides have been isolated from whole plants (Helsper et al., 1986b; Morse et al., 1987a; Sandelius and Morré, 1987) and tissue culture cells (Boss and Massel, 1985; Heim and Wagner, 1986; Strasser et al., 1986; Ettlinger and Lehle, 1988). These reports were based on identification of the lipids by co-migration with known standards of either the phospholipid or the hydrolyzed head groups and by both [³²P] and [³H]inositol labeling. Recently, more precise analysis was done of the phosphoinositides from carrot suspension culture cells using fast atom bombardment mass spectrometry (van Breeman et al, unpublished results). The inositol phospholipids were loaded onto a silicic acid column; washed with $CHCl_3$:MeOH(1:1); eluted with $CHCl_3$:MeOH:NH_4OH:H_2O (90:90: 7:22); dried (in vacuo); and analyzed with a JEOL HX 110 double-focusing mass spectrometer. The molecular ions for PI, LPI, PIP, and PIP_2 were identified. The major fatty acids were palmitic (16:0) and linoleic (18:2). These fatty acids, as well as linolenic acid, were found in PI from pollen (Helsper et al., 1987) and soybean (Mudd, 1980). There was no evidence for the presence of arachidonic acid. Although inositol sphingolipids have been reported in plants (Carter and Koob, 1969), all of the inositol lipids recovered from the carrot suspension culture cells were readily hydrolyzed in base, indicating that they were not sphingolipids.

The methods for extraction of polyphosphoinositides are, in general, based on a modification of the method of Folch et al. (1957) or Schacht (1981). The lipid headgroups are negatively charged; therefore, an acidic solvent is required to enhance their partitioning into the organic phase. Since the phosphoinositides seem particularly susceptible to hydrolysis, careful extraction is essential. Variations of this extraction include the addition of 2 M KCl (Watson et al., 1984) or the use of 10 mM tetrabutylammonium sulfate (Grove et al., 1981) rather than HCl. In addition, antioxidants such as creosol may be added (Sandelius and Sommarin, 1986). It has been our experience that the HCl extraction with the addition of EDTA to chelate metal ions gives consistent recovery of the phosphoinositides without generating lysolipids. As with all lipid extractions, the ratio of aqueous to organic phase is important. For each 0.1 ml of membranes or 0.05 g of cells, the following solutions are added in sequence with vortexing: 1.5 ml of $CHCl_3$:MeOH (1:2, v/v); 0.5 ml of 2.4 N HCl; 0.5 ml 1 mM EDTA; 0.5 ml of $CHCl_3$ (Boss and Massel, 1985). The extract is centrifuged at approximately $1,700 \times g$ for 5 min, and the lower, organic phase is removed. The aqueous phase is re-extracted twice with 0.5 ml $CHCl_3$. The combined organic phases are extracted twice with 2 ml MeOH:1 N HCl (1:1, v/v). The aqueous phase is removed, and the organic layer is dried in vacuo and stored frozen under nitrogen. It is essential to remove all of the acidic MeOH upper phase before drying the sample. If this is not done, the inositol lipids will be hydrolyzed by the residual acid.

The precise separation and identification of the phosphoinositides is critical for accurate interpretation of the data. Careful scrutiny of many published reports of plant phosphoinositides will reveal that separation of lysolipids from the phospholipids has not always been achieved. This will lead to falsely high values for the PIP and PIP$_2$ concentrations. The importance of separating and correctly identifying the phosphoinositides is demonstrated in Figure 4. [^{3}H]Inositol-labeled lipids were extracted from protoplasts isolated from carrot suspension culture cells and were chromatographed on LK5D plates presoaked in 1% oxalate for 80 sec and dried at 110°C for 2 hr. Identical samples were spotted on two separate plates: one was developed in solvent A, CHCl$_3$:MeOH:NH$_4$OH:H$_2$O (90:90:7:22, v/v; Schacht, 1978) and the other in solvent B, CHCl$_3$:CH$_3$COCH$_3$:MeOH:HOAc:H$_2$O (40:15:13:12:8, v/v; Jolles et al., 1979). Note the following: 1) In solvent A, PIP migrated below LPI, whereas in solvent B, PIP migrated slightly above LPI. 2) LPIP was well resolved from PIP and PIP$_2$ in solvent A, whereas in solvent B, a peak that appears to

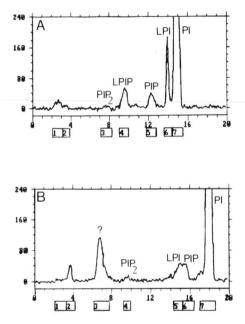

Fig. 4. Thin-layer analysis of [^{3}H]inositol-labeled lipids extracted from protoplasts isolated from wild carrot cells that had been labeled for 18 hr with [^{3}H]inositol. Equal aliquots of lipid were spotted on identical LK5 plates presoaked in 1% oxalate for 80 sec and oven-dried (110°C) for 2 hr. Solvent **A** was CHCl$_3$:MeOH:NH$_4$OH:H$_2$O, 90:90:7:22 (v/v). Solvent **B** was CHCl$_3$:CH$_3$COCH$_3$:MeOH:HOAc:H$_2$O, 40:15:13:12:8 (v/v). Migration of known standards (Sigma) are noted. For solvent **A**, compounds have been identified by negative ion fast atom bombardment mass spectrometry. The ratio of LPI to PIP is typical of the lipids extracted from protoplasts and is high relative to that of cells where PIP is greater than LPI (Boss, 1987). Boxes denote areas of integration using a Bioscan 500 Imaging Scanner.

be LPIP, based on predicted retardation factor (R_f) and quantity, was not well resolved and in most instances co-migrated with PIP_2. 3) PIP_2 made up a very small proportion of the inositol-labeled lipid, and for this reason, LPIP could easily be mistaken for PIP_2 in either solvent system. In most plants studied thus far, the ratio of PIP to PIP_2 is on the order of 9 to 1. 4) The lysolipids, as well as the [^{3}H]inositol at the origin, were a greater percentage of the total recovered inositol when solvent B was used, relative to that when solvent A was used. This suggested increased breakdown of the lipids during the chromatography in the acidic solvent.

Another method of separating the phosphoinositides is to use neomycin affinity columns (Schacht, 1978). Neomycin is positively charged and binds to the negatively charged phosphoinositides. Satisfactory resolution of plant phosphoinositides using the neomycin columns has not yet been achieved (Boss, unpublished results). This may be due to the presence of lysolipids in the plant extracts.

Lysolipids can be produced in vivo by phospholipase A or during extraction by acid or base hydrolysis of the phospholipids. When lysolipids are produced during the drying of the lipids or on thin-layer plates, residual inositol-labeled spots just above the origin become prominent—a telltale sign of lipid lysis. Since phospholipase A may be activated during stress (Burgoyne et al., 1987; Leshem, 1987) and since the lysolipids do not always resolve well from the phospholipids, an increase in the lysophosphoinositides could be misinterpreted as an increase in the phosphoinositides. These are additional reasons why, when studying stimulus–response coupling, careful separation and identification of the phosphoinositides are essential.

LOCALIZATION OF THE PHOSPHOINOSITIDES AND THE ENZYMES INVOLVED IN THEIR METABOLISM

The phosphoinositides must be located in the plasma membrane if they are to play a role in transducing a stimulus from the external environment to the cytosol. Two approaches have been used to study the localization of the phosphoinositides. One was to identify the sites of lipid synthesis, that is, PI and PIP kinase activity (Sandelius and Sommarin, 1986; Sommarin and Sandelius 1988). The other was to isolate membranes from prelabeled cells and identify the membranes enriched in phosphoinositides (Wheeler and Boss, 1987). Both groups isolated plasma membranes by aqueous two-phase partitioning. These studies have demonstrated that plasma membranes from plant cells are enriched in PIP and PIP_2 and the enzymes to synthesize them.

In addition, they have led to new insights concerning plant phosphoinositide metabolism.

The membrane labeling studies were done with carrot cells grown in suspension culture. These studies revealed that the plasma membrane not only was enriched in PIP and PIP_2 but that it also was enriched in LPIP but not LPI (Figure 5, Wheeler and Boss, 1987). Only 1–2% of the total LPI recovered from the combined upper and lower phases was in the upper, plasma membrane-rich fraction (Table I). A high relative ratio of PIP to LPI correlated with the purity of the plasma membrane fraction. As LPI increased, contamination by endomembranes increased based on marker enzyme assays. While the LPI constituted < 5% of the plasma membrane phosphoinositides, LPIP made up 15–20% of the phosphoinositide recovered in the plasma membrane fraction. The LPIP was not simply generated during isolation, since the total LPIP recovered in the combined upper and lower phases was approximately the same percentage as that found in the

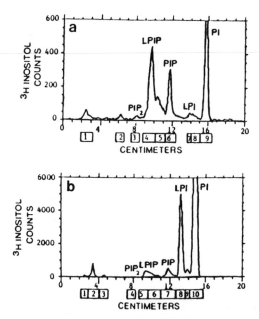

Fig. 5. Thin-layer analysis of [³H]inositol-labeled lipids extracted from membranes of fusogenic carrot cells. Membranes were isolated by aqueous two-phase partitioning: **a**: upper phase plasma membrane-rich; **b**: lower phase, microsomal. Note enrichment of PIP, LPIP, and PIP_2 in the plasma membrane-rich fraction and the high ratio of PIP to LPI in the plasma membrane-rich fraction relative to the microsomal fraction. Legend as per Figure 4. (Reproduced from Wheeler and Boss, 1987, with permission of the publisher.)

TABLE I. Distribution of [³H]Inositol-Labeled Lipids of Membranes From [³H]Inositol-Labeled Fusogenic Carrot Cells*

Cell type	PIP₂	LPIP	PIP	LPI	PI
Whole cells	0.6 ± 0.3	4.7 ± 1.3	7.0 ± 3.4	7.4 ± 1.6	76.0 ± 6.0
Plasma membrane (upper phase)	1.3 ± 0.1	18.9 ± 3.3	14.2 ± 1.5	4.5 ± 0.5	54.0 ± 4.5
Microsomes (lower phase)	0.3 ± 0.1	2.0 ± 0.6	2.7 ± 0.6	18.6 ± 1.4	73.5 ± 1.6
Nuclei isolated from protoplasts	3.6 ± 1.1	11.5 ± 0.9	5.0 ± 0.3	12.3 ± 0.1	55.7 ± 2.8

* Percent of total [³H]inositol lipid recovered ± SD from at least three separate experiments. PIP₂, phosphatidylinositol-4,5-bisphosphate; LPIP, lysoPIP; PIP, phosphatidylinositol-4-monophosphate; LPI, lysoPI; PI, phosphatidylinositol.

total cell extracts (4.3 ± 1% vs. 4.7 ± 1.3%, respectively). Furthermore, LPIP is synthesized not only by the hydrolysis of PIP but also by phosphorylation of LPI. Synthesis of ^{32}P-labeled LPIP from LPI and [γ^{32}P] ATP has been demonstrated using platelet microsomes (Thomas and Holub, 1987) and carrot plasma membranes (Wheeler and Boss, in preparation).

The carrot suspension culture cells used for these studies were from a fusogenic line of wild carrot cells, which readily incorporate inositol into phospholipids (Boss, 1987) and which yield protoplasts responsive to IP₃ in vivo (Rincon and Boss, 1987). While the high levels of LPIP found in the plasma membranes of the carrot cells correlate with the fusion potential of the protoplasts (Wheeler and Boss, in preparation), LPIP also has been found in the plasma membranes of nonfusogenic carrot cells and of soybean hypocotyls. The LPIP in nonfusogenic cells is predominantly the *sn* 2-linoleoyl analogue, whereas in the fusogenic cells it is the *sn* 1-palmitoyl analogue and has a slightly lower R_f in the ammonia chromatographic system (solvent A, Fig. 4; Wheeler and Boss, in preparation).

LPIP could play an important role in cellular physiology and signal transduction by directly enhancing cation transport. Hayashi et al. (1978) have shown that lysophosphoinositides will enhance K^+ transport in lipid bilayers and suggested that LPIP₂ facilitated monovalent cation transport in vivo by forming an ion-conducting channel.

Phosphoinositides and PI kinase activity have been reported to be associated with the nuclear membranes of animal cells (Smith and Wells, 1983;

Seyfred and Wells, 1984; Raval and Allan, 1985). When nuclei of [³H]inositol-labeled carrot cells were isolated according to Saxena et al. (1985), polyphosphoinositides were recovered (Hendrix and Boss, in preparation). A typical chromatograph is shown in Figure 6. In general, although the [³H]inositol incorporated per milligram of protein is relatively low (e.g., 250 cpm/mg protein compared with 20,000 cpm/mg protein for the plasma membrane), the percentage of PIP and PIP₂ is quite high, 5.0 and 3.6%, respectively (Table I). While the role of these lipids in the nuclear membrane is not completely understood, Smith and Wells (1983) demonstrated the PIP-enhanced nuclear envelope ATPase activity.

Another approach for determining the localization of the phosphoinositides is to study the enzymes involved in their synthesis and catabolism. Sandelius and Sommarin (1986) demonstrated that plasma membranes isolated from wheat seedlings were enriched in PI and PIP kinase activity. Kinase activity was greater in shoots than in roots. This means that if the phosphoinositides are involved in signal transduction in plants, the potential for cells to resynthesize the phosphoinositides and thus to receive and continually transmit stimulation may vary with organ type.

Careful analysis of the products of the kinase reaction, using [γ³²P]ATP and either endogenous substrates or exogenous PIP, indicated that phosphatidic acid (PA) was the primary radiolabeled product (Sandelius and Sommarin, 1986; Sommarin and Sandelius, 1988). These data suggested that there was a very active plasma membrane phospholipase C that hydrolyzed PIP to DAG and that DAG was subsequently phosphorylated by [γ³²P] ATP to PA or that DAG was already present in the plasma membrane.

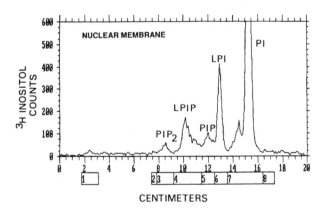

Fig. 6.   Thin-layer analysis of [³H]inositol-labeled lipids extracted from nuclei isolated from cells labeled for 22 hr with myo[2-³H]inositol. Legend as per Figure 4.

Sandelius and Sommarin went on to characterize the phospholipase C activity in the wheat seedlings (Melin et al., 1987). As with the kinase activity, the plasma membrane fraction was enriched in PIP and PIP_2 phospholipase C activity. Also, similar to the kinases, the specific activity of the polyphosphoinositide phospholipase C was greater in shoots than in roots (Table II). In addition, the specific activity of the plasma membrane enzyme from dark-grown shoots was higher than that from light-grown shoots. In contrast, the enzyme activity from the root plasma membrane was insensitive to the light regimen.

While phospholipase C activity was found in the cytosolic fraction, only the plasma membrane enzyme showed substrate specificity for PIP and PIP_2. The enzyme was activated by less than 1 μM calcium and inhibited by concentrations greater than 10 μM. At the lower calcium concentrations, PIP_2 was hydrolyzed more readily than PIP. In addition, at low lipid concentrations (less than 0.2 mM), PIP_2 was the preferred substrate (Melin et al., 1987). Thus, under normal resting conditions of low calcium and low substrate concentration, PIP_2 would be hydrolyzed preferentially, releasing DAG and IP_3 as second messengers. In contrast to animal cell phospholipase C, the enzyme in wheat was not activated by GTP or nonhydrolyzable analogues of GTP (Melin et al., 1987).

While low levels of calcium increased enzyme activity, increasing cytosolic calcium above 10 μM preferentially decreased hydrolysis of PIP_2. This is significant in that it indicates that the production of IP_3 would decrease as cytosolic levels of calcium increased above 10 μM. At the higher calcium concentrations, however, the other phosphoinositides should continue to be

TABLE II. **Phospholipase C Activity in Plasma Membranes Obtained From Shoots and Roots of Dark- and Light-Grown Wheat Seedlings**

Plant part	Phospholipase C activity (nmol/mg protein per min)		
	PIP_2	PIP	PI
Shoots Dark	147 ± 14 (6)	194 ± 14 (3)	10.4 ± 0.7 (3)
Light	100 ± 6 (3)	135 ± 6 (2)	9.9 (1)
Roots Dark	57.6 ± 7.8 (5)	79.5 ± 17.6 (3)	6.9 ± 1.7 (2)
Light	57.0 ± 7.0 (4)	85.3 ± 22.8 (3)	10.9 (1)

Each preparation was analyzed in triplicate. Results are given ± SD, or deviation of the mean when two preparations were analyzed. The number of preparations analyzed are indicated in parentheses. (From Melin et al., 1987, with permission.)
PIP_2, phosphatidylinositol-4,5-bisphosphate; PIP, phosphatidylinositol-4-monophosphate; PI, phosphatidylinositol.

metabolized, producing IP_2 and IP. For example, PI was hydrolyzed by plasma membrane phospholipase C, even at 1 mM calcium (Pfaffmann et al., 1987). When the plasma membrane PI-specific phospholipase C activity was characterized in bush bean and soybean (Pfaffman et al., 1987), the enzyme activity was stimulated by calcium with a broad optimum of 0.5 mM. As with animal PI-specific phospholipase C (Majerus et al., 1986), the bulk of the activity was soluble (cytosolic); however, the membrane-associated activity was specific to the plasma membrane. Phospholipase C activity, using PI as a substrate, also has been observed with cytosolic fractions of celery, cauliflower, and daffodils (Irvine et al., 1980) and with cytosolic and particulate fractions of lily pollen (Helsper et al., 1986a; 1987). PI phospholipase C activity was found to increase during pollen germination as the pollen tube began to elongate and membrane biosynthesis increased. As with the plasma membrane PI phospholipase C activity (Pfaffmann et al., 1987), the lily pollen enzyme also was stimulated by 1 mM calcium (Helsper et al., 1987).

In summary, the plasma membranes of plant cells are enriched in phosphoinositides and the enzymes involved in their biosynthesis and metabolism. Careful characterization of the plasma membrane phospholipase C indicates substrate specificity and regulation consistent with preferential hydrolysis of PIP_2 at low (1 µM) calcium. Furthermore, the presence of LPIP suggests that there is a phospholipase A_2 associated with the plasma membrane. In addition, other cellular membranes, especially the nuclear membrane, are potential sources of phosphoinositides.

INOSITOL PHOSPHATES AS SECOND MESSENGERS

If IP_3 is a second messenger in plants, it should mobilize intracellular calcium stores and cause an increase in cytosolic calcium. The first indication that inositol phosphates might affect calcium fluxes in plants was a report by Drøbak and Ferguson (1985). They showed a small transient change in $[^{45}Ca^{2+}]$ concentration when zucchini microsomes were treated with 20 µM IP_3. The in vitro response to IP_3 in plants was firmly established by Sze's laboratory using tonoplast vesicles (Schumaker and Sze, 1987). As with animal cell membranes, with the purified plant membranes the response was evident at low (1.5 µM) concentrations for IP_3, and no response was seen using IP_2 or IP at 10- and 100-fold higher concentrations, respectively. Unlike animal cells, however, the tonoplast or vacuolar membrane, rather than the ER, was the primary site for the IP_3-induced calcium efflux (Schumaker and Sze, 1987). These data have been confirmed by Ranjeva et al. and suggest that the vacuole might be a source of plant messengers (see Boudet and Ranjeva, Chapter 9).

If IP_3 is a second messenger, one should be able to add IP_3 to permeabilized cells and elicit a response without a stimulus. This has been demonstrated in pancreatic acinar cells, permeabilized with saponins (Streb et al., 1983). When IP_3 was added in the presence of $^{45}Ca^{2+}$, a transient net loss of $^{45}Ca^{2+}$ was observed. In the absence of exogenous calcium, IP_3 caused an increased $^{45}Ca^{2+}$ efflux. The efflux would be predicted if IP_3 had caused an increase in cytosolic calcium, which would have activated the plasma membrane calcium pumps causing a net efflux of calcium. A similar response was seen when protoplasts from fusogenic wild carrot cells were treated with IP_3 (Rincon and Boss, 1987). The response of the plant protoplasts was quite small compared with that of many animal cells. Only 17% of the accumulated calcium was lost vs. a 25–50% loss seen in many animal cells and EGTA rather than saponins was effective for permeabilizing the protoplasts.

The protoplasts used for the IP_3 studies were densely cytoplasmic and had only a few small vacuoles compared with the normally highly vacuolated plant cell. If the tonoplast is the primary source of calcium for the IP_3 response, one might expect a larger flux in calcium from more highly vacuolated protoplasts when IP_3 was added; however, the opposite was found. Vacuolated protoplasts did not respond to IP_3 (Rincon, unpublished results). The lack of response to IP_3 by the vacuolated carrot protoplasts could be due to differences in IP_3 metabolism (the vacuolated protoplasts may have higher phosphatase activity) or to a rapid reuptake of calcium into the vacuole rather than efflux from the cell. On the other hand, the lack of response may relate to the differences in calcium flux previously described for these two cell lines (Grimes and Boss, 1985). Reddy and Poovaiah (1987) were unable to detect a response to IP_3 with corn coleoptile protoplasts, which also are more vacuolated than the fusogenic carrot cells. Thus, while tonoplast vesicles and vacuoles have been shown to be sensitive to IP_3 in vitro, further studies are necessary to establish the mechanism of IP_3 response in vivo.

If IP_3 is a second messenger in plants, it should be present, albeit transiently, in detectable quantities. Separation of inositol phosphates by Dowex anion exchange chromatography is not an adequate method of analysis for plant inositol phosphates. As mentioned previously, plant cells metabolize [³H]inositol to [³H] compounds that co-elute with inositol phosphates when these techniques are used, (Coté et al., 1987; Rincon unpublished results, Fig. 3). In addition, glycerolphosphoinositides and nucleoside phosphates will co-elute with the inositol phosphates on the stepwise gradients. For these reasons, a method of high-pressure liquid chromatography (HPLC) analysis of plant inositol phosphates, which resolves the inositol phosphate from other [^{3}H] and [^{32}P] labeled compounds, was developed by Satter's laboratory (Coté et al., 1987). A less expensive and equally effective method of analysis is paper ionophoresis (Seiffert and Agranoff, 1965; Loewus, 1969; Rincon et al., unpublished results).

Curiously, there are few reports of IP$_3$ isolated from plant cells (Dillenschneider et al., 1986; Morse et al., 1987a; Ettlinger and Lehle, 1988). This may be due to the relatively low levels of PIP$_2$ and therefore IP$_3$ or to the fact that IP$_3$ is rapidly hydrolyzed by phosphatases. We have not been able to detect IP$_3$ in extracts from wild carrot cells (Rincon, unpublished results). Alternatively, other inositol phosphates may act as messengers in plants. For example, even though IP$_2$ did not affect the calcium transport of tonoplast vesicles (Schumaker and Sze, 1987), when microinjected into staminal hairs of *Setcreasea purpurea* IP$_2$ was equally as effective as IP$_3$ at inhibiting intercellular transport of carboxyfluorescein (Tucker, 1988). Whether or not this latter response involves an increase in cytosolic calcium remains to be shown.

More highly phosphorylated inositol phosphates (e.g., IP$_4$) also may be involved in regulating cytosolic calcium. Irvine and Moore (1986) suggest that IP$_3$ is phosphorylated to IP$_4$, which, in turn, activates calcium channels in the sea urchin eggs. IP$_6$ caused an efflux of calcium from carrot protoplasts; however, the metabolic fate of IP$_6$ was not followed in this system, and it was not determined whether IP$_6$ or its metabolites actually stimulated the calcium efflux (Rincon and Boss, 1987).

In liver and erythrocytes (Seyfred et al., 1984; Joseph and Williams, 1985; Downes et al., 1982), IP$_3$ phosphatase activity is associated with isolated plasma membranes. The working hypothesis for animal cells is that the ER or ER-like vesicles, which are sensitive to IP$_3$, are in close proximity to the plasma membrane, so that IP$_3$ will not be completely metabolized before reaching the ER receptor. While there have been careful studies of IP$_6$ and IP phosphatases in plants (Loewus and Loewus, 1983; Gumber et al., 1984), there is a dearth of information about IP$_3$ metabolism in plant cells. Considering the complexity of IP$_3$ metabolism in animal cells (Michell, 1986) and the potential for cyclic analogues as well as phosphorylation to IP$_4$, IP$_5$, and IP$_6$, it is necessary to determine precisely where and how IP$_3$ is metabolized in plant cells before the mechanism for the IP$_3$-stimulated calcium efflux can be delineated. Furthermore, to prove that IP$_3$ is involved in signal transduction in plants requires documentation of a transient increase in IP$_3$, which follows a given stimulus and precedes a transient increase in cytosolic calcium.

DIACYLGLYCEROL AS A SECOND MESSENGER

DAG is another product formed by phospholipase C hydrolysis of phospholipids. Several reports indicate that DAG or DAG analogues in concert with phosphatidylserine or phorbol esters will promote ATP-dependent protein phosphorylation in plants and *Neurospora* (Oláh and Kiss, 1986; Morré et al., 1984b; Elliott and Skinner, 1986; Schafer et al., 1985; Favre and Turian, 1987; Marmé, Chapter 3). Protein phosphorylation, which is phospholipid and calcium dependent, is considered to be due to a C-type protein kinase (Nishizuka, 1984). Unlike the C-type protein kinases of animals, the plant enzyme does not bind phorbol esters (Elliott and Skinner, 1986), suggesting that activation of protein phosphorylation by phorbol esters in plants may work by a mechanism different than that found for animal cells. Therefore, to demonstrate that a response involves protein kinase C activation, studies more rigorous than adding phorbol esters and observing a physiological response are necessary.

Other lipid activators of protein synthesis and proton pumps in plants are described both by Anderson (Chapter 8) and by Scherer (Chapter 7). There are no reports to date that DAG or DAG analogues stimulate protein kinase activity in vivo in plant cells. One could argue that since most plant cells have the ability to synthesize TAG from DAG, DAG is preferentially metabolized and does not activate protein phosphorylation in vivo or that the activation is very short-lived. Recent data (Morré et al., unpublished results) indicate that after exposure to 1 μM 2,4-dichlorophenoxyacetic acid, the DAG level of soybean hypocotyls does not change compared with controls. After cell fractionation, the DAG in the supernatant fraction was slightly increased, but the membrane fraction DAG was not. These findings are difficult to reconcile with postulated mechanisms for activation of a membrane-located protein kinase C. The increased DAG may simply reflect increased lipid turnover during auxin-induced growth or may reflect some as yet unidentified regulatory mechanism.

PHOSPHOINOSITIDE METABOLISM AND PLANT GROWTH

The first indication of changes in phospholipid metabolism in response to plant hormones was reported by Morré's laboratory in 1982 as a possible explanation for auxin-induced release of divalent cations from soybean membranes (Monroe et al., 1982; Buckhout et al., 1981; Morré et al., 1984a; see also Morré, Chapter 4). As with the first discovery of hormone-induced phosphatidylinositol metabolism in pancreatic cells (Hokin and Hokin, 1953), the significance of these data, for the most part, was to remain obscure until interest arose in the polyphosphoinositide signal transduction pathway. Also, like the animal response, some of the metabolic changes in phosphatidylinositol metabolism initially observed in plants have been shown to be associated not with the plasma membrane but with the endomembranes. For example, auxin will stimulate the inositol exchange reaction with the phosphatidylinositol found in ER membranes rather than that found in plasma membranes (Sandelius and Morré, 1987; Morré, Chapter 4).

Whether auxin or other plant growth regulators stimulate turnover of the polyphosphoinositides remains to be demonstrated. Zbell and Walter (1987) have shown that indoleacetic acid (IAA) affects the phosphorylation of membrane lipids. They used a plasma membrane-rich fraction isolated on Renografin gradients and preincubated with IAA 5 min. The IAA pretreatment caused a transient decrease in [^{32}P]-labeled phospholipids and a corresponding increase in water-soluble phosphates. While the data are suggestive of IAA-stimulated phosphoinositide metabolism, careful identification of the phospholipids and of the water-soluble products is needed before a specific effect on the polyphosphoinositide pathway can be considered to be demonstrated.

Similar problems with analysis have plagued in vivo studies of polyphosphoinositides. Ettlinger and Lehle (1988) reported that auxin caused a transient increase in IP$_2$ and IP$_3$ and a concomitant decrease followed by an increase in PIP and PIP$_2$ in *Catharanthus roseus* suspension culture cells, which had been arrested in G$_1$ phase and treated for 1 hr with 10 mM inositol. The purported IP$_2$ plus IP$_3$ recovered (391,256 dpm) far exceeded the amount of inositol-labeled PIP plus PIP$_2$ (223,725 dpm). This was true even though PIP$_2$ was reported to be 2.5% of the recovered inositol lipids. This value is more typical of LPIP, which was not identified on their chromatograms. Furthermore, if IP$_2$ and IP$_3$ were present in such large amounts and increased transiently 200–300% upon stimulation, one would expect to detect a large change in IP. This was not observed. These results may reflect some unique aspect of the cell cycle or culture conditions.

Elicitors produced by pathogens provide another type of stimulus for plant cells (see Boller, Chapter 10). To test whether phosphoinositide metabolism provides a link between the plasma membrane receptor for an elicitor and increased synthesis of phytoalexins, Strasser et al. (1986) used two cell cultures, from parsley and from soybeans, in which the amount and activity of the mRNA for phytoalexin had been shown to increase in response to elicitors. They found no significant change in phosphoinositides or inositol phosphates when elicitors were added. As they point out, the changes may have occurred in less than the 1–2 min sampling time that they used. The fact that no significant differences were observed in response to the elicitors suggests either that changes in phosphoinositide metabolism in response to elicitors were very rapid, with the cells returning to homeostasis within 1–2 min, or that polyphosphoinositide metabolism was not involved in the response to elicitors in these cells.

While elicitors may not alter phosphoinositide metabolism, the fungal enzymes used to digest cell walls, specifically Driselase, do alter plasma membrane protein phosphorylation. Treating cells for 1.5 min with Driselase resulted in changes in the calcium-dependent in vitro phosphorylation pattern of several plasma membrane proteins (Blowers et al., 1988). The response was not seen with boiled enzyme or with boiled "once-used" enzyme, which would contain cell wall fragments. Treatment of carrot cells with Driselase also caused a change in the phospholipids (Fig. 7). In both Driselase-treated and control cells, PIP_2 increased significantly by 2 min. After a 10 min treatment, the percentage of PIP_2 of the Driselase-treated cells was greater than the controls. This suggested either increased synthesis or decreased breakdown of PIP_2. There was no significant change in the PIP; however, LPI increased suggesting activation of a plasma membrane phospholipase A. Activation of plasma membrane phospholipase A by Driselase has been substantiated in in vitro assays using NBD-phosphatidylcholine as a substrate (Dengler, 1987). As with the work of Strasser et al. (1986), there was no indication from the carrot studies for activation of phospholipase C in response to fungal enzymes, fungal elicitors, or auxin.

The most promising evidence that phosphoinositides may play a role in signal transduction in plants comes from Dr. Satter's laboratory using *Samanea saman* pulvini which contain motor cells. In response to light, these cells generate ion gradients resulting in increased turgor in the cells on one side of the pulvinus. The cells enlarge, causing the leaflets to move either to an open or closed position, depending on whether or not the upper or lower cells of the pulvinus become turgid. Light can trigger the opening or closing of the leaflets in what is called a nastic or reversible movement. Morse et al. (1987b) found that after 30 sec of light, the IP_2 + IP_3/inositol-lipid recovered

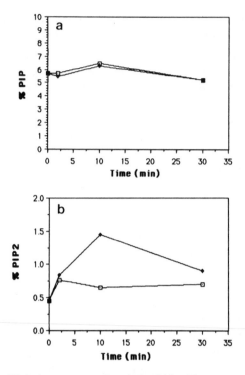

Fig. 7. The effects of Driselase on carrot phosphoinositides. The percentage of [³H]inositol-labeled PIP (a) and PIP₂ (b) extracted from cells placed in a 0.4 molal sorbitol osmoticum in a 2 mM MES buffer (□), pH 4.8 or cells placed in 0.4 sorbitol, 2 mM MES, pH 4.8 plus 2% (w/v) of the cell wall digestion enzymes, Driselase (♦). Cells were labeled for 18 hr with [³H]inositol prior to treatment.

increased to 30% of the control or resting value and the percentage of polyphosphoinositides decreased.

The *Samanea* pulvini are a good system to study signal transduction in that it is possible to give a single stimulus and observe a rapid response. Unfortunately, the pulvini are a mixture of cell types not all of which will respond to light by increased turgor. As Morse et al. (1987b) pointed out, chloroplasts, the site for fatty acid synthesis, also will respond to light. Indeed, since they reported inositol phosphates as a percentage of the inositol lipid recovered, the results would be influenced by a variation in the absolute amount of PI. Forthcoming experiments demonstrating a red/far-red light reversible response will define the role of inositol phosphates in light-induced signaling.

QUESTIONS AND CONTROVERSIES INVOLVING
PHOSPHOINOSITIDE METABOLISM

In order for PIP_2 to be involved in signal transduction according to the animal paradigm, hydrolysis of PIP_2 by phospholipase C followed, in turn, by an increase in IP_3 and DAG, must occur. These events should lead to a transient increase in cytosolic calcium and activation of protein kinase C. These criteria have not been met in a single plant system. In Figure 8 some alternative metabolic pathways for phosphoinositide metabolism in plants are suggested.

Many aspects of plant phosphoinositide metabolism are not well understood. For example, is there sufficient PIP_2 present in plants to produce sufficient IP_3 for stimulus–response coupling? While values for PIP_2 of $> 1\%$ of the total inositol lipid have been reported in unstimulated plants, if one carefully analyzes the data, in each instance the values can be attributed to the co-migration of LPIP with PIP_2. PIP_2 in plant cells is usually $< 1\%$ of the total inositol-labeled lipid recovered, and a 10% decrease in PIP_2 or increase in IP_3 would be near the limits of detection. By the same token, since the pool of PIP is relatively large (8- to 10-fold greater than PIP_2), a loss in PIP comparable to the PIP_2 pool also would be difficult to detect. These data reveal technical obstacles, and, more importantly, suggest a difference in regulation of phosphoinositide metabolism in plant cells relative to animal cells where the PIP to PIP_2 ratio can be on the order of 1 to 1 (Abdel-Latif et al., 1985).

Does a large pool of PIP afford plants a means of continually producing low levels of PIP_2 and therefore IP_3 and DAG in response to a constant

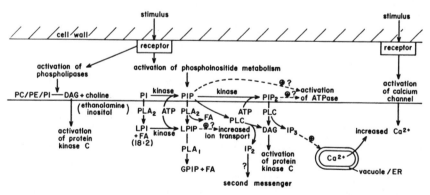

Fig. 8. A model depicting possible pathways for phosphoinositide metabolism involved in signal transduction in plant cells.

LPI, lysoPI; LPIP, lysoPIP; GPIP, glycerophosphoinositol phosphate; FA, fatty acid. Other abbreviations are as per Figure 2 legend.

stimulus? Berridge (1987) has suggested that continuous (several hours) stimulation and production of second messengers are needed to affect animal cell growth and development. This is in contrast to a short, rapid stimulus, which results in secretion. In plant cells, where growth responses usually involve irreversible cell expansion or cell division (the pulvinus being an exception), a continuous low level of second messenger may be required to alter growth. Such a mechanism would eliminate erratic growth in response to transient changes in environment. Predicted changes in the second messengers would be small but continuous, perhaps not unlike the pattern of sinusoidal changes in pH and potential measured in response to auxin (Felle, 1986; see also Felle, Chapter 6, and Morré, Chapter 4).

If plants do require continual pulses of a second messenger to alter growth and development, this may explain why phorbol esters, which continuously stimulate protein kinase C in animal cells, were so effective at mimicking cytokinin and stimulating bud development in moss protonema (Conrad and Hepler, 1986). In addition, in cells that respond rapidly but with nastic (reversible) growth such as the pulvini, a more pronounced and singular change in polyphosphoinositide metabolism would be predicted and has been found.

An alternative explanation for the large pool of PIP relative to PIP_2 is that PIP may be metabolized by other pathways. For example, PIP might be metabolized by phospholipase A_2-releasing fatty acid and LPIP. LPIP has been reported to be in the plasma membranes of plant cells. Does it act as a K^+ or H^+ channel or activate inositol phosphate phosphatase as reported for other lysolipids (Hayashi et al., 1978; Lim and Tate, 1971)? Does the released fatty acid act as a second messenger?

Reports of changes in PI and PIP kinase activity with growth conditions (Sommarin and Sandelius, 1988) are intriguing and suggest a mechanism for regulating phosphoinositide pools and thereby second messenger levels. Activation of phosphoinositide kinases could also lead to altered cell physiology. There is evidence for regulation of calcium transport in the sarcoplasmic reticulum by phosphorylation of PI to PIP and activation of the calcium ATPase (Varsanyi et al., 1983; Schafer et al., 1987). In these experiments, PIP was subsequently dephosphorylated to PI, and there was no evidence for involvement of phospholipases. In addition, PIP and PIP_2 have been shown to activate plasma membrane calcium ATPases of hepatocytes (Lin and Fain, 1984) and erythrocytes (Choquette et al., 1984) and nuclear membrane ATPases (Smith and Wells, 1983).

Another controversy that has arisen in plant phosphoinositide metabolism is the mechanism of action of IP_3. If the vacuole is the only site of IP_3-

sensitive calcium stores, how does IP$_3$ diffuse from the plasma membrane to the tonoplast without being metabolized? Are there IP$_3$ receptors on the tonoplast? How does IP$_3$ stimulate calcium efflux? Is there PIP present in tonoplast that would regulate calcium transport? What other metabolites of IP$_3$ occur in stimulated cells, and do any of these (particularly the more highly phosphorylated forms) affect cellular metabolism? For example, does IP$_4$ stimulate calcium transport? Specific inhibitors of inositol phosphate metabolism will aid in answering these questions.

Other questions concerning plant polyphosphoinositide metabolism need to be addressed. For example, is protein kinase C in plants activated by DAG in vivo? Is GTP involved in plant phosphoinositide metabolism in vivo? What is the role of the polyphosphoinositides in the nuclear membrane?

In summary, the phosphoinositides, the enzymes necessary to metabolize them, and a calcium-regulated amplification system responsive to IP$_3$, are present in plants. However, there are multifaceted pathways for inositol metabolism in plants, which are not prevalent in animal cells, and distinct differences in the relative amounts of the lipids and their sensitivity to stimuli and hormones are also evident. In addition, one should not overlook the potential for these negatively charged phospholipids to play an important role in cellular physiology by directly affecting membrane structure and activating membrane proteins. Clearly, the mechanisms by which the phosphoinositides regulate cellular metabolism remain to be established.

ACKNOWLEDGMENTS

Grants from the National Science Foundation, the U.S. Army Research Office, and the Department of Energy supported the author's research on phosphoinositide metabolism. I would like to thank Drs. Morré, Sandelius, Sommarin, and Tucker for providing manuscripts prior to publication.

REFERENCES

Abdel-Latif AA, JP Smith, RA Akhtar (1985): Polyphosphoinositides and muscarinic cholinergic and β-adrenergic receptors in the iris smooth muscle. In Bleasdale JE, J Eichberg, G Hauser (eds): "Inositol and Phosphoinositides: Metabolism and Regulation." Clifton, NJ: Humana Press, pp 275–298.

Akazawa T, T Mitsui (1985): Biosynthesis, intracellular transport, and secretion of amylase in rice seedlings. In Hill RD, L Munck (eds): "New Approaches to Research on Cereal Carbohydrates." Amsterdam: Elsevier Scientific, pp 129–137.

Asamizu T, A Nishi (1979): Biosynthesis of cell wall polysaccharides in cultured carrot cells. Planta 146:49–54.

Berridge MJ (1987): Inositol lipids and cell proliferation. Biochim Biophys Acta 907:33–45.

Berridge MJ, RMC Dawson, CP Downes, JP Heslop, RF Irvine (1983): Changes in the levels of inositol phosphates after agonist-dependent hydrolysis of membrane phosphoinositides. Biochem J 212:473–482.

Berridge MJ, RF Irvine (1984): Inositol trisphosphate, a novel second messenger in cellular signal transduction. Nature 312:315–321.

Blowers DP, WF Boss, AJ Trewavas (1988): Rapid changes in plasma membrane protein phosphorylation during initiation of cell wall digestion. Plant Physiol 86:505–509.

Boss WF (1987): Fusion permissive protoplasts: A plant system for studying cell fusion. In Sowers AE (ed): "Cell Fusion." New York: Plenum, pp 145–166.

Boss WF, MO Massel (1985): Polyphosphoinositides are present in plant tissue culture cells. Biochem Biophys Res Commun 132:1018–1023.

Brown EF, RP Newton (1981): Cyclic AMP and higher plants. Phytochem 20:2453–2463.

Buckhout TJ, KA Young, PS Low, DJ Morré (1981): In vitro promotion by auxins of divalent ion release from soybean membranes. Plant Physiol 68:512–515.

Burgoyne RD, TR Cheek, AJ O'Sullivan (1987): Receptor-activation of phospholipase A2 in cellular signalling. Trends Biol Sci 12:332–333.

Carter HE, JL Koob (1969): Sphingolipids in bean leaves (*Phaseolus vulgaris*). J Lipid Res 10:363–369.

Choquette D, G Hakim, AG Filotep, GA Plishker, JR Bostwick, JT Penniston (1984): Regulation of plasma membrane Ca^{2+} ATPases by lipids of the phosphatidylinositol cycle. Biochem Biophys Res Commun 125:908–915.

Cohen JD, RS Bandurski (1982): Chemistry and physiology of the bound auxins. Annu Rev Plant Physiol 33:403–430.

Conrad PA, PK Hepler (1986): The PI cycle and cytokinin-induced development in *Funaria*. Plant Physiol 80S:108.

Coté GG, MJ Morse, RC Crain, RL Satter (1987): Isolation of soluble metabolites of the phosphatidylinositol cycle from *Samanea saman*. Plant Cell Rep 6:352–355.

Dengler LA (1987): NBD-phosphatidylcholine: A tool to study endocytosis and phospholipase activity in carrot protoplasts. Master's thesis. North Carolina State University Raleigh, N.C.

Dillenschneider, A Hetherington, A Graziana, G Alibert, P Berta, J Haiech, R Ranjeva (1986): The formation of inositol phosphate derivatives by isolated membranes from *Acer pseudoplatanus* is stimulated by guanine nucleotides. Fed Eur Biochem Soc 208:413–417.

Downes CP, MC Mussat, RM Michell (1982): The inositol trisphosphate phosphomonoesterase of the human erythrocyte membrane. Biochem J 203:169–177.

Drøbak BK, IB Ferguson (1985): Release of Ca^{2+} from plant hypocotyl microsomes by inositol-1,4,5-trisphosphate. Biochem Biophys Res Commun 130:1241–1246.

Elliott DC, JD Skinner (1986): Calcium-dependent, phospholipid-activated protein kinase in plants. Phytochem 25:39–44.

Ettlinger C, L Lehle (1988): Auxin induces rapid changes in phosphatidylinositol metabolites. Nature 331:176–178.

Evans M (1985): Action of auxin on plant cell elongation. CRC Rev Plant Sci 2:317–367.

Fain JN, MJ Berridge (1979): Relationship between phosphatidylinositol synthesis and recovery of 5-hydroxytryptamine-responsive Ca^{++} flux in blowfly salivary glands. Biochem J 180:655–661.

Favre B, G Turian (1987): Identification of a calcium- and phospholipid-dependent protein kinase (protein kinase C) in *Neurospora crassa*. Plant Sci 49:15–21.

Felle H, B Brummer, A Bertl, RW Parish (1986): Indole-3-acetic acid and fusicoccin cause cytosolic acidification of corn coleoptile cells. Proc Natl Acad Sci USA 83:8992–8995.

Folch J, M Lees, GHS Stanley (1957): A simple method for the isolation and purification of total lipids from animal tissues. J Biol Chem 226:497–509.

Grimes HD, WF Boss (1985): Intracellular calcium and calmodulin involvement in protoplast fusion. Plant Physiol 79:253–258.

Grove RI, D Fitzpatrick, SD Schimmel (1981): Effect of Ca+ + on triphosphoinositide extraction in fusing myoblasts. Lipids 16:691–693.

Gumber SC, MW Loewus, FA Loewus (1984): Further studies on myo-inositol-1-phosphatase from the pollen of Lillium longiflorum Thunb. Plant Physiol 76:40–44.

Hallcher LM, WR Sherman (1980): The effects of lithium ion and other agents on the activity of myo-inositol-1-phosphatase from bovine brain. J Biol Chem 255:10896–10901.

Hanson JB (1984): The functions of calcium in plant nutrition. In Tinker PB, A Lauchli (eds): "Advances in Plant Nutrition," Vol 1. New York: Praeger pp 149–208.

Hayashi F, M Sokabe, M Takagi, K Hayashi, U Kishimoto (1978): Calcium-sensitive univalent cation channel formed by lysotriphosphoinositide in bilayer lipid membranes. Biochim Biophys Acta 510:305–315.

Heim S, KG Wagner (1986): Evidence of phosphorylated phosphatidylinositols in the growth cycle of suspension cultured plant cells. Biochem Biophys Res Commun 134:1175–1181.

Helsper JPFG, PFM De Groot, HF Linskens, JF Jackson (1986a): Phosphatidylinositol phospholipase C activity in pollen in Lilium longiflorum. Biochem J 25:2053–2055.

Helsper JPFG, PFM De Groot, HF Linskens, JF Jackson (1986b): Phosphatidylinositol monophosphate in Lilium pollen and turnover of phospholipid during pollen tube extension. Phytochem 25:2193–2199.

Helsper JPFG, JWM Heemskerk, JH Veerkamp (1987): Cytosolic and particulate phosphatidylinositol phospholipase C activities in pollen tubes of Lilium longiflorum. Physiol Planta 71:120–126.

Hepler PK, RO Wayne (1985): Calcium and plant development. Annu Rev Plant Physiol 36:397–439.

Hokin LE (1985): Receptors and phosphoinositide-generated second messengers Annu Rev Biochem 54:205–235.

Hokin MR, LE Hokin (1953): Enzyme secretion and the incorporation of ^{32}P into phospholipids of pancreas slices. J Biol Chem 203:967–977.

Irvine RF, AJ Letcher, RMC Dawson (1980): Phosphatidylinositol phosphodiesterase in higher plants. Biochem J 192:279–283.

Irvine RF, RM Moor (1986): Microinjection of inositol 1,3,4,5-tetrakisphosphate activates sea urchin eggs by a mechanism dependent on external Ca^{2+}. Biochem J 240:917–920.

Jolles J, KWA Wirtz, P Schotman, WH Gispen (1979): Pituitary hormones influenced polyphosphoinositide metabolism in rat brain. Fed Eur Biochem Soc Lett 105:110–114.

Jones RL, J Carbonell (1984): Regulation of the synthesis of barley aleurone amylase by gibberellic acid and calcium ions. Plant Physiol 76:213–218.

Joseph SK, RJ Williams (1985): Subcellular localization and some properties of the enzymes hydrolysing inositol polyphosphates in rat liver. Fed Eur Biochem Soc Lett 180:150–154.

Kaibuchi K, Y Takai, Y Nishizuka (1981): Cooperative roles of various membrane phospholipids in the activation of calcium-activated, phospholipid-dependent protein kinase. J Biol Chem 256:7146–7149.

Kauss H (1987): Some aspects of calcium-dependent regulation in plant metabolism. Annu Rev Physiol 38:47–72.

Leshem YY (1987): Membrane phospholipid catabolism and Ca^{2+} activity in control of senescence. Physiol Plantarum 69:551–559.

Lim PE, ME Tate (1971): The phytases. I. Lysolecithin-activated phytase from wheat brain. Biochim Biophys Acta 250:155–164.

Lin SH, JN Fain (1984): Calcium-magnesium ATPase in rat hepatocyte plasma membranes: Inhibition by vasopressin and purification of the enzyme. Prog Clin Biol Res 168:25–30.

Litosch I, C Wallis, JN Fain (1985): 5-Hydroxytryptamine stimulates inositol phosphate production in a cell-free system from blowfly salivary gland. J Biol Chem 260:5464–5471.

Loewus F (1969): Metabolism of inositol in higher plants. Ann NY Acad Sci 165:577–598.

Loewus FA, MW Loewus (1983): Myo-Inositol: Its biosynthesis and metabolism. Annu Rev Plant Physiol 34:137–161.

Maiti IB, FA Loewus (1978): Myo-Inositol metabolism in germinating wheat. Planta 142:55–60.

Majerus PW, TM Connolly, H Deckmyn, TS Ross, TE Bross, H Ishii, VS Bansal, DB Wilson (1986): The metabolism of phosphoinositide-derived messenger molecules. Science 234:1519–1526.

Marmé D, P Dieter (1983): Role of Ca^{2+} and calmodulin in plants. In Cheung WY (ed): "Calcium and Cell Function," Vol 4. New York: Academic Press, pp 263–311.

Melin PM, M Sommarin, AS Sandelius, B Jergil (1987): Identification of Ca^{2+} stimulated polyphosphoinositide phospholipase C in isolated plant plasma membranes. Fed Eur Biochem Soc Lett 223:87–91.

Michell RH (1975): Inositol phospholipids and cell surface receptor function. Biochim Biophys Acta 415:81–147.

Michell RH (1986): Inositol phosphates: Profusion and confusion. Nature 319:176–177.

Monroe A, B Gripshover, DJ Morré (1982): Labeled inositol turnover of soybean membranes stimulated by auxin in vitro. Plant Physiol 69:151.

Morré DJ, B Gripshover, A Monroe, JT Morré (1984a): Phosphatidylinositol turnover in isolated soybean membranes stimulated by the synthetic growth hormone 2,4-dichlorophenoxyacetic acid. J Biol Chem 259:15364–15368.

Morré DJ, JT Morré, RL Varnold (1984b): Phosphorylation of membrane-located proteins of soybean in vitro and response to auxin. Plant Physiol 75:265–268.

Morse MJ, RC Crain, RL Satter (1987a): Phosphatidylinositol cycle metabolites in *Samanea saman* pulvini. Plant Physiol 83:640–644.

Morse MJ, RC Crain, RL Satter (1987b): Light-stimulated phosphatidylinositol turnover in *Samanea saman* leaf pulvina. Proc Natl Acad Sci 84:7075–7078.

Mudd JB (1980): Phospholipid biosynthesis. In Stumpf PK, EE Conn (eds): "The Biochemistry of Plants: A Comprehensive Treatise," Vol 4. New York: Academic Press, pp 249–282.

Nishizuka Y (1984): Turnover of inositol phospholipids and signal transduction. Science 225:1365–1370.

Oläh Z, Z Kiss (1986): Occurrence of lipid and phorbol ester activated protein kinase in wheat cells. Fed Eur Biochem Soc Lett 195:33–37.

Oron Y, N Dascal, E Nadler, M Lupu (1985): Inositol 1,4,5-trisphosphate mimics muscarinic response in *Xenopus* oocytes. Nature 313:141–143.

Pfaffmann H, E Hartmann, AO Brightman, DJ Morré (1987): Phosphatidylinositol specific phospholipase C of plant stems: Membrane associated activity concentrated in plasma membranes. Plant Physiol 85:1151–1155.

Pickard GB (1985): Roles of hormones, protons, and calcium in geotropism. In Pharis RP, DM Reid (eds): "Encyclopedia of Plant Physiology," Vol 11. Berlin: Springer-Verlag, pp 193–281.

Poovaiah BW, ASN Reddy (1987): Calcium messenger system in plants. CRC Crit Rev Plant Sci 6:47–103.

Prentki M, TJ Biden, DD Janjic, RF Irvine, MJ Berridge, CB Wollheim (1984): Rapid mobilization of Ca^{2+} from rat insulinoma microsomes by inositol-1,4,5-trisphosphate. Nature 309:562–564.

Putney JW Jr (1986): Phosphoinositides and receptor mechanisms. In Venter JC, LC Harrison (eds): "Receptor Biochemistry and Methodology." Vol 7. New York: Alan R. Liss.

Raval PJ, D Allan (1985): Ca^{2+}-induced polyphosphoinositide breakdown due to phosphomonoesterase activity in chicken erythrocytes. Biochem J 231:179–183.

Reddy ASN, BW Poovaiah (1987): Inositol 1,4,5-trisphosphate induced calcium release from corn coleoptile microsomes. J Biochem 101:569–573.

Rincon M, WF Boss (1987): Myo-inositol trisphosphate mobilizes calcium from fusogenic carrot (Daucus carota L.) protoplasts. Plant Physiol 83:395–398.

Roux SJ, RD Slocum (1982): The role of calcium in mediating cellular functions important for growth and development in higher plants. In Cheung WY (ed): Calcium and Cell Function, Vol 3. New York: Academic Press, pp 409–453.

Roux SJ, RO Wayne, N Datta (1986): Role of calcium ions in phytochrome responses: An update. Physiol Plant 66:344–348.

Sandelius AS, DJ Morré (1987): Characteristics of a phosphatidylinositol exchange activity of soybean microsomes. Plant Physiol 84:1022–1027.

Sandelius AS, M Sommarin (1986): Phosphorylation of phosphatidylinositols in isolated plant membranes. Fed Eur Biochem Soc Lett 201:282–286.

Saxena PK, LC Fowke, J King (1985): An efficient procedure for isolation of nuclei from plant protoplasts. Protoplasma 128:184–189.

Schacht J (1978): Purification of polyphosphoinositides by chromatography on immobolized neomycin. J Lipid Res 19:1063–1067.

Schacht J (1981): Extraction and purification of polyphosphoinositides. Methods Enzymol 2:626–631.

Schafer M, G Behle, M Varsanyi, LMG Heilmeyer, Jr (1987): Ca^{2+} regulation of 1-(3-sn-phosphatidyl)-1D-myo-inositol 4-phosphate formation and hydrolysis on sacroplasmic-reticular Ca^{2+}-transport ATPase. Biochem J 247:579–587.

Schafer A, F Bygrave, S Matzenauer, D Marmé (1985): Identification of a calcium and phospholipid-dependent protein kinase in plant tissue. Fed Eur Biochem Soc Lett 187:25–28.

Schumaker KS, H Sze (1987): Inositol 1,4,5-trisphosphate releases Ca^{2+} from vacuolar membrane vesicles of oat roots. J Biol Chem 262:3944–3946.

Seiffert UB, BW Agranoff (1965): Isolation and separation of inositol phosphates from hydrolysates of rat tissues. Biochim Biophys Acta 98:574–581.

Sexton JC, TS Moore (1981): Phosphatidylinositol synthesis by an Mn^{2+}-dependent exchange enzyme in castor bean endosperm. Plant Physiol 68:18–22.

Seyfred MA, WW Wells (1984): Subcellular incorporation of ^{32}P into phosphoinositides and other phospholipids in isolated hepatocytes. J Biol Chem 259:7659–7665.

Seyfred MA, LE Farrell, WW Wells (1984): Characterization of D-myo-inositol 1,4,5-trisphosphate phosphatase in rat liver plasma membranes. J Biol Chem 259:13204–13208.

Smith CD, CC Cox, R Snyderman (1986): Receptor-coupled activation of phosphoinositide-specific phospholipase C by an N protein. Science 232:97-100.

Smith CD, WW Wells (1983): Phosphorylation of rat liver envelopes. II. Characterization of in vitro lipid phosphorylation. J Biol Chem 258:9368–9373.

Sommarin M, AS Sandelius (1988): Phosphatidylinositol and phosphatidylinositol phosphate kinases in plant plasma membranes. Biochim Biophys Acta 958:268–278.

Strasser H, C Hoffmann, H Grisebach, U Matern (1986): Are polyphosphoinositides involved in signal transduction of elicitor-induced phytoalexin synthesis in cultured plant cells? Z Naturforsch [C] 41:717–724.

56 / Boss

Streb H, RF Irvine, MJ Berridge, I Schulz (1983): Release of Ca^{++} from a non-mitochondrial intracellular store in pancreatic acinar cells by inositol-1,4,5-trisphosphate. Nature 306:67–69.

Thomas LM, BJ Holub (1987): The formation of lysophosphatidylinositol phosphate in human platelet microsomes. Lipids 22:144–147.

Trewavas AJ (1986): "Molecular and Cellular Aspects of Calcium in Plant Development." New York: Plenum.

Tucker EB (1988): Inositol bisphosphate and inositol trisphosphate inhibit cell-to-cell passage of carboxyfluorescein in staminal hairs of *Setcreasea purpurea*. Planta 174:358–363.

Varsanyi M, HG Tölle, LMG Heilmeyer, Jr, RMC Dawson, RF Irvine (1983): Activation of sarcoplasmic reticular Ca^{2+} transport ATPase by phosphorylation of an associated phosphatidylinositol. EMBO J 2:1543–1548.

Verma DC, DK Dougall (1979): Biosynthesis of myo-inositol and its role as a precursor of cell-wall polysaccharides in suspension cultures of wild-carrot cells. Planta 146:55–62.

Watson SP, RT McConnell, EG Lapetina (1984): The rapid formation of inositol phosphates in human platelets by thrombin is inhibited by prostacyclin. J Bio Chem 259:13199–13203.

Wheeler JJ, WF Boss (1987): Polyphosphoinositides are present in plasma membranes isolated from fusogenic carrot cells. Plant Physiol 85:389–392.

Wolniak SM, PK Hepler, WT Jackson (1983): Ionic changes in the mitotic apparatus at the metaphase/anaphase transition. J Cell Biol 96:169–173.

Yunghaus WN, DJ Morré (1977): Adenylate cyclase activity is not found in soybean hypocotyl and onion meristem. Plant Physiol 60:144–149.

Zbell B, C Walter (1987): About the search for the molecular action of high-affinity auxin-binding sites on membrane-localized rapid phosphoinositide metabolism in plant cells. In Klambt (ed): "Plant Hormone Receptors." Berlin: Springer-Verlag, pp 141–153.

Second Messengers in Plant Growth
and Development, pages 57–80
© 1989 Alan R. Liss, Inc.

3. The Role of Calcium and Calmodulin in Signal Transduction

Dieter Marmé

Institute of Biology III, University of Freiburg, 7800 Freiburg, Federal Republic of Germany

INTRODUCTION

Living organisms have established several mechanisms by which they convey extracellular signals into an intracellular language that can be understood by the biochemical machinery of the cell. One such mechanism has been discovered in animal cells about 20 years ago by Sutherland and associates (Robinson et al., 1971). It is called the second messenger principle, with cAMP as the intracellular second messenger.

Plant physiologists have tried to establish a similar role for cAMP in higher plants; they expected to gain insight into the molecular mechanism of the transduction of extracellular signals, such as (nonphotosynthetic) light, hormones, or gravity. There now is convincing evidence for the existence of cAMP in higher plants (Brown and Newton, 1981), but there is no physiological evidence for a role as second messenger similar to that in animal cells. In particular, no cAMP protein kinase has been demonstrated thus far (Salimath and Marmé, 1983). However, a large number of substrate peptides for the catalytic subunit of bovine heart cAMP protein kinase are present in plant extracts (Salimath and Marmé, 1983).

With the discovery of *calmodulin,* an intracellular Ca^{2+}-binding protein, by Cheung, Kakiuchi, and Wang (Dieter and Marmé, 1988), it became apparent that Ca^{2+} fulfills all the criteria for being a second messenger, as well. In plants, there are many physiological processes that had been reported to be under the control of Ca^{2+}: for example, cell elongation, cell

Address reprint requests to Dr. Dieter Marmé, Gödecke Research Institute, Mooswaldallee 1-9, 7800 Freiburg, Federal Republic of Germany.

division, protoplasmic streaming, enzyme secretion, hormone action, turgor regulation, phototactic and gravitropic movements, and enzyme activities (Roux and Slocum, 1982; Marmé and Dieter, 1983; Hepler and Wayne, 1985). However, in no example has the molecular mechanism of the Ca^{2+} regulation been completely elucidated. One reason was the difficulty to measure the cytoplasmic content of Ca^{2+}.

Only within the last few years some of the techniques that had been successfully used with animal cells have also been applied to plant cells (Williamson and Ashley, 1982; Gilroy et al., 1986; Bush and Jones, 1987). The other reason was the lack of evidence for the existence of calmodulin. It was in 1978 that Anderson and Cormier presented the first evidence for a heat-stable factor that could stimulate the calmodulin-deficient cAMP phosphodiesterase from bovine brain in a Ca^{2+}-dependent manner (Anderson and Cormier, 1978). Shortly after its discovery, this heat stable factor was identified as calmodulin (Charbonneau and Cormier, 1979). Since then, considerable progress has been made in identifying the sequential steps in plant signal transduction that involve Ca^{2+} as a second messenger.

This chapter reviews the current information regarding biochemistry, protein structure, and subcellular distribution of plant calmodulin and emphasizes the role of plant calmodulin in regulating cellular function. Furthermore, the available evidence for the various steps in signal transduction, as illustrated in Figure 1, will be discussed. A number of reviews on plant Ca^{2+} and calmodulin have appeared. The reader is referred to these for background information or discussion of topics such as the molecular biology of calmodulin or the role of Ca^{2+} in gravitropism and photomorphogenesis, which are beyond the scope of this chapter (Roux and Slocum, 1982; Marmé and Dieter, 1983; Hepler and Wayne, 1985; Marmé, 1983; Roberts et al., 1988; Marmé, 1985).

CALMODULIN

Since the discovery of troponin C and calmodulin as intracellular receptors of Ca^{2+} and the elucidation of their molecular mode of action, it became evident that cellular Ca^{2+}-binding proteins play an important role in conveying the Ca^{2+} message to various cellular functions. Over the last years several Ca^{2+}-binding proteins, in addition to troponin C and calmodulin, have been discovered (Dieter and Marmé, 1986). Thus far in plants calmodulin has been identified as the major Ca^{2+}-binding protein. In addition, two other Ca^{2+}-binding proteins have been identified: calcimedin in pea seedlings (Dauwalder et al., 1986) and the 63-kDa subunit of the quinate:NAD oxidoreductase (Graziana et al., 1984).

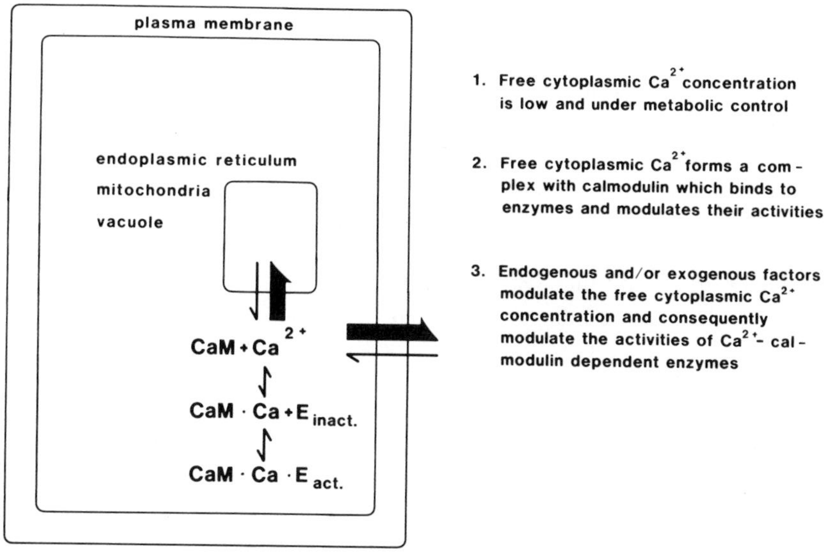

Fig. 1. Proposed scheme for the role of calcium and calmodulin in the regulation of cellular enzymes. The three statements summarize the basic requirements for Ca^{2+} as a second messenger in plant cells. CaM, calmodulin; $E_{inact.}$, inactive form of enzyme; $E_{act.}$, active form of enzyme.

Biochemistry

Calmodulin has been identified, purified, and characterized from a large variety of higher plants, including zucchini (Dieter and Marmé, 1980d), spinach (Watterson et al., 1980), peanut (Charbonneau and Cormier, 1979), barley (Schleicher et al., 1983; Grand et al., 1980), oat (Biro et al., 1984), and wheat (Toda et al., 1985). In addition, calmodulin has been purified from the green algae *Chlamydomonas* (Gitelman and Witman, 1980) and *Mougeotia* (Wagner et al., 1984), the fungi *Neurospora* (Ortega-Perez et al., 1981) and *Blastocladiella* (Gomes et al., 1979), and the yeasts *Candida* and *Saccharomyces* (Hubbard et al., 1982).

Purification of plant calmodulin was mostly achieved by a combination of several chromatography methods. After homogenization in an EDTA-containing buffer, the homogenate was cleared by centrifugation and either a DEAE-Sephacel column or batch ion-exchange step preceded the affinity chromatography on chlorpromazine- or fluphenazine-Sepharose 4B. In some instances further purification was achieved by semipreparative high-pressure liquid chromatography (HPLC) (Lukas et al., 1984).

A calcium-dependent change in electrophoretic mobility, which is typical for animal calmodulins has also been demonstrated for purified calmodulin from zucchini (Marmé and Dieter, 1983). Zucchini calmodulin moves further down into a 12.5% sodium dodecylsulfate polyacrylamide gel (SDS-PAGE) in the presence of Ca^{2+} compared with the absence of Ca^{2+}. The apparent molecular weight of zucchini calmodulin estimated by SDS-PAGE is 14,500 in the presence of Ca^{2+} compared with 16,000 for bovine brain calmodulin and 17,000 in the presence of EGTA compared with 19,000 for bovine brain calmodulin. The apparent molecular weights of two other plant and one algal calmodulin as estimated by SDS-PAGE in the presence of EGTA are 17,300 for peanuts (Anderson et al., 1980), 17,000–19,000 for spinach (Watterson et al., 1980), and 16,500 for *Mougeotia* (Wagner et al., 1984). Zucchini calmodulin does not appear as a single band on SDS-PAGE either in the presence or in the absence of calcium (Marmé and Dieter, 1983). Aside from the main band, there are other visible bands with an apparently higher molecular weight. Similar results for other plant calmodulins have been reported (Watterson et al., 1980; Anderson et al., 1980; Charbonneau et al., 1980).

Protein Structure

The amino acid compositions of various plant calmodulins are almost indistinguishable, and they are similar to those obtained for animal calmodulins (Marmé and Dieter, 1983). The similarities between bovine brain and the five plant calmodulins—peanut, spinach, barley, corn, and zucchini—include the presence of two proline residues and one trimethylysine residue, a high content of negatively charged amino acids, and a lack of tryptophan. Of interest is the presence of one cysteine residue in peanut (Anderson et al., 1980), zucchini (Görlach et al., 1985), spinach (Watterson et al., 1980), and wheat calmodulin (Toda et al., 1985), which has not been found in calmodulin from animals. Another significant difference is the high phenylalanine/tyrosine ratio. All plant calmodulins contain one tyrosine residue, whereas bovine brain calmodulin contains two such residues. The presence of only one tyrosine in plant calmodulin is also confirmed by the $E_{276}^{1\%}$ of 0.9 compared with 1.8 for bovine brain calmodulin (Anderson et al., 1980).

Recently, the amino acid sequences of calmodulin from spinach (Lukas et al., 1984), *Chalmydomonas* (Lukas et al., 1985), and wheat germ (Toda et al., 1985) have been reported. The amino acid sequence from spinach calmodulin differs from that of bovine brain calmodulin at 13 positions. Similar to other calmodulins, most of the sequence changes occur in the third and fourth structural domains. Interestingly, 12 of the 13 substitutions occur either within the Ca^{2+}-binding loops or in the flanking alpha-helical regions. Some substitutions are conservative and possess hydrophobic or charge properties similar to the amino acid residues at equivalent positions in other calmodulins. Of particular interest is the cysteine substitution for threonine-26, which occurs at a position that is a Ca^{2+}-binding ligand in the first domain (Roberts et al., 1988).

Calmodulin from the green alga *Chlamydomonas* differs from spinach calmodulin at 22 positions compared with only 19 positions with respect to vertebrate calmodulin. *Chlamydomonas* calmodulin possesses at some positions sequence characteristics of vertebrate and at other positions of plant calmodulin. Certain changes such as the histidine-81 substitution for serine or threonine, the glycine-148 substitution for lysine or arginine, and the carboxyl terminal amino acid extension by 11 amino acids represent significant alterations compared with other known calmodulin structures (Roberts et al., 1988). Wheat germ calmodulin is a peptide of 149 amino acids (Toda et al., 1985). A comparison with bovine brain calmodulin exhibits 11 amino acid substitutions other than amide assignments. Two are insertions and one is a deletion. From amino acid sequences we see that plant and animal calmodulins show a high degree of similarity and may be one of the most highly conserved proteins known. This also may imply that the molecular mechanisms of Ca^{2+}-mediated processes in plant cells may be very similar to those observed in animals. But it should be noted that plant-specific differences exist, for example, the light-dependent regulation of the (Ca^{2+} + Mg^{2+})ATPase and the species-dependent localization of the NAD kinase.

Plant calmodulin has been investigated for its Ca^{2+}-binding properties (Lukas et al., 1985). By gel filtration using ^{45}Ca^{2+} it could be shown that 3.3 mol of Ca^{2+} was bound for each mole of purified zucchini calmodulin. This is comparable to bovine brain calmodulin where, under similar conditions, the same value was obtained (Wolff et al., 1977). The binding constants for Ca^{2+} binding to plant calmodulin have been shown to be similar as those obtained for scallop calmodulin (Minowa and Yagi, 1984) and reported to be 0.255, 0.250, 0.130, and 0.165 μM^{-1}. The far- and near-UV circular dichroic spectra of the Ca^{2+}- and Mg^{2+}-saturated, as well as of the metal-free forms of zucchini calmodulin, reveal that upon Ca^{2+} binding the alpha-helix content increases (Dieter et al., 1980). A comparison with the spectra obtained with bovine brain calmodulin indicates that both calmodulins have similar secondary structures and similar Ca^{2+}-induced conformational changes.

Subcellular Localization

Analyses of the subcellular localization of calmodulin in various animal cells show that calmodulin is predominantly associated with the freely soluble fraction but also that variable amounts appear to be associated with subcellular fractions. Utilizing monospecific antibodies against native calmodulin, Brady et al. (1985) have localized calmodulin in cultured cells via indirect immunofluorescence. They show that in interphase cells the stain exhibits a more or less diffuse pattern, indicating that calmodulin is distributed throughout the cytoplasm. Mitotic cells, however, exhibit striking calmodulin fluorescence within the mitotic spindle apparatus. In metaphase the pole-to-kinetochore fibers of the mitotic spindle are stained and in the anaphase only the spindle poles are stained. In telophase calmodulin fluorescence is observed within the midbody. These results indicate that, while homogeneously distributed throughout the cytoplasm during interphase, calmodulin becomes highly concentrated within the spindle apparatus during mitosis.

Using rabbit antibodies to purified spinach calmodulin, Wick et al. (1985) have investigated the subcellular distribution of calmodulin at all stages of the cell cycle in onion and pea root meristematic cells, and they have compared this distribution with that of tubulin, using monoclonal antitubulin. They also show that calmodulin from prophase onward is associated with microtubular arrays of the mitotic spindle. At the time when spindle formation is underway, the nuclear surface is surrounded by anticalmodulin fluorescence. After nuclear envelope breakdown, most of the cell's reactive calmodulin is associated with microtubules of the spindle, as seen with double labeling by anticalmodulin and antitubulin. However, no association of calmodulin with microtubular arrays could be detected during interphase.

Localization of calmodulin in plant cellular organelles is not well documented. Most authors cannot rule out that the small amounts they find in mitochondria and chloroplasts are due to contamination by calmodulin from the cytoplasm. Muto (1982), by using the calmodulin-dependent NAD kinase or a radioimmunoassay to quantify calmodulin content, reported that 5–9% of the cellular calmodulin of wheat leaf is associated with mitochondria, and about 1–2% is associated with the chloroplasts. However, the criteria for the purity of the organelle fractions were poor, and contamination of the fractions could not be excluded. Simon et al. (1982) have calculated that, on a mg protein basis, the soluble cytoplasmic fraction from spinach contains about 2,000 times more calmodulin than the stromal fraction of spinach chloroplasts. Based on their calculations, they argue that the low amount of calmodulin detected in the stroma is due to contamination. Using a calmodulin radioimmunoassay, Biro et al. (1984) investigated a variety of particulate structures and purified organelles from etiolated oat seedlings. Immunoreactive material was found associated with mitochondria, etioplasts, and nuclei. However, contamination by cytoplasmic calmodulin cannot be excluded here. Matsumoto et al. (1983) reported on the existence of a heat-stable, calcium-dependent activator of NAD kinase and cAMP phosphodiesterase associated with chromatin from pea tissue.

CELLULAR Ca²⁺ TRANSPORT

The precise control of the free cytoplasmic Ca^{2+} concentrations is a prerequisite for the role of Ca^{2+} as a second messenger in animal cells. The low concentration of about 10^{-7}M free Ca^{2+} in the cytoplasm of unstimulated, resting cells is guaranteed by a calmodulin-dependent $(Ca^{2+} + Mg^{2+})$ATPase, which is located in the plasma membrane. The steep Ca^{2+} gradient across the plasma membrane, 1,000 μM outside compared with 0.1 μM inside has to be maintained against the Ca^{2+} leaking into the cytoplasm by unspecific permeation. The fact that this is achieved by the expenditure of less than 1% of the basal cellular energy utilization is mainly due to the very low permeability of the plasma membrane for Ca^{2+} (Borle, 1981) and because the Ca^{2+}-pumping $(Ca^{2+} + Mg^{2+})$ATPase is turned on by a $Ca^{2+}/$ calmodulin-dependent mechanism only when the free cytoplasmic Ca^{2+} concentration has reached a critical concentration of about 1 μM (Schatzmann, 1985).

Intracellular compartments and organelles also take part in the regulation of cytoplasmic Ca^{2+}: endoplasmic reticulum (ER), sarcoplasmic reticulum (SR), and mitochondria. The ER and SR have important functions in the physiological control of cellular Ca^{2+}-dependent processes, as it is obvious for the skeletal and cardiac muscle cells but also in nonmuscular cells. This became apparent when Streb et al. (1983) reported that inositol 1,4,5-trisphosphate, a receptor-mediated breakdown product of phosphatidylinositol 4,5-bisphosphate, caused a release from nonmitochondrial intracellular Ca^{2+} stores, presumably those in the ER. The mitochondria seem to function as a temporal buffer system for Ca^{2+} during cell stimulation (Rasmussen, 1985). Beside the receptor-mediated intracellular Ca^{2+} release by inositol 1,4,5-trisphosphate, another mechanism is predominantly involved in the exitation/response coupling: the voltage-gated Ca^{2+} channel (Reuter, 1983; Reuter, 1984).

For plant cells much less is known about the mechanisms that control excitation and those responsible for elimination of Ca^{2+} from the cytoplasm after cell stimulation. It appears at present that plants control the low Ca^{2+} level in the resting cell by pumping out the Ca^{2+} through the plasma membrane by means of a calmodulin-dependent $(Ca^{2+} + Mg^{2+})$ATPase and by sequestering Ca^{2+} into the ER, the vacuole, and the mitochondria. Cellular stimulation by light seems to be achieved by a slow, light-dependent change of the regulatory and kinetic properties of the $(Ca^{2+} + Mg^{2+})$ATPase and by a change of the sequestration into the mitochondria. There is also same evidence for the existence of Ca^{2+} channels (Andrejauskas et al., 1985).

Calmodulin-Stimulated $(Ca^{2+} + Mg^{2+})$ATPase

Gross and Marmé (1978) presented evidence for an active, ATP-dependent transport system in a microsomal fraction isolated from different plant species. Dieter and Marmé (1980d) showed that Ca^{2+} accumulation into a membrane vesicle fraction, most probably derived from the plasma membrane and having an inside-out orientation, could be stimulated by the addition of either purified plant or animal calmodulin. In subsequent work with corn it was demonstrated that the calmodulin-stimulated Ca^{2+} transport is mainly due to the calmodulin activation of a $(Ca^{2+} + Mg^{2+})$ATPase (Dieter and Marmé, 1981b). By affinity chromatography of a solubilized membrane fraction on calmodulin-Sepharose, it was possible to partially purify the $(Ca^{2+} + Mg^{2+})$ATPase and to demonstrate a threefold stimulation of the enzyme activity by calmodulin. The same preparation exhibited essentially one band on SDS-PAGE at about relative molecular weight (M_r) 130,000 (Dieter, unpublished), which compares with M_r 135,000 found for the $(Ca^{2+} + Mg^{2+})$ATPase purified by a similar procedure from bovine aortic smooth muscle plasma membrane (Furukawa and Nakamura, 1984).

It should be mentioned that the demonstration of membrane-bound Ca^{2+}/calmodulin-activated enzymes is sometimes difficult to assess because there may be contaminating calmodulin present in the membrane preparations. Therefore, extensive washing of the membranes in the absence of Ca^{2+} or, even better, with additional millimolar EDTA in the medium is required. The application of calmodulin antagonists, such as calmidazolium, trifluoperazine, or chlorpromazine, does not always reveal the participation of calmodulin in the activation of membrane-bound enzymes because these agents are highly lipophilic and can cause perturbations of the lipid bilayer and thus of the lipid microenvironment of the enzymes. Furthermore, the use of EDTA/EGTA buffers for Ca^{2+} has been shown to be absolutely necessary for micromolar range investigation of the Ca^{2+} dependence of Ca^{2+}/calmodulin-activated enzymes.

The calmodulin-dependent Ca^{2+} transport system prepared from dark-grown corn coleoptiles has been further investigated for its kinetic properties (Dieter and Marmé, 1983). It was shown that the addition of calmodulin caused a slight decrease of the apparent K_m and a twofold increase of V_{max}. When membrane vesicles were obtained from light-grown plants instead of from dark-grown plants, the calmodulin stimulation of both the Ca^{2+} transport and the $(Ca^{2+} + Mg^{2+})$ATPase was lost in the absence of calmodulin (Dieter and Marmé, 1981a). The apparent affinity for Ca^{2+} and the V_{max} of the transport were slightly decreased due to the irradiation, whereas addition of calmodulin to the membrane vesicles derived from light-treated plants did not cause an increase of the affinity for Ca^{2+} nor an enhancement of the V_{max} as observed for membrane vesicles prepared from dark-grown plants (Dieter and Marmé, 1983). The kinetics of Ca^{2+} accumulation were almost identical both in the presence and in the absence of calmodulin (Dieter and Marmé, 1981a). From these results, one may conclude that light treatment of dark-grown plants causes an increase of cytoplasmic Ca^{2+} due to the loss of the $(Ca^{2+} + Mg^{2+})$ATPase stimulation by calmodulin. In support of this hypothesis, increased uptake of $^{45}Ca^{2+}$ into intact plant segments has been observed when segments were incubated in the light instead of the dark (Marmé, unpublished).

In addition to light, a possible role of plant hormones in the regulation of calcium transport has been proposed by Oláh et al. (1983) based on their work with cytokinins. They found that growth of wheat plants in the presence of 6-benzylaminopurine, a synthetic cytokinin, resulted in a threefold enhancement of the activity of the $(Ca^{2+} + Mg^{2+})$ATPase from plasma-membrane-enriched microsomes. This indicates that Ca^{2+} extrusion might be increased when the plants were treated with this synthetic cytokinin. Furthermore, Kubowicz et al. (1982) demonstrated that a 2-hr treatment of excised segments from the elongation zone of soybean hypocotyls with growth-promoting concentrations of indole-3-acetic acid resulted in an increase of the Ca^{2+} accumulation of up to 100% in isolated membrane vesicles. Conversely, a similar pretreatment with zeatin was inhibitory. In the meristematic and maturing zones of the soybean hypocotyl, zeatin has the effect of promoting Ca^{2+} transport. These authors could not demonstrate any effect of the hormones on the Ca^{2+} transport system in vitro. Similarly, it could not be shown that light had any effect on Ca^{2+} transport in vitro (Dieter and Marmé, 1981a).

Ca^{2+} Transport Into the Endoplasmic Reticulum

Ca^{2+} transport into the endoplasmic reticulum would result in a decrease of cytoplasmic Ca^{2+}. In addition, fusion of the ER vesicles with the plasma membrane, as has been suggested by Griffing and Ray (1979), for ER vesicles loaded with cell wall material, would result in a further export of Ca^{2+} out of the cell. Gross (1982) provided the first evidence for an ATP-dependent Ca^{2+} accumulation in membrane vesicles derived from the ER. Through sucrose density gradient analysis, he showed that, on loading the vesicles with Ca^{2+} in the presence of oxalate, the Ca^{2+}-containing vesicles shifted, together with the NADH-cytochrome c reductase (a marker enzyme for plant ER), toward higher sucrose densities. Similar results have been obtained by Bush and Sze (1986). These authors presented evidence for a Ca^{2+} transport system isolated from cultured carrot cells which comigrated on a linear sucrose density gradient with the antimycin-A-insensitive NADH-cytochrome c reductase. This Ca^{2+} transport activity was inhibited by vanadate with an $IC_{50} = 12$ µM and was insensitive to the proton ionophore carbonyl-cyanide-m-chlorophenylhydrazone. The transport exhibits cooperative $Mg^{2+} - ATP$-dependent kinetics and has an apparent affinity for Ca^{2+} between 0.7 and 2 µM.

Ca^{2+} Transport Into the Vacuole

Plant cells contain large vacuoles that occupy up to 90% of the cell volume. These vacuoles contain large amounts of Ca^{2+}, often deposited as calcium oxalate. Only a few data are available on the mechanism(s) by which Ca^{2+} is taken up into and released from the vacuoles. However, because of the large amounts of Ca^{2+} present in this organelle, a regulatory function of the vacuoles may be anticipated. Gross (1982) presented the first evidence that vesicles derived from vacuoles of zucchini hypocotyls take up Ca^{2+} in the presence of ATP and oxalate. Bush and Sze (1986) also reported on a Mg^{2+}- and ATP-requiring Ca^{2+} transport system in membrane vesicles derived from isolated vacuoles originating from cultured carrot cells. Ca^{2+} transport was insensitive to vanadate but was strongly inhibited by the proton ionophore carbonyl-cyanide-m-chlorophenylhydrazone. The apparent K_m for Ca^{2+} was found to be about 20 µM. From these data and those reported for the plasma membrane, it may be concluded that in plant cells at least two different ATP-requiring Ca^{2+} transport systems exist: one type located in the plasma membrane and the ER, which operates via direct translocation of Ca^{2+} by means of a $(Ca^{2+} + Mg^{2+})$ATPase, and a second type located at the tonoplast and representing a Ca^{2+}/H^+ antiport. Evidence for two such different mechanisms has also been reported for pea (Rasi-

Caldogno et al., 1982) and for corn (Zocchi and Hanson, 1983). However, no attempt was made to determine the origin of either system (Zocchi and Hanson, 1983).

Recently the release of Ca^{2+} from isolated vacuoles by inositol-1,4,5 trisphosphate has been demonstrated (for further discussion, see Boudet and Ranjeva, Chapter 9). This finding implies a role of the vacuole in the regulation of the free, cytoplasmic Ca^{2+} concentration.

Ca^{2+} Transport Into the Mitochondria

Intact and well-coupled plant mitochondria take up large amounts of Ca^{2+} in the presence of phosphate (Hodges and Hanson, 1965). Oxidizable substrates such as succinate or ATP can be used as an energy source. The Ca^{2+} accumulated in the presence of phosphate can be readily released by addition of the divalent cationophore A23187 (Dieter and Marmé, 1980b). Na^{+}-induced Ca^{2+} release, as shown for rat heart mitochondria (Crompton et al., 1976), could not be demonstrated for corn mitochondria (Dieter and Marmé, 1980b). Plant mitochondrial Ca^{2+} accumulation cannot be enhanced by calmodulin (Dieter and Marmé, 1980c). It is interesting to note that the Ca^{2+} transport activity of corn mitochondria depended on the developmental stage of the corn seedlings (Dieter and Marmé, 1980b). When corn seedlings are transferred from darkness to (nonphotosynthetic) far-red light, a decrease of the Ca^{2+} transport activity can be observed. This effect has a slow onset and can be observed only after about 6-hr irradiation (Dieter and Marmé, 1981a).

Antimicrotubular herbicides and fungicides affect Ca^{2+} transport into plant (Hertel et al., 1980) and animal (Hertel et al., 1981) mitochondria. Hertel and Marmé (1983) showed that these substances interfere directly with the Ca^{2+} uptake mechanism. Electron transport and ADP phosphorylation are affected only at concentrations higher than those needed for Ca^{2+} transport inhibition. Based on these data, it is suggested that deregulation of the cytoplasmic Ca^{2+} concentration by the specific inhibition of mitochondrial Ca^{2+} accumulation may represent the mechanism of action for certain antimicrotubular or antimitotic substances.

The kinetic parameters of mitochondrial Ca^{2+} transport (Dieter and Marmé, 1983), possess a relatively low affinity, apparent K$_m$ (250 μM) and a relatively high V$_{max}$ (63 nmol × min^{-1} × mg^{-1}). Comparison with the corresponding values of the microsomal Ca^{2+} transport system shows that the V$_{max}$ of mitochondria is about 10–20 times higher but that the affinity is only about one-tenth that of microsomes. Therefore, it may be assumed that, similar to animal cells, the low Ca^{2+} concentration in an unstimulated plant cell is maintained by a plasma-membrane-located, calmodulin-dependent $(Ca^{2+} + Mg^{2+})$ATPase and/or by an ER-located $(Ca^{2+} + Mg^{2+})$ATPase. It may be further assumed that mitochondria may be involved in temporary Ca^+ buffering after massive cellular Ca^{2+} influx.

Ca^{2+}/CALMODULIN-STIMULATED ENZYMES

Aside from calmodulin-dependent $(Ca^{2+} + Mg^{2+})$ATPase, only a few more Ca^{2+}/calmodulin-regulated enzyme activities have been demonstrated: a soluble NAD kinase in a cytoplasmic fraction from zucchini (Dieter and Marmé, 1980a); a particulate NAD kinase at the outer mitochondrial membrane of corn (Dieter and Marmé, 1984a); a particulate NAD kinase located at the chloroplast envelope of light-grown peas (Simon et al., 1984); and a number of protein kinases, either soluble or membrane bound.

NAD Kinases

Oxidized and reduced nicotinamide adenine nucleotides play an important role as co-factors of many enzymatic reactions in all living organisms. Most of the pyridine nucleotide-dependent enzymes have a clear preference for NAD or NADP. Therefore, the NAD kinase (ATP: NAD 2'-phosphotransferase: EC 2.7.1.23), an enzyme that catalyzes the phosphorylation of NAD in the presence of ATP, is a favorite target for many regulatory signals.

Muto et al. (1981) have shown that light induces the conversion of NAD to NADP in intact leaves from wheat and peas. The same effect also was observed in mesophyll protoplasts of wheat leaves and in chloroplasts prepared from these protoplasts. They presented evidence that the NAD kinase is localized partly in the chloroplasts and partly in the cytoplasm. The relative amounts of the enzyme activity in both fractions changes from one plant to the other (e.g., in wheat 95% of the total NAD kinase activity is associated with the chloroplasts, whereas in peas 95% is found to be soluble and is most probably located in the cytoplasm). Tezuka and Yamamoto (1972) reported on the photoregulation of the NAD kinase in cell-free extracts from peas. However, the involvement of phytochrome, the photoreceptor for plant photomorphogenesis, could not be verified by others (Hopkins and Briggs, 1973).

Muto and Miyachi (1977) presented the first evidence for a protein factor involved in the activation of the plant NAD kinase. Attempts to purify the enzyme from crude extracts of peas on a DEAE-cellulose column always resulted in a great—sometimes total—loss of the enzyme activity. However, when they added the boiled crude extract to the fractions eluted from the ion-exchange column, the activity could be reconstituted. The reason for the loss of enzyme activity during DEAE-cellulose chromatography was the separation of the NAD kinase from a protein activator. The activator was found to exist in pea, spinach, Chinese cabbage, rice, corn, and *Chlorella*. In their discussion, the authors suggest that this protein activator is similar to calmodulin. The protein activator of the plant NAD kinase, which was reported originally by Muto and Miyachi (1977), has been identified and characterized by Anderson and Cormier (1978) and Anderson et al. (1980) as calmodulin.

Shortly after the discovery of the heat stable and highly acidic protein activator of plant NAD kinase, Anderson and Cormier (1978) showed that the activation occurred only in the presence of Ca^{2+}. The fact that Muto and Miyachi (1977) observed an activation without exogenously added Ca^{2+} can be explained by the presence of sufficient amounts of Ca^{2+} in their protein preparations and in their enzyme assay. Anderson and Cormier (1978) further demonstrated that a calmodulin-deficient cAMP phosphodiesterase from porcine brain could be stimulated by an activator protein from various plant sources. From these data, it was already obvious that animal and plant calmodulin have similar functional properties. NAD kinase has been highly purified from yeast (Apps, 1970) and from rat liver (Oka and Field, 1968), both sources of calmodulin. During the purification process, which included DEAE-cellulose chromatography, no loss of enzyme activity could be detected, indicating that yeast and liver do not have a calmodulin-dependent NAD kinase. The only other nonplant source for a calmodulin-dependent NAD kinase has been reported to be the sea urchin egg (Epel et al., 1981). After fertilization of the sea urchin egg, the intracellular Ca^{2+} concentration increases (Steinhardt et al., 1977) concomitantly with NAD kinase activity (Epel, 1964). Calmodulin offers a unique possibility for the transfer of information from Ca^{2+} to this NAD kinase.

The physiological implications of the calmodulin dependence of the plant NAD kinase have not yet been elucidated in detail. However, it is possible that in plant cells extracellular signals (e.g., light, hormones, and other physiologically active substances), which are known to influence a large number of cellular processes, make use of the second messenger Ca^{2+} to regulate an enzyme of great importance for the cellular metabolism. As a consequence of turning on or off such a central process, a large number of biochemical reactions could be affected.

It appears from the available data that NAD kinase activities are located in different cellular compartments, depending on the plant species.

Soluble NAD kinase was found in a cytoplasmic fraction obtained from dark-grown zucchini (Dieter and Marmé, 1980a). The enzyme has been purified partially by calmodulin-Sepharose affinity chromatography and could be stimulated about eightfold by calmodulin in a calcium-dependent manner (Fig. 2). The stimulation could be achieved by either bovine brain or zucchini calmodulin at concentrations for one-half saturation of about 0.1 μM. At optimal calmodulin concentrations, one-half saturation occurs at a calcium concentration of about 4 μM. The M_r of the NAD kinase from zucchini has been estimated by Sephadex G-100 chromatography to be about 50 kDa.

NAD kinase activity from dark-grown corn coleoptiles was shown to be almost totally dependent on Ca^{2+} and calmodulin. Nearly all of the enzyme activity was found in a particulate fraction (Dieter and Marmé, 1984a). On differential and density-gradient centrifugation, the NAD kinase activity co-migrated with the mitochondrial cytochrome c oxidase, whereas marker activities for nuclei, etioplasts, ER, and microbodies were separated well,

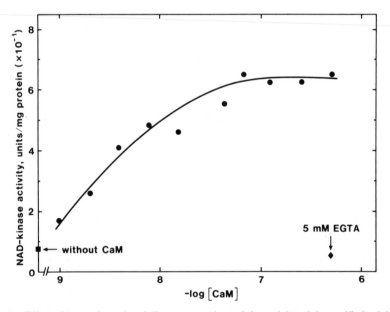

Fig. 2. Effect of increasing calmodulin concentrations of the activity of the purified soluble nicotinamide adenine dinucleotide (NAD) kinase from zucchini hypocotyls. The free Ca^{2+} concentration was 10 μM. Calmodulin (CaM) was purified to homogeneity from the same source.

indicating that the NAD kinase was associated with mitochondria. This NAD kinase, associated with intact mitochondria, was activated by exogenously added Ca^{2+} and calmodulin. To investigate the submitochondrial localization of the NAD kinase, the organelles were ruptured by osmotic treatment and sonication, and the submitochondrial fractions were separated again by density-gradient centrifugation. The NAD kinase activity exhibited the same density pattern as the antimycin-A-insensitive, NADH-dependent cytochrome c reductase, a marker enzyme of the outer mitochondrial membrane. Marker enzymes for the mitochondrial matrix and the inner mitochondrial membrane indicated that the Ca^{2+}/calmodulin-dependent NAD kinase from coleoptiles of dark-grown corn seedlings was located at the outer mitochondrial membrane.

In homogenates from light-grown pea seedlings more than one-half of a Ca^{2+}, calmodulin-dependent activity and most of a Ca^{2+}/calmodulin-independent activity of the homogenate were associated with chloroplasts (Simon et al., 1984). The Ca^{2+}/calmodulin-dependent activity could be detected by adding Ca^{2+} and calmodulin to the incubation medium containing purified, intact chloroplasts. This activity could not be separated from the chloroplasts by successive washes or by phase partition in aqueous two-polymer-phase systems. After chloroplast fractionation, the Ca^{2+}/calmodulin-dependent NAD kinase activity was localized at the envelope, and a Ca^{2+}/calmodulin-independent activity was recovered from the stroma.

In all three examples, the NAD kinase was able to sense changes of the free Ca^{2+} concentration in the cytoplasm, a prerequisite of the Ca^{2+} messenger concept. Furthermore, the NAD kinase was of special interest because of its light dependence. We showed that in corn coleoptiles, when exposed to far-red light, the NADP(H)/NAD(H) ratio increased (Dieter and Marmé, 1986). This reflected the activation of the NAD kinase, located in the outer mitochondrial membrane, by an increase of the free, cytoplasmic Ca^{2+} concentration due to the light inhibition of the Ca^{2+} transport activity (Dieter and Marmé, 1981a). The effect of far-red light was mimicked by incubating segments of corn coleoptiles at high or low Ca^{2+} concentrations in the presence of the divalent cationophore A23187 (Dieter and Marmé, 1984b). Similar data were obtained with segments from zucchini hypocotyls. These results suggested that the Ca^{2+}-dependent, regulatory mechanism was the same for the soluble, cytoplasmic NAD kinase in zucchini and the outer mitochondrial membrane-bound NAD kinase from corn.

Protein Kinases

Although protein kinases are well documented to occur in higher plants (Trewavas, 1976; Lin and Key, 1980; Ranjeva and Boudet, 1987; Blowers and Trewavas, Chapter 1), very little is known about the mechanisms by which their activities are controlled. Our laboratory failed to discover any protein kinase activity in plant homogenates that was stimulated by either cAMP or cGMP (Salimath and Marmé, 1983). However, addition of the catalytic subunit of the bovine heart cAMP protein kinase resulted in a pronounced phosphorylation of many plant proteins (Salimath and Marmé, 1983).

Ca^{2+}-dependent protein kinase activity had been shown to occur in membranes from pea shoots (Hetherington and Trewavas, 1982) and a $Ca^{2+}/$ calmodulin-dependent protein kinase activity could be demonstrated, together with its substrate peptides, in membranes from zucchini (Salimath and Marmé, 1983). Several Ca^{2+}-dependent and $Ca^{2+}/$calmodulin-dependent protein kinases have been partially purified from wheat embryo (Polya and Davies, 1982; Polya et al., 1983; Polya and Micucci, 1984). Schäfer et al. (1985) identified and partially purified a Ca^{2+} and phospholipid-dependent protein kinase from zucchini. This enzyme is of special interest because it may link the plant Ca^{2+} messenger system to another regulatory system that involves breakdown of polyphosphoinositides, as is known for the protein kinase C (Nishizuka, 1984). A similar stimulation by both phospholipid and Ca^{2+} has been reported also by Muto and Shimogawara to occur at the chloroplast envelope from spinach chloroplasts (Muto and Shimogawara, 1985). However, all these Ca^{2+}, $Ca^{2+}/$calmodulin-, and $Ca^{2+}/$phospholipid-dependent protein kinase activities, were investigated in homogenates from various plant tissues; and nothing is known about their properties, that is, the Ca^{2+}-dependent regulation in situ. The other important lack of information is that nothing is known about the biochemical or physiological function of the phosphorylated substrates.

There seems to be one exception: the quinate : NAD^+ oxidoreductase (QORase) kinase from light-grown carrot cells. QORase reversibly converts dehydroquinate into quinate, a by-product of the shikimate pathway, which accumulates at high concentrations in many plants (Boudet, 1973). Initial evidence that the enzyme is regulated by a phosphorylation/dephosphorylation mechanism came from work of Refeno et al. (1982). Furthermore, it was shown that activation of the inactivated QORase occurred only when, in addition to ATP, calmodulin and Ca^{2+} were present in the incubation medium (Ranjeva et al., 1983). By manipulating the cellular free Ca^{2+} concentration in light-grown carrot cells, Graziana et al. (1983b) were able to demonstrate that variations of cellular Ca^{2+} content resulted in changes of the degree of phosphorylation and activity of the QORase. These data are in agreement with the hypothesis that an increase of the free cytoplasmic Ca^{2+} content causes the activation of a $Ca^{2+}/$calmodulin-dependent protein kinase that phosphorylates the QORase and thus yields an activated enzyme. Further work with this promising system will be needed to evaluate the exact molecular mechanisms and to verify the above-postulated chain of events.

The modulation of the activity of QORase becomes more complex if the effects of light are considered. Graziana et al. (1983a; 1984) reported that transfer of carrot cell-suspension cultures from light to darkness considerably changes the regulatory and structural properties of the QORase. The enzyme prepared from dark-grown carrot cells became directly activatible by Ca^{2+}. This activation was reversible: addition of EGTA, sufficient to chelate the free Ca^{2+}, causes a decrease in QORase activity; subsequent addition of more Ca^{2+} again results in an activation of the enzyme. The activity could not be further increased by addition of calmodulin and was not inhibited when fluphenazine, a calmodulin antagonist, was added to the enzyme assay medium. The authors reported also that the apparent M_r shifted from 42,000 to 110,000 on transfer of the cell cultures from light to darkness. Analysis of the enzyme, purified to apparent homogeneity from dark-grown cells, revealed two subunits on SDS polyacrylamide gels. In addition to the 42,000 peptide, the only peptide obtained for the "light" enzyme, a new peptide of about 60,000 Mr appeared. This 60K peptide binds Ca^{2+} but has no catalytic properties, whereas the 42K protein is the catalytic subunit. It could further be shown that the 42K catalytic subunit is a phosphoprotein that is not susceptible to exogenous or endogenous phosphatases in a complex with the 60K Ca^{2+}-binding protein but could undergo phosphatase-catalyzed dephosphorylation after removal of the 60K subunit.

Based on these investigations, it was proposed that diurnal variations in the regulation of quinate : NAD oxidoreductase occurs. In light-grown cells it is thought that enzyme activity is modulated through reversible phosphorylation/dephosphorylation by a Ca^{2+}/calmodulin-dependent protein kinase and a phosphatase. It is suggested that, in dark-grown cells, the enzyme is maintained in an active state by the association of an additional subunit that prevents dephosphorylation of the catalytic subunit. This subunit also can activate the enzyme further in the presence of Ca^{2+}. Although certain aspects of this model remain to be substantiated through additional experimentation, it remains an attractive model for the study of Ca^{2+} and calmodulin regulation in plants.

STIMULUS–RESPONSE COUPLING

The importance of Ca^{2+} for many plant physiological processes is well established (Marmé, 1983). The presence of calmodulin and of calmodulin-dependent enzymes has been demonstrated. Several Ca^{2+} transport mechanisms have been identified. Less is known about the mechanisms by which primary signals (e.g., light, hormones) are translated into changes of the cytoplasmic free Ca^{2+} concentration. The effect of light and hormones on the microsomal $(Ca^{2+} + Mg^{2+})$ATPase has been described in some detail. In the case of the nonphotosynthetic light-dependent regulation of the NAD kinase in the cytoplasm of zucchini hypocotyl cells and in the outer mitochondrial membrane of corn coleoptile cells, the chain of events has been traced from the effect of light on the calmodulin-dependent $(Ca^{2+} + Mg^{2+})$ATPase and subsequent increase of cytosolic Ca^{2+} to the calmodulin-mediated NAD kinase activation.

Recently, Miller and Sanders (1987) have shown that Ca^{2+}-sensitive microelectrodes can be used to report photosynthetically related changes of the free cytoplasmic Ca^{2+} concentration in the characean alga *Nitellopsis*. They found that illumination caused a decrease of the cytoplasmic Ca^{2+} concentration from 350 nM to about 50 nM. They proposed that the light-induced depletion of Ca^{2+} from the cytoplasm constitutes a fundamental signal that causes the rate of extrachloroplastic metabolism to be geared to photosynthetic processes in the chloroplast.

Felle demonstrated, also using Ca^{2+}-sensitive microelectrodes, that the cytoplasmic Ca^{2+} concentration increased on the addition of auxin (Felle, Chapter 6). In both instances the underlying mechanisms are still unknown. However, these examples clearly demonstrate that, by using the appropriate techniques, signal-dependent changes of the free cytoplasmic Ca^{2+} concentration can be detected in intact plant cells.

Thus far we have only very preliminary evidence for the existence of Ca^{2+} channels in plant plasma membranes (Williamson and Ashley, 1982; Hayama et al., 1979; Kikuyama and Tazawa, 1983). However, in a few cases the effects of Ca^{2+} channel blockers such as verapamil have been demonstrated to cause specific inhibition of Ca^{2+}-dependent processes. Verapamil has been reported to inhibit cytokinin-stimulated bud formation in the moss *Funaria* (Saunders and Hepler, 1983), cytoplasmic streaming in the alga *Micrasterias* (Lektonen, 1984), and phototaxis in the alga *Chlamydomonas* (Nultsch et al., 1986). Andrejauskas et al. (1985) recently demonstrated specific binding of the Ca^{2+} channel blocker [^{3}H]verapamil to a membrane fraction from plants. Binding was saturable and reversible. Sucrose-density fractionation of the membrane preparations revealed that the specific verapamil binding sites are located primarily at the plasma membrane.

The same authors (Andrejauskas et al., 1986) found that 3,4,5-triiodobenzoic acid was able to enhance verapamil binding to plant membranes. 3,4,5-Triiodobenzoic acid is a transport inhibitor of the plant growth hormone auxin. They also showed that 3,4,5-triiodebenzoic acid enhanced verapamil binding to membranes from rabbit skeletal muscles; relaxed arterial smooth muscles contracted either by K$^+$ depolarization or by norepinephrine. It might be hypothesized that the effects demonstrated for 3,4,5-triiodobenzoic acid link together hormone transport and Ca^{2+} channels. This idea is further substantiated by the fact that the polar, basipetal auxin transport is accompanied by a polar, acropetal Ca^{2+} transport (DeGuzman and DelaFuente, 1984) and that this Ca^{2+} transport is abolished when the auxin transport is inhibited by triiodobenzoic acid. However, the precise mechanism is still unclear and needs further experimentation.

Thus it seems likely that plants possess mechanisms similar to those of animals to control the Ca^{2+} fluxes into and out of the cytoplasm and that the activities of these processes are regulated by extracellular stimuli such as light and hormones. Calmodulin exerts its function as a cellular Ca^{2+} receptor in plants to convey the information stored in the free cytoplasmic Ca^{2+} to the biochemical machinery of the cell in order to regulate plant growth and development.

REFERENCES

Anderson JM, H Charbonneau, HP Jones, RO McCann, MJ Cormier (1980): Characterization of the plant nicotinamide adenine dinucleotide kinase activator protein and its identification as calmodulin. Biochemistry 19:3113–3120.

Anderson JM, MJ Cormier (1978): Calcium-dependent regulator of NAD kinase. Biochem Biophys Res Commun 84:595–602.

Andrejauskas E, R Hertel, D Marmé (1985): Specific binding of the calcium antagonist [^{3}H]verapamil to membrane fractions from plants. J Biol Chem 260:5411–5414.

Andrejauskas E, R Hertel, D Marmé (1986): 3,4,5-Triiodobenzoic acid effects [^{3}H]verapamil binding to plant and animal membrane fractions and smooth muscle contraction. Biochem Biophys Res Commun 18:1269–1275.

Apps DK (1970): The NAD kinases of Saccharomyces cervesiae. Eur J Biochem 13:223–230.

Biro RL, S Daye, BS Serlin, ME Terry, N Datta, SK Sopory, SJ Roux (1984): Characterization of oat calmodulin and radioimmunoassay of its subcelluar distribution. Plant Physiol 75:382–386.

Borle AB (1981): Control, modulation and regulation of cell calcium. Rev Physiol Biochem Pharmacol 90:14–153.

Boudet AM (1973): Les acides quinique et shikimique chez les angiospermes arborescentes. Phytochem 12:363–370.

Brady RC, FR Cabral, MJ Schibler, JR Dedman (1985): Cellular localization of calmodulin and calmodulin-acceptor proteins. In Marmé D (ed): "Calcium and Cell Physiology." Heidelberg: Springer, pp 140–147.

Brown EG, RP Newton (1981): Cyclic AMP and higher plants. Phytochem 20:2453–2463.

Bush DS, RL Jones (1987): Measurement of cytoplasmic Ca^{2+} in aleurone protoplasts using indo-1 and fura-2. Cell Calcium 8:455–472.

Bush DR, H Sze (1986): Calcium transport in tonoplast and endoplasmic reticulum vesicles isolated from cultured carrot cells. Plant Physiol 80:549–555.

Charbonneau H, MJ Cormier (1979): Purification of plant calmodulin by fluphenazine-sepharose affinity chromatography. Biochem Biophys Res Commun 90:1039–1047.
Charbonneau H, HW Jarrett, RO McCann, MJ Cormier (1980): Calmodulin in plants and fungi. In Siegel FL, E Carafoli, RH Kretsinger, DH MacLennan, Wasserman RH (eds): "Calcium Binding Proteins: Structure and Function." New York: Elsevier/North-Holland: pp 155–164.
Crompton M, MI Capano, E Carafoli (1976): The sodium-induced efflux of calcium from rat heart mitochondria: A possible mechanism for the regulation of mitochondrial calcium. Eur J Biochem 69:453–462.
Dauwalder M, SJ Roux, LK Hardison, JR Redman (1986): Localization of calcimedin in pea seedlings by immunocytochemistry. J Cell Biol 103:453a.
DeGuzman CC, RK DelaFuente (1984): Polar flux in sunflower hypocotyl segments. Plant Physiol 76:347–357.
Dieter P, JA Cox, D Marmé (1980): Partial purification of plant NAD kinase by calmodulin-Sepharose affinity chromatography. Cell Calcium 1:279–286.
Dieter P, D Marmé (1980a): Calmodulin-activated plant microsomal Ca^{2+} uptake and purification of plant NAD kinase and other proteins by calmodulin-sepharose chromatography. Ann NY Acad Sci 356:371–373.
Dieter P, D Marmé (1980b): Ca^{2+} Transport in mitochondrial and microsomal fractions from higher plants. Planta 150:1–8.
Dieter P, D Marmé (1980c): Calmodulin activation of plant microsomal Ca^{2+} uptake. Proc Natl Acad Sci USA 77:7311–7314.
Dieter P, D Marmé (1981a): Far-red light irradiation of intact corn seedlings affects mitochondrial and calmodulin-dependent microsomal Ca^{2+} transport. Biochem Biophys Res Commun 101:749–755.
Dieter P, D Marmé (1981b): A Calmodulin-dependent, microsomal ATPase from corn (*Zea mays* L.). FEBS Lett 125:245–248.
Dieter P, D Marmé (1983): The effect of calmodulin and far-red light on the kinetic properties of the mitochondrial and microsomal calcium-ion transport system from corn. Planta 159:277–281.
Dieter P, D Marmé (1984a): A Ca^{2+}, calmodulin-dependent NAD kinase from corn is located in the outer mitochondrial membrane. J Biol Chem 259:184–189.
Dieter P, D Marmé (1984b): The role of calcium and calmodulin in higher plants. In Boudet AM, G Alibert, G Marigo, PJ Lea (eds): "Twenty-Fourth Annual Proceedings of the Phytochemical Society of Europe." Oxford: Oxford University Press, pp 213–229.
Dieter P, D Marmé (1986): NAD kinase in corn: Regulation by far-red light is mediated by Ca^{2+} and calmodulin. Plant Cell Physiol 27:1327–1333.
Dieter P, D Marmé (1988): The history of calcium binding proteins. In MP Thompson (ed): "Calcium Binding Proteins." Boca Raton, FL: CRC Press pp 1–9.
Epel D (1964): A primary metabolic change of fertilization: Interconversion of pyridine nucleotides. Biochem Biophys Res Commun 17:62–68.
Epel D, C Patton, RW Wallace, WY Cheung (1981): Calmodulin activates NAD kinase of sea urchin eggs: An early event of fertilization. Cell 23:543–549.
Furukawa KI, H Nakamura (1984): Characterization of the (Ca^{2+} + Mg^{2+})ATPase purified by calmodulin-affinity chromatography from bovine aortic smooth muscle. J Biochem 96:1343–1350.
Gilroy S, WA Hughes, AJ Trewavas (1986): The measurement of intracellular calcium levels in protoplasts from higher plant cells. FEBS Lett 199:217–221.
Gitelman SE, GB Witman (1980): Purification of calmodulin from *Chlamydomonas:* Calmodulin occurs in cell bodies and flagella. J Cell Biol 87:764–770.

Gomes SL, L Mennucci, JC Maia (1979): A calcium-dependent protein activator of mammalian cyclic nucleotide phosphodiesterase from *Blastocladiella emersonii*. FEBS Lett 99:39–42.

Görlach M, P Dieter, HH Seydewitz, C Kaiser, I Witt, D Marmé (1985): Characterisation of calmodulin from *Drosophila* heads. Biochim Biophys Acta 832:228–232.

Grand RJA, AC Nairn, SV Perry (1980): The preparation of calmodulins from barley (*Hordeum* sp.) and basidiomycete fungi. Biochem J 185:755–760.

Graziana A, M Dillenschneider, R Ranjeva (1984): A calcium-binding protein is a regulatory subunit of quinate:NAD⁺ oxidoreductase from dark-grown carrot cells. Biochem Biophys Res Commun 125:774–783.

Graziana A, R Ranjeva, AM Boudet (1983b): Provoked changes in cellular calcium controlled protein phosphorylation and activity of quinate:NAD⁺ oxidoreductase in carrot cells. FEBS Lett 156:325–328.

Graziana A, R Ranjeva, BP Salimath, AM Boudet (1983a): The reversible association of quinate:NAD⁺ oxidoreductase from carrot cells with a putative regulatory subunit depends on light conditions. FEBS Lett 163:306–310.

Griffing LR, PM Ray (1979): Dependence of cell wall secretion on calcium. Plant Physiol 63:51.

Gross J (1982): Oxalate-enhanced active calcium uptake in membrane functions from zucchini squash. In Marmé D, E Marré, R Hertel (eds): "Plasmalemma and Tonoplast: Their Functions in the Plant Cell." Amsterdam: Elsevier, pp 369–376.

Gross J, D Marmé (1978): ATP-dependent Ca²⁺ uptake into plant membrane vesicles. Proc Natl Acad Sci USA 75:1232–1236.

Hayama T, T Shimmen, M Tazawa (1979): Participation of Ca²⁺ in cessation of cytoplasmic streaming induced by membrane excitation in *Characeae* internodal cells. Protoplasma 99:305–321.

Hepler PK, WO Wayne (1985): Calcium and plant development. Annu Rev Plant Physiol 36:397–439.

Hertel C, D Marmé (1983): Herbicides and fungicides inhibit Ca²⁺ uptake by plant mitochondria: A possible mechanism of action. Pestic Biochem Physiol 19:282–290.

Hertel C, H Quader, DG Robinson, D Marmé (1980): Anti-microtubular herbicides and fungicides affect Ca²⁺ transport in plant mitochondria. Planta 149:336–340.

Hertel C, H Quader, DG Robinson, I Roos, E Carafoli, D Marmé (1981): Herbicides and fungicides stimulate Ca²⁺ efflux from rat liver mitochondria. FEBS Lett 127:37–39.

Hetherington A, A Trewavas (1982): Calcium-dependent protein kinase in pea shoot membranes. FEBS Lett 145:67–71.

Hodges TK, JB Hanson (1965): Calcium accumulation by maize mitochondria. Plant Physiol 40:101–108.

Hopkins DW, WR Briggs (1973): Phytochrome and NAD kinase: A re-examination. Plant Physiol 51:52.

Hubbard M, M Bradley, P Sullivan, M Shepherd, I Forrester (1982): Evidence for the occurrence of calmodulin in the yeasts *Candida albicans* and *Saccharomyces cerevisiae*. FEBS Lett 137:85–88.

Kikuyama M, M Tazawa (1983): Transient increase of intracellular Ca²⁺ during excitation of tonoplast-free *Chara* cells. Protoplasma 117:62–67.

Kubowicz BD, LN Vanderhoff, JB Hanson (1982): ATP-dependent calcium transport in plasmalemma preparations from soybean hypocotyls. Plant Physiol 69:187–191.

Lektonen J (1984): The significance for Ca²⁺ in the morphogenesis of *Micrasterias* studied with EGTA, verapamil, LaCi₃ and ionophore A 23187. Plant Sci Lett 33:53–60.

Lin PPC, JL Key (1980): Histone kinase from soybean hypocotyls: Purification, properties, and substrate specifities. Plant Physiol 66:360–367.

Lukas TJ, DB Iverson, M Schleicher, DM Watterson (1984): Structural characterization of a higher plant calmodulin. Plant Physiol 75:788–795.

Lukas TJ, ME Wiggins, DM Watterson (1985): Amino acid sequence of a novel calmodulin from the unicellular alga *Chlamydomonas*. Plant Physiol 78:477–483.

Marmé D (1983): Calcium transport and function. In Läuchli A, RL Bieleski (eds): "Inorganic Plant Nutrition: Encyclopedia of Plant Physiology," New Series, Vol 15. Heidelberg: Springer, pp 599–625.

Marmé D (1985): The role of calcium in the cellular regulation of plant metabolism. Physiol Veg 23:945–953.

Marmé D, P Dieter (1982): Calcium and calmodulin-dependent enzyme regulation in higher plants. In Marmé D, E Marrè, R Hertel (eds): Plasmalemma and Tonoplast: Their Functions in the Plant Cell. Amsterdam: Elsevier, North Holland, pp 111–118.

Marmé D, P Dieter (1983): Role of Ca^{2+} and calmodulin in plants. In Cheung WY (ed): "Calcium and Cell Functions," Vol 4. New York: Academic Press, pp 263–311.

Matsumoto H, M Tanigawa, T Yamaya (1983): Calmodulin-like activity associated with chromatin from pea buds. Plant Cell Physiol 24:593–602.

Miller AJ, D Sanders (1987): Depletion of cytosolic free calcium induced by photosynthesis. Nature 326:397–400.

Minowa O, K Yagi (1984): Calcium binding to tryptic fragments of calmodulin. J Biochem 96:1175–1182.

Muto S (1982): Distribution of calmodulin within wheat leaf cells. FEBS Lett 147:161–164.

Muto S, S Miyachi (1977): Properties of a protein activator of NAD kinase from plants. Plant Physiol 59:55–60.

Muto S, S Miyachi, H Usuda, GE Edwards, JA Bassham (1981): Light-induced conversion of nicotinamide adenine dinucleotide to nicotinamide adenine dinucleotide phosphate in higher plant leaves. Plant Physiol 68:324–328.

Muto S, K Shimogawara (1985): Calcium-and phospholipid-dependent phosphorylation of ribulose-1,5-bisphosphate carboxylase/-oxygenase small subunit by a chloroplast envelope-bound protein kinase in situ. FEBS Lett 193:88–92.

Nishizuka Y (1984): The role of protein kinase C in cell surface signal transduction and tumor promotion. Nature 308:693–698.

Nultsch W, J Pfau, R Dolle (1986): Effects of calcium channel blockers on phototaxis and motility of *Chlamydomonas reinhardtii*. Arch Microbiol 144:393–397.

Oka H, JB Field (1968): Inhibition of rat liver nicotinamide adenine dinucleotide kinase by reduced nicotinamide dinucleotide phosphate. J Biol Chem 242:815–819.

Oláh Z, A Bérczi, L Erdei (1983): Benzylaminopurine-induced coupling between calmodulin and Ca-ATPase in wheat root microsomal membranes. FEBS Lett 154:395–399.

Ortega-Perez R, D Van Tuinen, D Marmé, JA Cox, G Turian (1981): Purification and identification of calmodulin from *Neurospora crassa*. FEBS Lett 133:205–208.

Polya GM, JR Davies (1982): Resolution of Ca^{2+}-calmodulin-activated protein kinase from wheat germ. FEBS Lett 150:167–171.

Polya GM, JR Davies, V Micucci (1983): Properties of a calmodulin activated Ca^{2+}-dependent protein kinase from wheat germ. Biochim Biophys Acta 761:1–12.

Polya GM, V Micucci (1984): Partial purification and properties of a second calmodulin-activated Ca^{2+}-dependent protein kinase from wheat germ. Biochim Biophys Acta 785:68–74.

Ranjeva R, AM Boudet (1987): Phosphorylation of proteins in plants: Regulatory effects and potential involvement in stimulus-response coupling. Annu Rev Plant Physiol 38:73–93.

Ranjeva R, G Refeno, AM Boudet, D Marmé (1983): Activation of plant quinate:NAD^+ 3-oxidoreductase by Ca^{2+} and calmodulin. Proc Natl Acad Sci USA 80:5222–5224.

Rasi-Caldogno F, MI DeMichelis, MC Pugliarello (1982): Active transport of Ca^{2+} in membrane vesicles from pea: Evidence for a H^+/Ca^{2+} antiport. Biochim Biophys Acta 693:287–295.

Rasmussen H, W Zawalich, I Kojima (1985): Ca^{2+} and cAMP in the regulation of cell function. In Marmé D (ed): "Calcium and Cell Physiology." Heidelberg: Springer, pp 1–18.

Refeno G, R Ranjeva, AM Boudet (1982): Modulation of quinate:NAD^+ oxidoreductase activity through reversible phosphorylation in carrot cells suspensions. Planta 154:193–198.

Reuter H (1983): Calcium channel modulation by neurotransmitters, enzymes and drugs. Nature 301:569–574.

Reuter H (1984): Ion channels in cardiac cell membranes. Annu Rev Physiol 46:473–484.

Roberts DM, TJ Lukas, DM Watterson (1988): "Structure, Function and Mechanism of Action of Calmodulin." Boca Raton, FL: CRC Press (in press).

Robinson GA, RW Butcher, EW Sutherland (1971): "Cyclic AMP." New York: Academic Press.

Roux SJ, RD Slocum (1982): Role of calcium in mediating cellular functions important for growth and development in higher plants. In Cheung WY (ed): "Calcium and Cell Function," Vol 3. New York: Academic Press, pp 409–453.

Salimath BP, D Marmé (1983): Protein phosphorylation and its regulation by calcium and calmodulin in membrane fractions from zucchini hypocotyls. Planta 158:560–568.

Saunders MJ, PK Hepler (1983): Calcium antagonists and calmodulin inhibitors block cytokinin-induced bud formation in Funaria. Dev Biol 99:41–49.

Schäfer A, F Bygrave, S Matzenauer, D Marmé (1985): Identification of a calcium- and phospholipid-dependent protein kinase in plant tissue. FEBS Lett 187:25–28.

Schatzmann HJ (1985): Calcium extrusion across the plasma membrane by the calcium-pump and the $Ca^{2+}-Na^+$ exchange system. In Marmé D (ed): "Calcium and Cell Physiology." Heidelberg: Springer, pp 19–52.

Schleicher M, TJ Lukas, DM Watterson (1983): Further characterization of calmodulin from the monocotyledon barley (Hordeum vulgare). Plant Physiol 73:666–670.

Simon P, M Bonzon, H Greppin, D Marmé (1984): Subchloroplastic localization of NAD kinase activity: Evidence for Ca^{2+}, calmodulin-dependent activity at the envelope and for a Ca^{2+}, calmodulin-independent activity in the stroma of pea chloroplasts. FEBS Lett 167:332–338.

Simon P, P Dieter, M Bonzon, H Greppin, D Marmé (1982): Calmodulin-dependent and independent NAD kinase activities from cytoplasmic and chloroplastic fractions of spinach (Spinacia oleracea L.) Plant Cell Rep 1:119–222.

Steinhardt RA, R Zucker, F Schatten (1977): Intracellular calcium release at fertilization in the sea urchin egg. Dev Biol 58:185–196.

Streb H, RF Irvine, MJ Berridge, I Schulz (1983): Release of Ca^{2+} from a nonmitochondrial intracellular store in pancreatic acinar cells by inositol-1,4,5-trisphosphate. Nature 306:67–69.

Tezuka T, Y Yamamoto (1972): Photoregulation of nicotinamide adenine dinucleotide kinase activity in cell-free extracts. Plant Physiol 50:458–462.

Trewavas A (1976): Post-translational modification of proteins by phosphorylation. Annu Rev Plant Physiol 27:349–374.

Toda H, M Yazawa, F Sakiyama, K Yagi (1985): Amino acid sequence of calmodulin from wheat germ. J Biochem 98:833–842.

Wagner G, P Valentin, P Dieter, D Marmé (1984): Identification of calmodulin in the green alga Mougeotia and its possible function in chloroplast reorientational movement. Planta 162:62–67.

Watterson DM, DB Iverson, LJ Van Eldik (1980): Spinach calmodulin: Isolation, characterization, and comparison with vertebrate calmodulins. Biochemistry 19:5762–5768.

Wick SM, S Muto, J Duniec (1985): Double immunofluorescence labeling of calmodulin and tubulin in dividing plant cells. Protoplasma 126:198–206.

Williamson RE, CC Ashley (1982): Free Ca^{2+} and cytoplasmic streaming in the alga *Chara*. Nature 296:647–651.

Wolff DJ, PG Pairier, CO Brostrom, MA Brostrom (1977): Divalent cation binding properties of bovine brain Ca^{2+}-dependent regulator protein. J Biol Chem 252:4108–4117.

Zocchi G, JB Hanson (1983): Calcium transport and ATPase activity in a microsomal vesicle fraction from corn roots. Plant Cell Environ 6:203–210.

Second Messengers in Plant Growth
and Development, pages 81–114
© 1989 Alan R. Liss, Inc.

4. Stimulus–Response Coupling in Auxin Regulation of Plant Cell Elongation

D. James Morré

Department of Medicinal Chemistry and Pharmacognosy, Purdue University, West Lafayette, Indiana 47907

INTRODUCTION

Controlled growth is essential to the normal development of all plant, animal, and fungal cells. Growth is subject to regulation by hormones, growth factors, and a variety of environmental conditions, but mechanistic details are poorly understood. Even in animal cells where the molecular biology of growth factors, growth factor receptors, oncogenes, and other growth control elements has advanced (Neumann, 1985; Nishizuka, 1984; Berridge and Irvine, 1984), detailed mechanisms whereby the normal function of these growth control elements are translated into actual growth are still missing from our information.

Growth consists of two basic processes: an initial increase in cell number and a subsequent increase in cell size. In mammalian cells, these two processes are much more difficult to separate than in plants, where in growing stems and roots, for example, zones of cell division and of cell elongation are spatially separated and readily available for independent investigation.

The elongation growth in plants, which is the subject of this chapter, is the growth that results exclusively from an increase in cell size. Small increases in cell number may accompany cell elongation, but without expansion growth these increases do not in themselves give rise to appreciable increases in tissue volume. Because the expanding cell wall leaves a permanent

Address reprint requests to Dr. D. James Morré, Department of Medicinal Chemistry and Pharmacognosy, Purdue University, West Lafayette, IN 47907.

record of cell elongation, plants offer an experimental advantage over mammalian cells in that irreversible changes in cell dimensions can be readily monitored and recorded.

Numerous factors and conditions are known to modulate the rate and degree of polarity of plant expansion growth. Experimentally, growth regulators of the auxin type (Table I) have been used classically to modulate expansion rates of excised tissue segments floated on solutions. Most responsive have been segments taken from shoots (hypocotyl or epicotyl) of etiolated seedlings of dicots or the modified seed leaf (coleoptile) of grasses.

The auxins include a natural plant hormone, indole-3-acetic acid (IAA), plus numerous synthetic analogues such as 2,4-dichlorophenoxyacetic acid (2,4-D) and α-naphthyleneacetic acid (α-NAA). IAA and 2,4-D, at least, appear to regulate growth by similar, if not identical, modes of action. This chapter will be restricted to an analysis of messenger formation and function in the signal–response cascade involved in the control of expansion growth by regulators of the auxin type.

BASIC ELEMENTS OF THE SIGNAL–RESPONSE CASCADE INITIATED BY AUXIN

Effector (hormone)-responsive systems usually are considered to include at least three essential elements: 1) a receptor site to recognize and bind the extracellular signal (hormone); 2) some form of transduction mechanism

TABLE I. Effect of Auxins and Nonauxins on Membrane Thickness*

Treatment	No. of determinations	Membrane thickness (Å)
None	10	109 ± 4
Water	20	105 ± 4
IAA	10	93 ± 4
2,4-D	4	93 ± 6
α-NAA	3	93 ± 4
Picloram	3	97 ± 5
Benzoic acid	3	104 ± 2
2,3-D	3	103 ± 2
2,5-D	3	103 ± 2

* Auxins and nonauxins were added to isolated membrane vesicles at a final concentration of 1 μM and pH 6.0. Incubations were for 20 min. Measurements were on vesicles contrasted by reaction with phosphotungstic acid at low pH (Roland et al., 1972) to identify the reactive vesicles as plasma membrane. (Reproduced from Morré and Bracker, 1976, with permission of the publisher.)
IAA, indole-3-acetic acid; 2,4-D, 2,4-dichlorophenoxyacetic acid; 2,3-D, 2,3-dichlorophenoxyacetic acid; 2,5-D, 2,5-dichlorophenoxyacetic acid; α-NAA, α-naphthyleneacetic acid.

that recognizes a change in configuration or conformation of the receptor; and 3) an amplifier that translates the received message into an increase (or decrease) in the intracellular concentration of a chemical species capable of exerting control over intracellular processes (Michell et al., 1979).

Auxins elicit very rapid responses (within minutes or seconds) as well as longer-term responses requiring new macromolecular synthesis. In the sections that follow, each of these various considerations will be analyzed in detail and the evidence that serves to establish the potential operation of a signal–response cascade for auxin regulation of elongation growth in plants will be reviewed.

The Auxin Receptor

There is now general agreement that membrane fragments, including plasma membrane, bind the auxin regulators in a manner that is both reversible and specific. Additionally, cytoplasmic (or even nuclear) receptors, which might function to mediate the inductive responses involving macromolecular synthesis, have been sought (Venis, 1986).

The binding to membranes of radioactive IAA or NAA is saturable, reversible, and reaches equilibrium rapidly even at $0°C$ (Hertel et al., 1972). Bound radioactive auxins (e.g., $[^{14}C]$ NAA) were displaced by unlabeled auxins (IAA or 2,4-D). Average dissociation constants (K_d) of $1–2 \times 10^{-6}$ and $3–4 \times 10^{-6}$ M were estimated for NAA and IAA, respectively. High- and low-affinity binding sites were described subsequently (Batt et al., 1976; Ray et al., 1977a), with a K_d for high-affinity binding of NAA in the range $1–7 \times 10^{-7}$ M.

A variety of different analyses demonstrate that auxin-binding sites are located on more than one subcellular component. While an early correlation between plasma membrane content and auxin binding to soybean fractions was obtained (Fig. 1), other fractions bound disproportionally more auxin than expected on the basis of their plasma membrane contents, suggesting binding to membrane components other than plasma membrane.

A second major site of auxin-binding activity has been suggested to be endoplasmic reticulum (ER) (Ray, 1977; Dohrmann et al., 1978; Batt and Venis, 1976). Other possible locations include Golgi apparatus (Batt and Venis, 1976; Williamson et al., 1977) and tonoplast (Dohrmann et al., 1978). Mitochondria apparently do not bind auxins, but peroxisomes, etioplasts, and nuclear membranes have not been examined in this regard.

Kinetic and specificity differences that may distinguish the auxin binding at these different sites are less clear, and clarification must await studies with fractions of better defined composition. The ER sites may have a somewhat lower affinity for the synthetic auxin, 2,4-D, than expected on the basis of

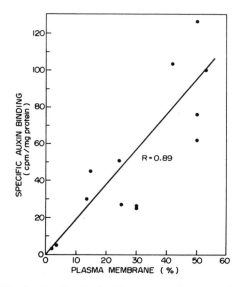

Fig. 1. Regression line showing the relationship between plasma membrane content and the specific activity of indole-3-acetic acid (IAA) binding to cell fractions prepared by sucrose gradient centrifugation of soybean homogenates. (Reproduced from Williamson et al., 1977, with permission of the publisher.)

growth activity (Ray et al., 1977b), whereas other fractions, tentatively identified as plasma membrane (site III), preferentially bind 2,4-D.

Auxin-binding activities are retained in Triton X-100 solubilized preparations (Batt et al., 1976; Ray et al., 1977a; Cross and Briggs, 1978; Cross et al., 1978). Subsequently, maize membranes have yielded auxin-binding preparations with protein bands at 40,000 and 46,000 Da (Venis, 1980). Löbler and Klämbt (1985) indicate only one band of 20,000 Da on sodium dodecylsulfate polyacrylamide gel electrophoresis (SDS-PAGE) following immunoaffinity purification. These apparent discrepancies may be reconciled by the assignment of the 40,000- and 46,000-Da bands as dimeric structures and by the enrichment of perhaps two, rather than one, monomer bands (Venis, 1986). Most recently, a 22,000-Da polypeptide monomer, glycosylated and capable of binding concanavalin A, has been purified as the monocot receptor (Venis, 1987).

Soluble proteins that bind auxins have been reported from tobacco leaves (Vreugdenhil et al., 1979) and cultured pith callus (Oostram et al., 1980). The latter show a high binding affinity for IAA ($K_d = 10^{-8}$ M) and are of apparent high molecular weight (150,000–300,000 Da). The concentration is low (0.01 pmol/mg protein). Unlike auxin binding to maize membranes, the binding in tobacco is time and temperature dependent and becomes optimal at 30°C and 30 min.

High-affinity binding is a necessary but insufficient condition for a binding molecule to achieve receptor status. A second necessary condition is that the binding interaction results in the initiation of a response cascade that generates a meaningful physiological response. Fulfillment of this second criterion for plasma membranes is complicated by the possibility that both transport (Lomax et al., 1985) and growth control receptors may be present. Either or both would be regarded as fulfilling a receptor function. Unequivocal functional assignments of receptor activity to auxin-binding molecules must await a more complete characterization of purified molecules solubilized from active membrane (particulate) or supernatant fractions.

Evidence for Conformational Alterations

While conformational alterations in response to ligand binding for membrane-located receptors are regarded widely as part of the signal–response cascade, experimental documentation of such changes generally is lacking. The auxin–response cascade in plants is no exception.

Ultrastructural evidence from alteration of plant plasma membranes induced by auxins was observed both in isolated and in situ plasma membranes of etiolated hypocotyls of soybean (Morré and Bracker, 1976). In these studies, fixed and embedded preparations were stained by phosphotungstic acid at low pH (Roland et al., 1972). Isolated plasma membrane vesicles treated with physiological concentrations of auxin were 10–15% thinner than the untreated vesicles. The response was auxin specific (Tables I, II), temperature and time dependent, and reversible. The response was rapid ($t_{1/2}$ of 2–3 min) and exhibited a concentration dependency that paralleled the growth response to an optimum at 1 μM. Supraoptimal concentrations were increasingly less effective in promoting both growth and membrane thinning (Fig. 2). Subsequently, it was shown for isolated fractions of the plasma membrane vesicles that 1 μM IAA resulted in an increase of 25% in the fluorescence polarization of a membrane-bound probe, N-phe-

TABLE II. Thickness of Plasma Membranes of Tissue Explants of Soybean Hypocotyls Incubated In Situ With and Without Auxin*

Treatment (1 hr)	Membrane thickness (Å)
Water	101 ± 3
IAA (1 μM)	87 ± 2

* Plasma membranes were contrasted by reaction with phosphotungstic acid at low pH as for Table I. (Reproduced from Morré and Bracker, 1976, with permission of publisher.) IAA, indole-3-acetic acid.

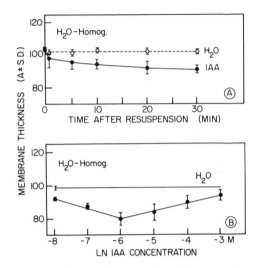

Fig. 2.   Membrane thickness of isolated plasma membrane fragments; **A:** as a function of time of auxin treatment with and without 1 μM indole-3-acetic acid (IAA); **B:** as a function of IAA concentration. Incubated for 20 min. (Redrawn from Morré and Bracker, 1976, with permission of the publisher.)

nylnaphthylamine, with no measurable change in the fluorescence life times (Helgerson et al., 1976). The amplitude of the polarization increase was maximal in the temperature range 12–22°C and confirmed that the membranes were capable of responding to auxin in a cell-free environment, as indicated from membrane measurements.

Further evidence for a conformation change associated with auxin treatment of isolated plasma membrane vesicles has come from infrared spectroscopy. Highly purified plasma membrane vesicles, prepared by aqueous two-phase partition according to Kjellbom and Larsson's (1984) variation of the general method (Albertsson et al., 1982), were examined (Morré et al., 1987). Spectra in the amide I and amide II regions showed a consistent pattern of change in response to the presence of 2,4-D (Fig. 3). In the absence of 2,4-D, the integrated intensity of the amide I band was about twice that of the amide II. With the addition of 1 μM 2,4-D, the amide I band broadened appreciably, and the relative ratios of amide II to amide I were reduced (Fig. 3). The broadening was proportional to the negative logarithm of the concentration of auxin over the range 10^{-8}–10^{-6} M, and the decline of relative ratios of amide II and amide I was proportional over the entire range of 10^{-8}–10^{-3} M.

From the infrared analyses, there was little effect seen in the hydrocarbon chain region of the plasma membrane between 10^{-6} and 10^{-4} M 2,4-D,

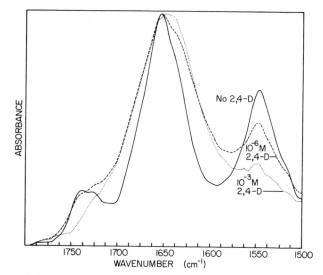

Fig. 3. The amide band region of the infrared difference spectrum for fully hydrated plasma membrane vesicles ($\sim$ 1 mg protein) of soybean with and without exposure to 1 μM or 1 mM 2,4-dichlorophenoxyacetic acid (2,4-D). Band broadening and a reduced ratio of amide peak II to amide peak I were seen. Broadening was maximal at 1 μM, and the peak ratios were log linear with increasing auxin concentration.

although the vibrational frequency of the choline stretch at 1,000 cm^{-1} declined in parallel to the relative ratios of amide II and amide I. There may be some relationship of the latter to the upfield chemical shifts observed by Jones et al. (1984) and Jones and Paleg (1984) with proton nuclear magnetic resonance for the protons of the choline head group of artificial phospholipid vesicles. However, the latter effects were observed at the much higher molar ratios of IAA/phosphatidylcholine of 0.1–1 and were greatest below pH 5, rather than at the physiological conditions of the infrared studies. Moreover, 2,4-D induced only a small chemical shift with proton nuclear magnetic resonance compared with IAA (Jones and Paleg, 1984), whereas the relative growth responses from the two compounds were very similar.

Thus, available evidence supports the interpretation that with isolated plasma membrane vesicles, a conformational response to auxins occurs and may extend beyond the interaction of proteins with the phospholipid head groups at the lipid/protein phase boundary. With isolated membranes, and at least at physiological auxin concentrations, something more than a simple interaction with choline-containing head groups of phospholipids appears to be involved. A number of enzymatic alterations in isolated membranes in response to auxin have been observed. The first of these was a change in the solubility characteristics of the products of glucan synthetase II (Vander-woude et al., 1972). Some, like the glucan synthetase effect, seem quite indirect (ie, a calcium response. See, for example, Kauss, 1987.) Others, like the response of NADH oxidase (Morré et al., 1986; Brightman et al., 1988) or that of the activity of the Ca^{2+} ATPase (Kubowicz et al., 1982), may be more closely coupled. If the dimension change observed in the electron microscope is a manifestation of this conformational response to auxin, then it would seem to occur in vivo as well as in vitro, since the auxin-induced thickening seen with PTA-stained membranes was observed in vivo with tissue segments, as well as in vitro with isolated membrane vesicles (Morré and Bracker, 1976).

Protein Phosphorylation-Dephosphorylation

A useful model for receptor function is that of the receptor for epidermal growth factor. Here a tyrosine protein kinase, structured as part of the receptor, is activated by the interaction with growth factors, so that a phosphorylated, "activated" receptor is the result (Cohen, 1982). In the absence of stimulus, the receptor dephosphorylates via a phosphoprotein phosphatase, thus affording a mechanism for relaxation of the transduction process when the stimulus is removed. Autophosphorylation has been sug-gested, for example, as a means of prolonging the effects of a transient calcium stimulus (Blowers and Trewavas, Chapter 1).

The elements necessary for protein–receptor phosphorylation–dephosphorylation are present in the plant plasma membrane. Isolated membranes, incubated with γ-[^{32}P] ATP, catalyze the phosphorylation of several proteins (Morré et al., 1984; Ranjeva and Boudet, 1987; Blowers and Trewavas, Chapter 1). Further incubation in the absence of ATP results in a decrease in phosphorylated proteins indicative of phosphoprotein phospha-tase activity. In the presence of 1 μM auxin (2,4-D), the phosphorylation of two peptide bands of 45,000 and 50,000 Da increased during very short (0.2 min) incubations of soybean microsomes with γ-[^{32}P]-ATP. Overall in-creases in membrane phosphorylation observed, as well as increased label-ing of low-molecular-weight constituents, were apparently due in part to increased turnover of phospholipids, e.g., phosphatidic acid (Morré et al., 1984).

The protein phosphorylation stimulated by 2,4-D is not stimulated by 2,3-dichlorophenoxyacetic acid (2,3-D), an inactive synthetic auxin analogue, is blocked by calcium and has not been observed with plasma membranes purified from total microsomes. In situ, at steady state, 2,4-D-stimulated labeling of microsomal phosphoproteins may be reduced, rather than in-creased (Varnold and Morré, 1985; Veluthambi and Poovaiah, 1986). A 37,000-Da peptide that co-purified with the auxin-stimulated NADH oxidase from soybean membrane (Brightman et al., 1988) appears to be phosphorylated, although this phosphorylation is not enhanced in the isolated membranes by auxin (Brightman, unpublished results). At present, there is no evidence for receptor phosphorylation–dephosphorylation associated with auxin action at the plasma membrane. This, however, would not preclude changes in phosphorylation in other membranes such as ER in response to auxin or of receptor phosphorylation-dephosphorylation still playing a major regulatory role in signal–response coupling in plants. Control by phosphorylation has been implicated for the fusicoccin receptor (Aducci et al., 1984). While phosphorylation sites other than tyrosine might well be involved, tyrosine phosphate (Elliott and Geytenbeek, 1985) and tyrosine phosphorylation (Torruella et al., 1986) have been reported for plants. The tyrosine protein kinase of Torruella et al. (1986) is located in the nucleus and thus is probably not important for plasma membrane stimulus–response.

Signal-Transducing Proteins

While there is an apparent need for proteins to transduce the receptor–auxin interaction into a membrane conformational change and for a response cascade of enzymatic activities, such proteins from plants are thus far unknown (Pall, 1985). A GTP-binding protein appears to fulfill such a role in animal cells. GTP-binding proteins have been extracted from *Lemna* (Hasunuma and Funadera, 1987) and etiolated pea epicotyls (Hasunuma et al., 1987). No GTP binding protein, however, has been found associated with the plant plasma membrane, and no guanosine nucleotide requirement or guanosine nucleotide enhancement of a plasma membrane activity been observed.

PHOSPHOLIPASE ACTIVATION AND LIPID CHANGES

Changes in lipid composition of membranes (Morré et al., 1984) have been offered as an explanation for the concentration-dependent release of bound calcium and other divalent ions from microsomes incubated with auxin in vitro (Buckhout et al., 1981a,b). The calcium-release experiments were patterned after membrane responses to local anesthetics (Low et al., 1979). Interactions of these agents with membranes resulted in release of $^{45}Ca^{2+}$ (equilibrated in the presence of ionophore A23187) from calcium-binding sites within the membrane (as opposed to luminal Ca^{2+}). The release of membrane-bound calcium showed a response to the logarithm of the auxin concentration that was maximal at 1 μM auxin and paralleled the dose–response relationship of growth to auxin. IAA and 2,4-D, but not 2,3-D, produced the response. When calcium was replaced by magnesium or

90 / Morré

TABLE III. Scatchard Analysis of Calcium Binding to Soybean Hypocotyl Plasma
Membranes*

| | nmole Ca^{2+} sites/mg protein | |
Treatment	Low affinity	High affinity
No auxin	226	47
1 μM auxin	187	28

* (Reproduced from Buckhout et al., 1981a, with permission of the publisher.)

manganese, release also was observed. Scatchard analyses revealed that the 2,4-D effect represented an actual loss of high-affinity binding sites for divalent ions, rather than a change in the binding affinity (Table III). This observation afforded an opportunity to determine a change in membrane composition resulting from incubation in vitro with auxins.

With isolated membranes incubated in vitro with auxin, phospholipid compositional changes were evidenced primarily by an accelerated breakdown of acidic phospholipids, phosphatidylinositol (PI), phosphatidylglycerol, and phosphatidic acid, as well as loss of phosphatidylcholine (Morré et al., 1984). Phospholipid changes in vitro were reported previously by Price-Jones and Harwood (1983) for the pea stem. Thus the hydrolysis of the acidic phospholipids was suggested as a mechanism to explain auxin-induced calcium release. The stoichiometry was such that the sum of the acidic, Ca^{2+}-binding phospholipids lost on auxin treatment was approximately equivalent to the number of high-affinity calcium sites also lost from the membranes in response to auxin.

Phosphatidylinositol-Specific Phospholipase C

Phosphoinositide-specific phospholipase C (Majerus et al., 1986) cleaves phophoinositides to yield 1,2-diacylglycerol and inositol phosphates. There have been several reports of soluble (cytoplasmic) phospholipase C activities in plants (Helsper et al., 1986a,b; Irvine, 1982). Additionally, breakdown of membrane-associated PI has been demonstrated, based on analyses using radiolabeled PI (Morré et al., 1984; Helsper et al., 1986a,b; Connett and Hanke, 1986). In soybeans, a PI-specific phospholipase C activity is present both in the soluble fraction and associated with membranes (Pfaffmann et al., 1987). Moreover, the membrane-associated form of the activity is located predominantly in the plasma membrane. Plasma membrane phospholipase C activity is specific for PI (Pfaffmann et al., 1987) and polyphosphoinositides (Melin et al., 1987), with relative activities dependent on calcium concentration. The membrane-associated activity is much higher toward exogenous phosphatidylinositol 4,5-bisphosphate (PIP_2) and phosphatidylinositol-4-monophosphate (PIP) than to exogenous PI, whereas the reverse is true for the soluble activity (Melin et al., 1987). The activities depend on the addition of both calcium and a detergent (Table IV). Other phospholipids are not hydrolyzed, and calmodulin is without effect. Most of the PI-specific phospholipase activity nevertheless is cytosolic, even in animal cells, although its substrates are in a membrane bilayer.

Products of the soybean enzyme from exogenous PI analyzed by differential elution from ion-exchange columns were inositol phosphate 85% and free inositol 15%; optimum activity was at pH 6.6 (Pfaffmann et al., 1987). With PI as substrate, there was an insignificant ($\sim$ 3%) stimulation of the membrane-associated activity on incubation of isolated plasma membrane vesicles with 2,4-D and no effect on the soluble activity (Table IV). A similar small auxin stimulation of about 4% also was obtained with the polyphosphoinositide-specific activity (M. Sommarin and A.S. Sandelius, private communication).

With a synthetic water-soluble phospholipase C substrate, p-nitrophenylphosphorylcholine (pNPP-choline), the auxin response with soybean membranes was a 10–20% (sometimes less) increase in hydrolysis rate. With EDTA-washed membranes, calcium was required for auxin stimulation. Also, stimulation was given by calcium plus calmodulin. The auxin-induced component of pNPP-choline hydrolysis exhibited a broad pH optimum at about 7.0 and an optimum stimulation by auxin at 1 µM (Morré and Drobes, 1987).

Phospholipase D

When isolated soybean microsomes were incubated with and without 2,4-D, reductions in both PI and in phosphatidylcholine were noted to result

TABLE IV. Response of a Phosphatidylinositol-Specific Phospholipase C to Auxin and Ca^{2+} When Assayed in the Presence of Deoxycholate

Enzyme source	Additions	Specific activity[a] (nmoles/hr · mg protein)
Cytosol	None	1.6 ± 0.2
	+ 0.5 mM CaCl₂	5.2 ± 0.3
	+ 0.5 mM CaCl₂ + 1 µM 2,4-D	4.4 ± 0.3
Plasma membrane	None	7.4 ± 0.4
	+ 0.5 mM CaCl₂	24.0 ± 0.5
	+ 0.5 mM CaCl₂ + 1 µM 2,4-D	24.7 ± 0.5

[a] This activity was about 85% cytosolic and 15% membrane associated. The latter was associated dominantly, if not exclusively, with the plasma membrane. Using exogenous phosphatidylinositol as substrate, optimal activity was dependent upon additions of both Ca^{2+} and detergent. Calmodulin (1 µM) was without effect. The products analyzed by differential elution from ion-exchange columns were 85% inositol phosphate and 15% free inositol. 2,4-D, 2,4-dichlorophenoxyacetic acid.
From a study with H. Pfaffman and E. Hartmann, University of Mainz, Mainz, FRG.

from the 2,4-D addition (Morré et al., 1984). To estimate phosphatidylcholine breakdown, tissue sections were preincubated with [^{14}C] choline for 4 hr. A 6,000g_{max} (10 min) supernatant of the filtered homogenate was centrifuged for 20 min at 45,000g max to prepare heavy microsomes. The freshly isolated membranes were resuspended in distilled water, with and without various additions and incubated for 15 min at 25°C. Reactions were stopped with 10% trichloroacetic acid (TCA); the membranes were pelleted for 5 min at 10,000 g_{max}, at 4°C; and the pellets were washed twice with 1-ml portions of cold distilled water.

The hydrolysis of the in vivo-labeled choline-containing phospholipids of isolated soybean membrane vesicles was rapid and reached a plateau within the 15 min incubation (Fig. 4). Radioactivity lost from the membrane pellet was recovered as free choline in the supernatant (Table V). The loss was accelerated by 1 μM 2,4-D, calcium, or calmodulin (Table V). Losses were greatest with the combination of 2,4-D, calcium and calmodulin (Table V, Fig. 4).

Since C-type phospholipases are not known to hydrolyze phosphatidylcholine, the breakdown is assumed to occur via a D-type phospholipase,

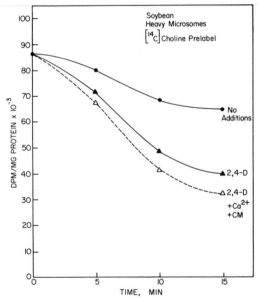

Fig. 4. Kinetics of loss of radioactivity from [^{14}C]choline-prelabeled soybean membranes as a function of time of incubation with or without 1 μM 2,4-dichlorophenoxyacetic acid (2,4-D) alone or in the presence of 1 μM Ca^{2+} + 1 μM calmodulin. (Reproduced from Morré and Drobes, 1987, with permission of the publisher.)

TABLE V.   Loss of [^{14}C]Choline From Membranes Prelabeled In Situ*

Additions	Membrane	Lost from membrane	Supernatant
		Cpm/mg protein	
Initial	27,025		
Incubated (15 min)	24,597	2,428	3,881
+ 1 μM 2,4-D	21,893	5,132	9,627
+ 1 μM Ca^{2+}	20,768	6,257	7,728
+ 1 μM CM	19,219	7,806	5,828
+ 2,4-D + Ca^{2+}	21,754	5,271	5,305
+ 2,4-D + CM	18,581	8,445	7,728
+ CM + Ca^{2+}	19,208	7,817	9,812
+ 2,4-D + Ca^{2+} + CM	17,076	9,949	12,715

* Membranes were prelabeled in intact hypocotyl segments with 1 μM 2,4-D and incubated 15 min under the conditions indicated. The radioactivity lost from the membrane and that entering the supernatant were approximately the same suggestive of a hormone-stimulated hydrolysis of phosphatidylcholine.
2,4-D, 2,4-dichlorophenoxyacetic acid; CM, calmodulin.

with the major products being phosphatidic acid and choline (see Martin et al., 1987). However, since soybean membranes contain both choline phosphate phosphatase and inositol phosphate phosphatase as well as phosphatidic acid phosphatase (Table VI), the products of phosphatidylcholine cleavage might be diacylglycerol, inorganic phosphate, and choline from either a D- or a C-type primary cleavage. Soybean membranes incubated with or without 2,4-D previously were shown to undergo a concentration-dependent release of inorganic phosphate from endogenous membrane components (Scherer and Morré et al., 1978, Fig. 5). The latter might derive from phospholipid hydrolysis, described above, through the subsequent action of membrane phosphatases.

Incorporation of [^{14}C]choline or [^{3}H]inositol by excised hypocotyl segments of soybean over 4 hr of incubation is accelerated 25–30%, suggestive of accelerated phospholipid biosynthesis or turnover during auxin growth. In endothelial cells, diacylglycerol has recently been shown to be released from

TABLE VI.   Phosphatase Activities of Soybean Membranes*

Substrate	μmoles/hr · mg protein
Phosphatidic acid	0.05
Myo-inositol-1-monophosphate	0.35
Choline phosphate	0.15

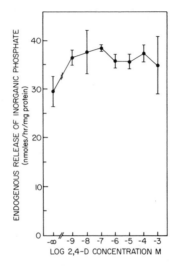

Fig. 5. Stimulation of endogenous phosphate release by 2,4-dichlorophenoxyacetic acid (2,4-D) from soybean membranes incubated for 1–2 hr with varying concentrations of 2,4-D. (Reproduced from Scherer and Morré, 1978, with permission of the publisher.)

phosphatidylcholine by a novel phospholipase C that does not require Ca^{2+} (Martin et al., 1987).

Polyphosphoinositide Turnover

Of major interest in terms of the possible role of membrane lipid constituents in transmembrane signaling are the polyphosphoinositides, PIP and PIP_2 (Fig. 6). These are cleaved by a C-type phospholipase to give rise to diacylglycerol (DAG), inositol-1,4-bisphosphate (IP_2), and inositol-1,4,5-trisphospate (IP_3), respectively, all three of which may serve important roles in signal transduction (Boss, Chapter 2). PIP and PIP_2 have now been identified unequivocally from carrot cells by fast atom bombardment mass spectroscopy (Wheeler and Boss, unpublished results).

An accelerated turnover of both PIP and PIP_2 in response to 2,4-D treatment was reported by Zbell and Walter (1987) for isolated plasma membrane vesicles from carrot labeled in vitro with $\gamma[^{32}P]ATP$. The kinase responsible for the phosphorylation of PI in plants is localized in the plasma membrane (Sandelius and Sommarin, 1986; Sommarin and Sandelius, 1988; Sandelius et al., in preparation). Additionally, Ettlinger and Lehle (1988) report rapid transient changes in IP_3 and IP_2 when *Catharanthus roseus* cultures arrested in G_1 were stimulated by auxin additions.

With isolated plasma membrane preparations from elongating soybean hypocotyls, we find that the PI kinases are not stimulated by auxin treatment

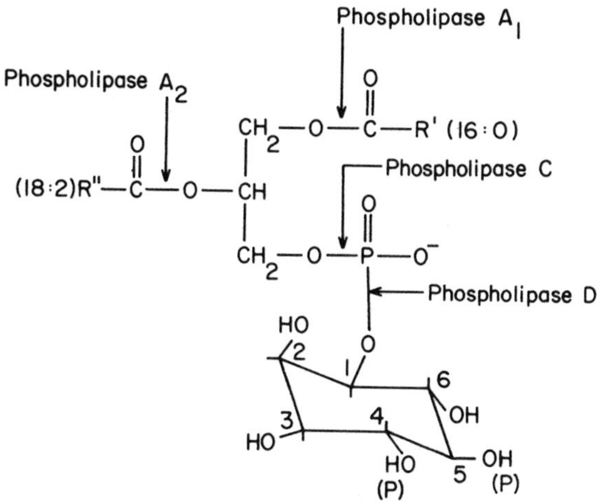

Fig. 6. Structure of phosphatidylinositol. The polyphosphoinositides contain a second phosphate in position 4 or second and third phosphates in positions 4 and 5 (Majerus et al., 1986).

TABLE VII. Effect of Auxin on Phosphorylation of Phospholipids of Plasma Membranes From Soybean*

Labeled constituent	Expt	Phosphate incorporation, (nmoles/mg protein)			
		Without 2,4-D		With 2,4-D	
		4 min	10 min	4 min	10 min
Phosphatidylinositol phosphate (PIP)	I	1.7 ± 0.1	2.8 ± 0.2	1.9 ± 0.2	3.1 ± 0.1
	II	2.9 ± 0.1	4.6 ± 0.5	2.4 ± 0.1	4.1 ± 0.04
Phosphatidylinositol bisphosphate (PIP₂)	I	0.15 ± 0.02	0.35 ± 0.2	0.17 ± 0.02	0.38 ± 0.02
	II	0.2 ± 0.003	0.5 ± 0.003	0.2 ± 0.004	0.5 ± 0.03
Phosphatidic acid (PA)	I	3.5 ± 0.4	6.8 ± 0.3	3.6 ± 0.2	7.4 ± 0.2
	II	6.8 ± 0.4	10.4 ± 0.1	6.0 ± 0.7	11.4 ± 0.1

* Plasma membranes were isolated from etiolated hypocotyls of soybean by aqueous two-phase partition and incubated (40 μg protein for 10 min with 75 nmol $\gamma[^{32}P]ATP$ (250–450 cpm/pmol), 5 μl HEPES/KOH, pH 7.0, containing 15 mM Mg^{2+}, 1.5 mM ATP, and 1 mM DTT, pH 7.0 in a final volume of 100 μl, according to Sommarin and Sandelius (1988) v/v with and without 1 μM 2,4-D. The reactions were stopped with $CHCl_3$:methanol:H_2O 1:2:0.5 and the lipids extracted and chromatographed as described by Sandelius and Sommarin (1986), after which radioactivity was determined. Unpublished results of F.E. Wilkinson, Purdue University, West Lafayette, Indiana.

of isolated plasma membrane vesicles (Table VII). However, the hydrolysis of the PI and PIP, once formed in short-term labeling with [^{3}H]inositol, is stimulated by 2,4-D treatment, but there appears to be no obvious accelerated formation of either IP$_2$ or IP$_3$ compared with IP (Tables VIII and IX). Thus, in the experiments of Tables VIII and IX, there is evidence of auxin-accelerated lipid breakdown with the isolated membranes but not necessarily a breakdown specifically involving polyphosphoinositides. With [^{32}P] incorporation from γ[^{32}P]ATP, a variable auxin stimulation of radioactivity into phosphatidic acid was seen, suggestive of auxin-accelerated formation of diacylglyceride and its subsequent phosphorylation (Table VII). Similar accelerated labeling of low-molecular-weight constituents, possibly phosphatidic acid, was seen earlier (Morré et al., 1974) with membrane labeled with γ[^{32}P]ATP with and without auxin. With the breakdown of [^{3}H]-labeled inositides, a major effect was on the loss of PI as reported previously (Table VIII).

While control cycles operating in animal cells may provide useful parallels to guide plant studies, plants also may be expected to have unique characteristics. For example, even though cyclic nucleotides may be found in plant extracts, cyclic-nucleotide-responsive proteins (e.g., protein kinases) are not (Hepler and Wayne, 1985). Auxin responsive soybean tissues and membranes lack both detectable cyclic AMP and adenylate cyclase (Yunghans and Morré, 1977) as determined by several sensitive techniques. This and other evidence suggests that both a cyclic AMP and a cyclic GMP involvement may be safely excluded from the auxin-response cascade.

TABLE VIII. Effect of Auxin on Breakdown of Inositol-Labeled Phosphatidylinositol Metabolites of Soybean Plasma Membranes*

| | | Dpm/100 µg plasma membrane protein × 10^{-2} | |
| | | Incubated 10 min | |
Inositol metabolite	Initial	Without 2,4-D	With 1 µM 2,4-D
Phosphatidylinositol (PI)	128	112	76
Phosphatidylinositol phosphate (PIP)	7.2	4.3	3.4
Phosphatidylinositol bisphosphate (PIP$_2$)	0.3	0.2	0.2

* Sections, 1 cm in length, of etiolated hypocotyls of soybean were incubated 4 hr with 100 µCi [^{3}H]inositol (10–20 Ci/mmol). Plasma membranes were then isolated by aqueous two-phase partition, resuspended in water and incubated for 0 or 10 min with and without 1 µM 2,4-D. The labeled phosphatidylinositols were then extracted with CHCl$_3$:1.2 N HCl, 1:1, v/v and chromatographed as described by Schacht (1978). Radioactivity was determined by liquid scintillation methods.

TABLE IX. Effect of Auxin on Breakdown of γ[³²P]ATP-Labeled Phosphatidylinositol Metabolites of Soybean Membranes*

	Dpm/100 μg plasma membrane protein × 10⁻²		
		Incubated 10 min	
Inositol metabolite	Initial	Without 2,4-D	With 1 μM 2,4-D
Phosphatidylinositol bisphosphate (PIP₂)	389 ± 35	350 ± 26	288 ± 33
Phosphatidylinositol phosphate (PIP)	3,610 ± 572	3,396 ± 214	2,948 ± 586
Phosphatidic acid (PA)	11,622 ± 1,171	11,089 ± 876	8,910 ± 1,711

* Soybean (*Glycine max* L.) hypocotyl plasma membranes purified by aqueous two-phase partition were phosphorylated with γ[³²P]ATP according to Sandelius and Sommarin (1986) for 15 min at 25°C. The membranes were collected by centrifugation at 12,000g for 10 min and washed twice with 180 mM Tris maleate containing 18 μM CaCl₂ and resuspended in the same buffer. The phosphorylated membranes were incubated for 0 and 10 min either with or without 1 μM 2,4-D in a total volume of 200 μl. The reactions were stopped by the addition of 600 μl of chloroform/methanol/water (1:2:0.5, v/v/v). The lipids were extracted, chromatographed, and their radioactivity determined as described (Sandelius and Sommarin, 1986). Unpublished results of FE Wilkinson, Purdue University, West Lafayette, Indiana.

Diacylglycerols (DAG): Formation In Vivo

If phospholipid breakdown by any of the mechanisms listed above and studied with isolated plasma membrane vesicles were taking place in situ, and if degradation exceeded the rate of resynthesis, the lipid hydrolysis should result in an accumulation of DAG in the membrane. The pool of DAG normally is very small, ~ 4% of the total lipid, so that small increases may be detected and quantitated.

With excised hypocotyl segments, incubated with and without 2,4-D, steady-state levels of DAG declined over 4 hr in the absence of auxin supplementation (Morré et al., in preparation). In the presence of 1 μM 2,4-D, in some experiments DAG levels declined less than those of control segments but the differences were small (5% of total). In those experiments were DAG were altered by 2,4-D (or IAA), the differences were primarily associated with a supernatant fraction and not with membranes.

Lipid Breakdown and Second Messenger Formation

Any of the lipid constituents whose breakdown in isolated vesicles is accelerated by auxin could serve as second messengers. This is especially so for the polyphosphoinositols, where the released inositol phosphates may evoke specific physiological responses. IP₃ has been reported to stimulate calcium efflux from microsomes (Drobak and Ferguson, 1985; Reddy and

Poovaiah, 1987) and tonoplast (Schumaker and Sze, 1987) vesicles in vitro and from protoplasts in vivo (Rincon and Boss, 1987). Hydrolysis of acidic phospholipids, in general, would generate, additionally, the release of bound calcium and other divalent ions. Even a D-type cleavage of phosphatidylcholine, for example, would produce DAG if coupled to phosphatidic acid phosphatase. The DAG are thought normally to remain with the membrane, but may enter the soluble fraction as well. DAG have been described as activators of C-type protein kinases (Nishizuka, 1984).

A regulatory role for DAG in plants remains to be determined. A relatively unspecific overall stimulation by DAG and Ca^{2+} was reported for in vitro phosphorylation of membrane proteins of soybean (Morré et al., 1984). Subsequently, Schäfer et al. (1985) partially purified a Ca^{2+}- and DAG-stimulated protein kinase from zucchini. The activity was phospholipid and Ca^{2+}-dependent, and the phospholipid requirements were met by phosphatidylserine (PS) and phosphatidylethanolamine but not by PI. Also a putative C-type protein kinase was isolated from wheat by Òláh and Kiss (1986). It was stimulated by a mixture of PS and either phorbol ester or 1,2-diolein, but a requirement for DAG was not well documented. A similar protein kinase was reported from *Ameranthus tricolor* (Elliott and Skinner, 1986) but was not phorbol ester activated. Neither was the C-type protein kinase from *Neurospora* (Favre and Turian, 1987) phorbol ester activated.

Activation of protein kinase C is a consequence of the increase by DAG of the affinity of the enzyme for Ca^{2+}. This reduces the amount of Ca^{2+} needed for activity (Donnelly and Jensen, 1983; Niedel and Blackshear, 1986; O'Brien et al., 1987). Tumor-promoting phorbol esters can substitute for DAG by binding to the same site on the enzyme (Ashendel, 1985; Berridge, 1984; Niedel and Blackshear, 1986).

A DAG stimulation of membrane-bound kinase activity may be difficult to demonstrate in lipid-containing membrane preparations. Amounts of DAG present or arising endogenously during isolation in the membrane may be sufficient to give maximal DAG-stimulated activity, even though levels of DAG in vivo might be rate limiting. While DAG are minor membrane constituents (4% of the total lipid), PIP_2 is even less (1% of the total phosphoinositides or between 0.2 and 0.3% of the total lipid). Thus the amounts of DAG coming even from complete hydrolysis of PIP_2 would be difficult to detect against the normal background of membrane DAG present and might even be physiologically insignificant.

ROLE OF INTERNAL ENDOMEMBRANES

ER (as well as Golgi apparatus) membranes are characterized by high-affinity auxin-binding sites (Batt and Venis, 1976; Ray, 1977). Similar distributions have been found for mammalian receptors, for example, the insulin receptor (Bergeron et al., 1973; Schilling et al., 1979), where one interpretation has been that these intracellular receptors may be newly synthesized en route to the cell surface or internalized receptors resulting from surface down regulation. A subcellular function for such intracellular hormone receptors, if any, is unknown.

In plants, evidence for a function of intracellular auxin receptors is scant. Morré et al. (1984) reported an auxin-responsive inositol-exchange reaction (as distinct from hydrolysis of PI) of microsomes from soybean in which free inositol is exchanged with inositol in PI by a reaction requiring cytidine nucleosides and Mn^{2+} first described by Sexton and Moore (1981). This activity is not found with purified plasma membrane fractions and has been assigned to endoplasmic reticulum based on cell fractionation studies (Sandelius and Morré, 1987). It is clearly an exchange and appears not to involve the major population of polyphosphatidylinositols located at the plasma membrane (Sandelius and Morré, 1987).

Recent developments suggest an involvement of ER in calcium-release phenomena. In view of such findings, it may be possible that the auxin-stimulated calcium release observed by Buckhout et al. (1981a,b) was influenced at least in part by release of plasma membrane-bound IP_3. Its interaction with ER or tonoplast vesicles present in the microsomes might ultimately release calcium sequestered within the lumens of such vesicles. Such a series of events with a requirement for two different kinds of membranes, while speculative, would explain, for example, why the auxin-stimulated release of calcium was more difficult to reproduce with purified cell fractions.

An interaction with soluble auxin-binding proteins, IAA, and nuclei to activate an RNA polymerase II activity has been reported (Van der Linde et al., 1984). Previously, Hardin et al. (1972) observed that some factor released from plasma membranes in the presence of auxin also could stimulate RNA polymerase II. The nature of the Hardin et al. factor was not determined although it was shown to be nonproteinaceous and of low molecular weight (Clark et al., 1976) (IP_3 or Ca^{2+}?).

PLASMA MEMBRANE REDOX: AN ESSENTIAL ROLE IN AUXIN-INDUCED GROWTH

Plasma membranes of all cells are characterized by an electron-transport chain capable of transferring electrons and protons and of generating a positive inside membrane potential (Crane et al., 1979, 1985, Chapter 5). In animal cells a role in growth had been inferred from the hormone responsiveness of the system, that is, responsiveness of the plasma membrane redox system of liver and fat cells to insulin or glucagon (Goldenberg et al., 1979). Stimulation of animal cell growth by external oxidants such as ferric chloride (Basset et al., 1986), diferric transferrin (Barnes and Sato, 1980), or ferric isopyridoxylnicotinate (Landschulz et al., 1984; Ponca et al., 1984) has been reported. However, even more compelling evidence came from the observation of Ellem and Kay (1983) that impermeable ferricyanide could stimulate the growth of serum-deficient melanoma cells. The natural donor and acceptor for the hormone-responsive redox system have been described as NADH (Navas et al., 1986) and transferrin-bound Fe^{3+} (Crane et al., 1985; Löw et al., 1986), respectively, for animal cells.

Evidence for a role for plasma membrane redox in plant growth was initially sought using carrot cells and external ferricyanide (Barr et al., 1985). Two additional sources have been provided from plant stems, where a correlation between inhibition of auxin growth and plasma membrane redox has been obtained with certain inhibitors and where a direct effect of auxin has been demonstrated.

Several drugs, well established as antitumor agents, are known to inhibit both growth and plasma membrane redox activities in mammalian cells (Sun et al., 1984; Crane et al., 1985). These same drugs when applied to apical segments cut from etiolated hypocotyls of soybean showed a strong correlation between inhibition of plasma membrane redox and inhibition of elongation growth. Among the drugs examined were the antiproliferative agents *cis* platin and Adriamycin, as well as the inhibitor of plasma membrane redox *p*-nitrophenylacetate (Barr et al., 1986) (Fig. 7). These drugs appeared to preferentially inhibit auxin growth (50–80%) at concentrations where control growth was much less inhibited (10–30%). The less active platin analogue, *trans* platin, was much less effective in inhibition of growth and plasma membrane electron transport in plants.

There are a number of redox activities associated with the plant plasma membrane that have been implicated at various times in proton efflux, potassium uptake, iron reduction, and membrane polarization (Crane et al., 1985; Møller and Lin, 1986). The inhibitor data were determined with a NADH-ferricyanide oxidoreductase activity and plasma membrane vesicles prepared by aqueous two-phase partitioning (Morré et al., 1988).

There is also an external oxidase associated with the plant plasma membrane. This oxidase has the interesting property of being activated by auxins (Table X). IAA and the active synthetic analogues, 2,4-D and α-NAA, all promote growth and stimulate plasma membrane redox, whereas the inac-

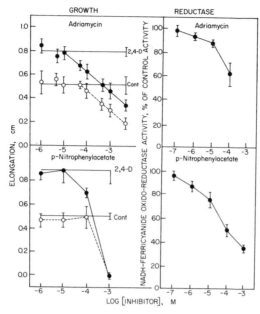

Fig. 7. Inhibition of growth and plasma membrane redox activity by an antiproliferative agent, Adriamycin, and by the inhibitor of plasma membrane electron transport, p-nitrophenylacetate (redrawn from Morré et al., 1988).

TABLE X. Stimulation of NADH Oxidase of Isolated Plasma Membranes of Soybean Hypocotyls Comparing Growth-Active (2,4-D, α-NAA) and Inactive (2,3-D, β-NAA) Synthetic Auxins With Indole-3-Acetic (IAA) Acid and Benzoic Acid*

Compound	Growth (% of no auxin) (M)	NADH oxidase (% of no auxin) (M)
Indole-3-acetic acid (IAA)	166 (10^{-4})	222 (10^{-5})
Benzoic acid	102 (10^{-5})	83 (10^{-6})
2,4-dichlorophenoxyacetic acid (2,4-D)	160 (10^{-5})	250 (10^{-6})
2,3-dichlorophenoxyacetic acid (2,3-D)	109 (10^{-5})	111 (10^{-6})
1-Naphthaleneacetic acid (α-NAA)	164 (10^{-6})	200 (10^{-7})
2-Naphthaleneacetic acid (β-NAA)	116 (10^{-6})	117 (10^{-7})

* Values reported are for compounds tested at the molar concentrations given in parentheses.

tive but chemically related auxin analogues 2,3-D and β-NAA do not. The concentration dependency of tissue growth and NADH oxidase activity of isolated plasma membrane vesicles are closely correlated for both IAA and 2,4-D, except that maximum stimulations of NADH oxidase with the iso-

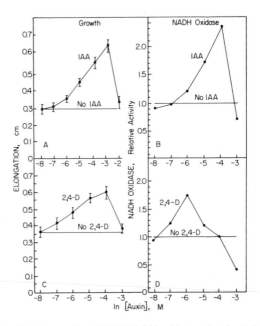

Fig. 8. Stimulation of **A, C**: growth and **B, D**: NADH oxidase activty by **A, B**: the natural auxin indole-3-acetic acid (IAA) and by **C, D**: the synthetic auxin 2,4-dichlorophenoxyacetic acid (2,4-D.) (Reproduced from Morré et al., in press.)

lated vesicles are achieved at approximately tenfold lower concentrations of auxin than for optimum tissue growth (Fig. 8). As with membrane thinning, calcium release and other responses of membranes to auxin, the stimulation of NADH oxidase gives an ascending response to an optimum and a descending response as the optimum is exceeded. This response parallels the well-established growth responses to auxin, where similar optimum curves are recorded (Fig. 8).

Unlike other direct responses of membrane-associated activities to auxin, the NADH oxidase response is given by detergent-solubilized membranes suggestive of an opportunity to purify the auxin-stimulated activity (Fig. 9). Using CHAPS (3-([3-cholamidopropyl]dimethylammonio)-1-propane-sulfonate) solubilized extracts, the plasma membrane NADH oxidase activity was purified some 2,000-fold relative to the total homogenate and the purified activity retained its auxin responsiveness (Fig. 10).

One of the most potent inhibitors of plant growth, actinomycin D (Key, 1964), is also a potent inhibitor of NADH oxidase (Brightman et al., 1988). With the purified enzyme, it is inhibitory at 0.01 μM and gives 100%

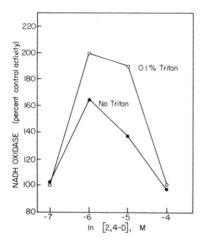

Fig. 9. Concentration-dependent stimulation by 2,4-D of NADH oxidase of plasma membrane in the presence or absence of the non-ionic detergent Triton X-100. Vesicles were incubated for 10 min both with detergent and hormone and then assayed for NADH oxidation. (Reproduced from Brightman et al., 1988, with permission of the publisher.)

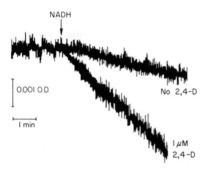

Fig. 10. Spectrophotometric tracing of purified, NADH oxidase activity. The sample was incubated with or without 1 μM 2,4-D for 3 min prior to addition of NADH. (Reproduced from Brightman et al., 1988, with permission of the publisher.)

inhibition of activity at 0.05 μM. Other plasma membrane electron transport inhibitors such as *cis* platin and *p*-nitrophenylacetate, while inhibiting auxin-induced growth, do not inhibit NADH oxidase. Moreover, while all plant cells so far examined have exhibited plasma membrane redox activity, the activity is greatest in regions of most rapid elongation (Qui et al., 1985). This is in keeping with our findings (Brightman et al., 1988) that the NADH oxidase of mature portions of the soybean hypocotyl is no longer auxin responsive. Thus, a working hypothesis derived from these and other data

(Fig. 11) and depicted in the model of Figure 12 is that cytoplasmic NADH is oxidized via a transmembrane electron transport (NADH: external oxidant oxidoreductase) in a manner coupled to O_2 via the NADH oxidase, which would be a rate-limiting terminal oxidase. The auxin signal–response cascade, with its subsequent conformational alteration in the membrane, might be linked somehow to the NADH oxidase to stimulate activity. How such increased electron transport might be translated into an increased growth rate is the subject of the next section.

Early indications of an auxin effect on the redox state of the cytoplasm were reported for soybean membranes and soluble proteins (Morré, 1970). These findings were initially of interest as an explanation of the work of Galston and Kaur (1963), who showed an increase with auxin in the heat coagulability of cytoplasmic protein in response to auxin. Both the sulfhydryl (-SH) content of cytoplasmic proteins and the heat coagulability showed a concentration dependency in response to auxin that paralleled the growth response and the auxin effect on the NADH oxidase of Figure 9 (Morré, 1970).

The NADH oxidase stimulated by auxin requires oxygen and is inhibited by an argon atmosphere. A mitochondrial location is eliminated on the basis of a lack of sensitivity to azide. Finally, the auxin-stimulated NADH oxidase has been purified from the plasma membrane as a complex of three proteins of molecular weight 36,000, 52,000, and 72,000 Da. The 36,000-Da component appears to be phosphorylated.

The proposed linkage between NADH oxidase and transmembrane electron transport is based on the auxin sensitivity of the oxidase and the sensitivity of NADH-external oxidant oxidoreductase (but not NADH oxidase) to the antiproliferative agents cis platin and p-nitrophenylacetate. With the latter, auxin-induced growth and NADH-ferricyanide oxidoreductase activity, but not NADH oxidase, were inhibited. Therefore, the drugs must act at sites in the main part of the electron-transport chain. Additional evidence is provided by experiments with ferricyanide added as an external oxidant. In control cells, growth is stimulated presumably through increased electron transport. However, auxin-induced growth is inhibited, presumably by a mechanism where, with ferricyanide, electrons bypass the auxin-stimulated oxidase. Additionally, the electron transport across the plasma membrane to oxygen may be related to the auxin-stimulated oxygen uptake (Polevoy and Salamatova, 1977; Novak and Ivankina, 1983; Sijmons et al., 1984a,b; Loppert, 1983) that is resistant to inhibition by cyanide and salicylhydroxamate (Table XI, Crane et al., 1985).

The NADH oxidase may represent a primary response site to auxin. An alternative would be that one of the peptides co-purified with the oxidase is a transducing protein that somehow alters the conformation of the membrane to activate not only the NADH oxidase but also to account for the responses to auxins of other plasma membrane enzymes and characteristics as well.

ROLE OF MEMBRANE POTENTIAL AND PROTON GRADIENTS

The trans plasma membrane electron transport stimulated by auxin in plants could have several important implications. One already suggested (Morré et al., 1986) would be to account for cytoplasmic acidification (Brummer and Parish, 1983; Felle et al., 1986; Fig 11) induced by auxin. This has, in turn, been implicated as a mechanism to stimulate the vanadate-inhibited ATP-driven proton pump of the plasma membrane (Morré et al., 1986; Rubinstein and Stern, 1986). The latter would be in keeping with the well-established lag of 5–7 min between addition and the onset of external proton secretion with stem or coleoptile segments (Cleland, 1976; Cleland and Rayle, 1978).

The explanation for the lag in growth and cell wall acidification remains a major unanswered question. Clarification may reside in work of Felle (1987b, Chapter 6) (Fig. 11), where both membrane potential and cytoplasmic pH were measured. The lag in the rise in membrane potential (Fig. 11, upper curve) parallels the lag in proton extrusion, and it may be assumed that the two are related. At the same time, there is a corresponding acidification of the cytoplasm (Fig. 11, lower curve).

The activation of proton release by plant plasma membranes finds a parallel in HeLa cells, for example, where cytoplasmic acidification is related to activation of a H^+/Na^+ antiport (Sun et al., 1987). A similar mechanism

TABLE XI. Effect of Plant Hormones on Residual Respiration of Carrot Cells*

| Hormone | Residual respiration[a] | | Increase (%) |
	Control	Hormone added	
Indoleacetic acid (1×10^{-8} M)	0.99 ± 0.17	2.21 ± 0.16	22
Gibberellic acid (5×10^{-7} M)	1.41 ± 0.06	1.79 ± 0.08	27

* Residual respiration is oxygen uptake in the presence of 1 mM KCN and salicylhydroxamic acid to inhibit mitochondrial respiration (Misra PC, FL Crane, unpublished results). Other auxin stimulations of respiration have been reviewed (Audus, 1960; Evans, 1976).
[a] nmol/min · mg dry weight.

Fig. 11.   Oscillations of cytosolic pH (pH_c) and membrane potential (ψ_m) following treatment of *Zea mays* coleoptiles with 1 µM indole-3-acetic acid (IAA) added to the medium, $pH_o = 6$, measured simultaneously with double-barreled pH microelectrodes fabricated as described by Felle (1987a). Dashed line indicates control-pH_c before IAA addition. Data courtesy of Dr. H. Felle, Botanisches Institut I, Justus Liebig University, Giessen, FRG.

could occur at the plant cell surface. As mentioned above, the source of protons for cytoplasmic acidification might very well be NADH oxidation at the inner surface of the plasma membrane (Ivankina and Novak, 1980; Crane, Chapter 5) (Fig. 12). This phenomenon has been used to explain the induction of proton extrusion in animal cells by external impermeant electron acceptors such as ferricyanide and diferric transferrin.

In plants, the cytoplasmic acidification and proton pumping (membrane potential) responses are sinusoidal (Fig. 11). They oscillate with the same period but are displaced from each other by λ/2. Thus, there exists experimental evidence for a highly coordinated and balanced relationship between cytoplasmic pH control and the mechanism that ultimately results in cell wall acidification. Both the necessity and the sufficiency of cell wall acidification for cell elongation to occur has been challenged, however (Kutschera and Schopfer, 1985).

CELL ELONGATION

The driving force of cell elongation is generally considered to be turgor. There is no question that both turgor and wall extensibility may influence elongation rate. What remains to be proven, however, is that high turgor and wall extensibility are sufficient for elongation to occur. For example, experimental conditions have been generated with actinomycin D or impermeant -SH reagents (Eisinger and Morré, 1968; Morré and Eisinger, 1968). Where turgor remained high, wall loosening induced by auxin was more than sufficient to support rapid elongation, and membrane permeability was retained. Yet the tissue sections did not elongate. These findings may serve as a reminder that plant growth, especially auxin-induced growth, may be mediated by active processes.

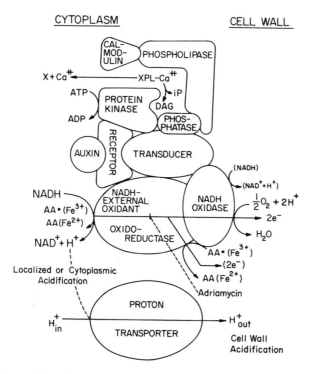

Fig. 12. Diagram illustrating the hypothetical arrangement of constituents leading to signal transduction in auxin regulation of elongation growth in plants. To reconcile the various findings, electron flow from NADH is regarded to be coupled to molecular oxygen via the external NADH oxidase. Impermeant electron acceptors such as ferricyanide or ascorbate radical (AA·) also are reduced by the transmembrane activity. Also indicated in the diagram is the ability of both the NADH oxidase and the external oxidant oxidoreductase to oxidize external NADH directly, which provides the normal biochemical assays of these activities. Other components of the signal response cascade are based on evidence discussed in the text. The role of calcium released from the membrane as a second messenger has been reviewed by Morré (1986). GTP is without effects on the auxin stimulated activities thus far examined with the isolated soybean membranes. The existence of a "transducer" protein is indicated only as conjectural. An auxin receptor protein has not been isolated from soybean plasma membranes. The designation "phosphatase" is to indicate the presence in the membrane of not only phosphatases capable of dephosphorylating phosphoproteins but phosphatases cleaving phosphorylinositol, phosphorylcholine, and phosphatidic acid (phosphatidic acid phosphatase) (Table VI). The latter activities result in the appearance with isolated membranes where phospholipid cleavage is stimulated by auxin of inorganic phosphate and the free phospholipid headgroup (X), as well as diacylglycerols (DAG) either from a phospholipase C or a phospholipase D type cleavage (Fig. 6). The expected auxin concentration dependent release of inorganic phosphate has been observed experimentally (Scherer and Morré, 1978; Fig. 5).

The nature of the active components of growth, even if they exist, is unknown. Explanations based on the initial conformational change in response to auxin or a direct coupling of growth to electron flow via the plasma membrane redox system are inconsistent with the 5–7 min lag between auxin addition and the initiation of growth. The lag appears not to be due to cell wall acidification. Rather, it may be the result of an enzyme system, perhaps activated by cytoplasmic acidification, for example. It would then function in parallel to proton extrusion as both a necessary and a sufficient requirement for expansion to proceed.

Recent findings by Kutschera and Briggs (1987a) remind us that auxin-induced cell elongation is primarily, perhaps exclusively, a function of epidermal cells. In these cells, incorporation of cell wall precursors occurs simultaneously with the onset of the auxin-induced growth (Kutschera and Briggs, 1987b). An increase in cell wall matrix components, particularly of epidermal cells, correlated with increased wall loosening (plastic extensibility), has been offered as an alternative to excretion of protons as a mechanism of cell wall loosening in auxin-induced growth.

The role of second messengers in the control of cell elongation and their formation in response to growth regulatory stimuli and/or auxin hormones is poorly understood at best, and the underlying mechanisms, not at all. While calcium levels, phospholipid metabolism, and protein phosphorylation may eventually prove to be important, definitive studies are lacking. Whether or not the animal paradigm based on polyphosphoinositide metabolism operates in plant growth control or in response to hormone addition is not clear. Phospholipid changes do appear to occur but not necessarily via an auxin or growth-regulated PIP_2-specific phospholipase C (Fig. 12). One must not lose sight of the possibility that plants may have evolved second messenger systems in parallel to but unique from those of animal cells. A major limitation derives from our overall lack of understanding of the growth mechanism itself. Once molecular details of the growth process are better understood, these regulatory mechanisms and second messenger functions unique to control of plant growth may be placed in correct perspective.

REFERENCES

Aducci P, A Ballio, L Fiorucci, E Simonetti (1984): Inactivation of solubilized fusicoccin-binding by endogenous plant hydrolases. Planta 160:422–427.
Albertsson PA, B Anderson, C Larsson, HE Akerlund (1982): Phase partition: A method for purification and analysis of cell organelles and membrane vesicles. Methods Biochem Anal 28:115–150.
Ashendel CL (1985): The phorbol ester receptor: A phospholipid-regulated protein kinase. Biochem Biophys Acta 822:219–242.

Audus LJ (1960): Effect of growth-regulating substances on respiration. In Ruhland W (ed): "Encyclopedia of Plant Physiology," Vol 12, Part B. Berlin: Springer-Verlag, pp 360–387.

Barnes D, G Sato (1980): Serum-free cell culture: A unifying approach. Cell 22:649–655.

Barr R, TA Craig, FL Crane (1984): Plant growth regulators affect transmembrane ferricyanide reduction in carrot cells. In Randall DD, DG Blevins, R Larson, BJ Rapp (eds): "Current Topics in Plant Biochemistry and Physiology." Columbia: University of Missouri Press, p 171.

Barr R, TA Craig, FL Crane (1985): Transmembrane ferricyanide reduction in carrot cells. Biochim Biophys Acta 812:49–52.

Barr R, AS Sandelius, FL Crane, DJ Morré (1986): Redox reactions of tonoplast and plasma membranes isolated from soybean hypocotyls by free-flow electrophoresis. Biochim Biophys Acta 852:254–261.

Basset P, Y Quesneau, J Zwiller (1986): Iron induced L1210 growth: Evidence of a transferrin independent iron transport. Cancer Res 46:1644–1647.

Batt S, MA Venis (1976): Separation and localization of two classes of auxin binding sites in corn coleoptile membranes. Planta 130:15–21.

Batt S, MB Wilkins, MA Venis (1976): Auxin binding to corn coleoptile membranes: Kinetics and specificity. Planta 130:7–13.

Bergeron JJM, WH Evans, II Geschwind (1973): Insulin binding to rat liver Golgi fractions. J Cell Biol 59:771–776.

Berridge MJ (1984): Inositol triphosphate and diacylglycerol as second messengers. Biochem J 220:345–360.

Berridge MJ, RF Irvine (1984): Inositol triphosphate: A novel second messenger in cellular signal transduction. Nature 312:315–321.

Brightman AO, R Barr, FL Crane, DJ Morré (1988): Auxin-stimulated NADH oxidase purified from plasma membrane of soybean. Plant Physiol 86:1264–1269

Brummer P, RW Parish (1983): Mechanisms of auxin-induced plant cell elongation. Fed Eur Biochem Soc Lett 161:9–13.

Buckhout TJ, KA Young, PS Low, DJ Morré (1981a): In vitro promotion by auxins of divalent ion release from soybean membranes. Plant Physiol 68:512–515.

Buckhout TJ, KA Young, PS Low, DJ Morré (1981b): Response of isolated plant membranes to auxins: Calcium release. Bot Gaz 141:418–421.

Clark JE, DJ Morré, H Cherry, WN Yunghans (1976): Enhancement of RNA polymerase activity by non-protein components from plasma membranes of soybean hypocotyls. Plant Sci Lett 7:233–238.

Cleland RE (1976): Kinetics of hormone-induced H^+ secretion. Plant Physiol 58:210–213.

Cleland RE, DL Rayle (1978): Auxin, H^+-excretion and cell elongation. Bot Mag Special Issue 1:1215–139.

Cohen P (1982): The role of protein phosphorylation in neutral and hormonal control of cellular activity. Nature 296:613–620.

Connett RJA, DE Hanke (1986): Breakdown of phosphatidylinositol in soybean cells. Planta 169:216–221.

Crane FL, H Goldenberg, DJ Morré (1979): Dehydrogenases of the plasma membrane. Subcell Biochem 6:345–399.

Crane FL, I Sun, MG Clark, G Grebing, L Löw (1985): Transplasma membrane redox systems in growth and development. Biochim Biophys Acta 811:233–264.

Cross JW, WR Briggs (1978): Properties of a solubilized auxin-binding protein from coleoptiles and primary leaves of Zea mays. Plant Physiol 62:152–157.

Cross JW, WR Briggs, UC Dohrmann, PM Ray (1978): Auxin receptors of maize coleoptile membranes do not have ATPase activity. Plant Physiol 61:581–584.

Dohrmann U, R Hertel, H Kowalik (1978): Properties of auxin binding sites in different subcellular fractions from maize coleoptiles. Planta 140:97–106.

Donnelly TE, R Jensen (1983): Effects of fluphenazine on the stimulation of calcium-sensitive, phospholipid-dependent protein kinase by 12-o-tetradecanoyl phorbol-13-acetate. Life Sci 33:2247–2253.

Drobak BK, IB Ferguson (1985): Release of Ca^{2+} from plant hypocotyl microsomes by inositol-1,4,5-triphosphate. Biochem Biophys Res Commun 130:1241–1246.

Eisinger WR, DJ Morré (1968): The effect of sulfhydryl inhibitors on plant cell elongation. Proc Indiana Acad Sci for 1967 77:136–143.

Ellem KAO, GF Kay (1983): Ferricyanide can replace pyruvate to stimulate growth and attachment of serum restricted human melanoma cells. Biochem Biophys Res Commun 112:183–190.

Elliott DC, M Geytenbeek (1985): Identification of products of phosphorylation in T37-transformed cells and comparison with normal cells. Biochem Biophys Res Commun 845:317–323.

Elliott DC, JD Skinner (1986): Calcium-dependent, phospholipid-activated protein kinase in plants. Phytochemistry 25:39–44.

Ettlinger C, L Lehle (1988): Auxin induces rapid changes in phosphatidylinositol metabolism. Nature 331:176–178.

Evans ML (1976): Rapid responses to plant hormones. Annu Rev Plant Physiol 25:195–223.

Favre B, G Turian (1987): Identification of a calcium and phospholipid-dependent protein kinase C in Neurospora crassa. Plant Sci 49:15–21.

Felle H (1987a): Proton transport and pH control in Sinapis alba root hairs: A study carried out with double-barrelled microelectrodes. J Exp Bot 38:340–354.

Felle H (1987b): Auxin causes oscillations of cytosolic free calcium and pH in Zea mays coleoptiles (submitted).

Felle H, B Brummer, A Bertl, RW Parish (1986): Indole-3-acetic acid and fusicoccin cause cytosolic acidification of corn coleoptile cells. Proc Natl Acad Sci USA 83:8992–8995.

Galston A, R Kaur (1963): An effect of auxins on the heat coagulability of the proteins of growing plant cells. Proc Natl Acad Sci USA 45:1587–1590.

Goldenberg H, FL Crane, DJ Morré (1979): NADH oxido-reductase of mouse liver plasma membranes. J Biol Chem 254:2491–2498.

Hardin JW, JH Cherry, DJ Morré, CA Lembi (1972): Enhancement of RNA polymerase activity by a factor released by auxin from plasma membrane. Proc Natl Acad Sci USA 63:3146–3150.

Hasunuma K, K Funadera (1987): GTP-binding protein(s) in green plant, Lemna Paucicostata. Biochem Biophys Res Commun 143:908–912.

Hasunuma K, K Furukawa, K Tomita, C. Mukai, T. Nakamura (1987): GTP-binding proteins in etiolated epicotyls of Pisum sativum (Alaska) seedlings. Biochem Biophys Res Commun 148:133–139.

Helgerson SL, WA Cramer, DJ Morré (1976): Evidence for an increase in microviscosity of plasma membranes from soybean hypocotyls induced by the plant hormone indole-3-acetic acid. Plant Physiol 58:548–551.

Helsper PFG, PFM de Groot, HF Linskens, JF Jackson (1986a): Phosphatidylinositol phospholipase C activity in pollen of Lilium longiflorum. Phytochem 25:2053–2055.

Helsper PFG, PFM de Groot, HF Linskens, JF Jackson (1986b): Phosphatidyl monophosphate in Lilium pollen and turnover of phospholipid during pollen tube extension. Phytochem 25:2193–2199.

Hepler PK, RO Wayne (1985): Calcium and plant development. Annu Rev Plant Physiol 36:397–439.

Hertel R, K-S Thomson, VEA Russo (1972): In vitro auxin binding to particulate cell fractions from corn coleoptiles. Planta 107:325–340.

Irvine RF (1982): The enzymology of stimulated lipid turnover. Cell Calcium 3:295–309.

Ivankina NG, VA Novak (1980): H^+-transport across plasmalemma: H^+-ATPase or redox-chain? In Spanswick RM, WJ Lucas, J Dainty (eds): "Plant Membrane Transport: Current Conceptual Issues." Amsterdam: Elsevier, pp 503–504.

Jones GP, A Marker, LG Paleg (1984): Complex formation between indole-3-acetic acid and phospholipid membrane components in aqueous media. I. Parameters of the systems. Biochemistry 23:1514-1520.

Jones GP, LG Paleg (1984): Complex formation between indole-3-acetic acid and phospholipid membrane components in liquid media. II. Interaction of auxins and related compounds with phosphatidylcholine membranes. Biochemistry 23:1521–1524.

Kauss H (1987): Some aspects of calcium-dependent regulation in plant metabolism. Annu Rev Plant Physiol 38:47–72.

Key J (1964): Ribonucleic acid and protein synthesis as essential processes for cell elongation. Plant Physiol 39:365–370.

Kjellbom P, C Larsson (1984): Preparation and polypeptide composition of chlorophyll-free plasma membranes from leaves of light-grown spinach and barley. Physiol Plant 62:501–509.

Kubowicz BD, LN Vanderhoef, JB Hanson (1982): ATP-dependent calcium transport in plasmalemma preparations from soybeans hypocotyls: Effects of hormone treatments. Plant Physiol 69:187–191.

Kutschera U, WR Briggs (1987a): Differential effect of auxin on in vivo extensibility of cortical cylinder and epidermis in pea internodes. Plant Physiol 84:1361–1366.

Kutschera U, WR Briggs (1987b): Rapid auxin-induced stimulation of cell wall synthesis in pea internodes. Proc Natl Acad Sci USA 84:2747–2751.

Kutschera U, P Schopfer (1985): Evidence against the acid growth theory of auxin action. Planta 163:483–493.

Landschulz W, I Theslaff, P Esblom (1984): A lipophilic iron chelator can replace transferrin as a stimulator of cell proliferation and differentiation. J Cell Biol 98:596–601.

Löbler M, D Klämbt (1985): Auxin-binding protein from coleoptile membranes of corn. I. Purification by immunological methods. J Biol Chem 260:9848–9853.

Lomax TL, RJ Mehlhorn, WR Briggs (1985): Active auxin uptake by zucchini membrane vesicles: Quantitation by ESR, volume and pH measurements. Proc Natl Acad Sci USA 82:6541–6545.

Loppert H (1983): Energy coupling for membrane hyperpolarization in Lemma: Respiration rate, ATP level and membrane potential at low oxygen concentrations. Planta 159:329–335.

Löw H, IL Sun, P Navas, C Grebing, FL Crane, DJ Morré (1986): Transplasmalemma electron transport is part of a diferric transferrin reductase system. Biochem Biophys Res Commun 139:1117–1123.

Low P, DH Lloyd, TM Stein, JA Rogers III (1979): Calcium displacement by local anesthetics: Dependence on pH and anesthetic charge. J Biol Chem 254:4119–4125.

Majerus PW, TM Connolly, H Deckmyn, TS Ross, TF Bross, H Ishii, VS Bansal, DB Wilson (1986): The metabolism of phosphoinositide-derived messenger molecules. Science 234:1519–1529.

Martin TW, RB Wysolmerski, D Lagunoff (1987): Phosphatidyl choline metabolism in endothelial cells: Evidence for phospholipase A and a novel Ca^{2+}-independent phospholipase C. Biochim Biophys Acta 917:296–307.

Melin PM, M Sommarin, AS Sandelius, B Jergil (1987): Identification of Ca^{2+}-stimulated polyphosphoinositide phospholipase C in isolated plant membranes. Fed Eur Biochim Soc Lett 223:87–91.

Michell RH, SS Jafferji, LM Jones (1977): The possible involvement of phosphatidylinositol breakdown in the mechanism of stimulus-response coupling at receptors which control cell-surface calcium gates. Adv Exp Med Biol 83:447–464.

Møller IM, Lin W (1986): Membrane-bound NAD(P)H dehydrogenases in higher plant cells. Annu Rev Plant Physiol 37:309–334.

Morré DJ (1970): Auxin effects on the aggregation and heat coagulability of cytoplasmic proteins and lipoproteins. Physiol Plant 23:38–50.

Morré DJ (1986): Calcium modulation of auxin-membrane interactions in plant cell elongation. In Trewavas AJ (ed): "Molecular and Cellular Aspects of Calcium in Plant Development." New York: Plenum, pp 293–300.

Morré DJ, CE Bracker (1976): Ultrastructural alteration of plant plasma membranes induced by auxin and calcium ions. Plant Physiol 58:544–547.

Morré DJ, FL Crane, R Barr, C Penel, LY Wu (1988): Inhibition of plasma membrane redox and elongation growth of soybean. Physiol Plant 72:236–240.

Morré DJ, JH Crowe, DM Morré, LM Crowe (1987): Infrared spectroscopic evidence for a conformational alteration of plant plasma membranes upon exposure to the growth hormone analog, 2-4-dichlorophenoxyacetic acid. Biochem Biophys Res Commun 147:506–512.

Morré DJ, B Drobes (1987): A membrane-located, calcium-/calmodulin-activated phospholipase stimulated by auxin. In Stumpf PK, JB Mudd, WD Nes (eds): "The Metabolism, Structure and Function of Plant Lipids." New York: Plenum, pp 229–231.

Morré DJ, Eisinger WR (1968): Cell wall extensibility: Its control by auxin and relationship to cell elongation. In Wightman F, G Setterfield, (eds): "Biochemistry and Physiology of Plant Growth Substances." Ottawa: Ringe Press, pp 625–645.

Morré DJ, B Gripshover, AL Monroe, JT Morré (1984): Phosphatidylinositol turnover in isolated soybean membranes stimulated by the synthetic growth hormone, 2,4-dichlorophenoxyacetic acid. J Biol Chem 259:15364–15368.

Morré DJ, JT Morré, RL Varnold (1984): Phosphorylation of membrane located proteins of soybean in vitro and response to auxin. Plant Physiol 75:265–268.

Morré DJ, P Navas, C Penel, CJ Castillo (1986): Auxin-stimulated NAPH oxidase (semi-dehydroascorbate reductase) of soybean plasma membrane: Role in acidification of cytoplasm. Protoplasma 133:195–197.

Navas P, IL Sun, DJ Morré, FL Crane (1986): Decrease of NADH in HeLa cells in the presence of transferrin or ferricyanide. Biochem Biophys Res Commun 135:110–115.

Neumann P (1985): Events at the surface of the cell. Nature 317:380.

Niedel JE, PJ Blackshear (1986): Protein kinase C. In Putwey JW Jr (ed): "Phosphoinositides and Receptor Mechanisms." New York: Alan R. Liss, pp 47–48.

Nishizuka Y (1984): Studies and perspective of protein kinase C. Science 225:1365–1368.

Novak VA, NG Ivankina (1983): Influence of nitro blue tetrazolium on the membrane potential and ion transport in water thyme. Soviet Plant Physiol 30:845–853.

O'Brien CA, WL Arthur, IB Weinstein (1987): The activation of protein kinase C by the polyphosphoinositol 4-phosphate. Fed Eur Biochem Soc Lett 214:339–342.

Òláh Z, Z Kiss (1986): Occurrence of lipid and phorbol ester activated protein kinase in wheat cells. Fed Eur Biochem Soc Lett 195:33–36.

Oostrom H, Z Kulescha, TB Van Vliet, RB Libbenga (1980): Characterization of a cytoplasmic auxin receptor from tobacco pith callus. Planta 140:44–47.

Pall ML (1985): GTP: A central regulator of cellular anabolism. Curr Top Cell Regul 25:1–20.

Pfaffmann H, E Hartmann, AO Brightman, DJ Morré (1987): Phosphatidylinositol specific phospholipase C of plant stems: Membrane associated activity concentrated in plasma membranes. Plant Physiol 85:1151–1155.

Polevoy VV, N Salamatova (1977): Auxin, proton pump and cell tropics. In Marré E, O Ciferri (eds): "Regulation of Cell Membrane Activities in Plants." Amsterdam: Elsevier, pp 209–216.

Ponca R, RW Brady, A Wilczyhska, HM Schylman (1984): The effect of various chelating agents on the mobilization of iron from reticulocytes in the presence and absence of pyridoxal isonicotinoyl hydrazone. Biochim Biophys Acta 802:477–489.

Price-Jones MJ, JL Harwood (1983): Hormonal regulation of phosphatidylcholine synthesis in plants: The inhibition of cytidyl-transferase activity by indole-3-ylacetic acid. Biochem J 216:627–631.

Qui Z-S, B Rubinstein, AI Stern (1985): Evidence for electron transport across the plasma membrane of Zea mays. Plant Physiol 80:805–811.

Ranjeva R, AM Boudet (1987): Phosphorylation of proteins in plants: Regulatory effects and potential involvement in stimulus/response coupling. Annu Rev Plant Physiol 38:73–94.

Ray PM (1977): Auxin binding sites of maize coleoptiles are localized on membranes of endoplasmic reticulum. Plant Physiol 59:594–59.

Ray PM, V Dohrmann, R Hertel (1977a): Characterization of naphthaleneacetic acid binding to receptor sites on cellular membranes of maize coleoptile tissue. Plant Physiol 59:357–364.

Ray PM, U Dohrmann, R Hertel (1977b): Specificity of auxin binding sites on maize coleoptile membranes as possible receptor sites for auxin action. Plant Physiol 60:585–591.

Reddy ASN, BW Poovaiah (1987): Inositol 1,4,5-triphosphate induced calcium release from corn coleoptile microsomes. J Biochem 101:569–573.

Rincon M, WF Boss (1987): Myo-inositol triphosphate mobilizes calcium from fusogenic carrot (Daucus carota L.) protoplasts. Plant Physiol 83:395–398.

Roland J-C, CA Lembi, DJ Morré (1972): Phosphotungstic acid-chromic acid as a selective electron-dense stain for plasma membrane of plant cells. Stain Technol 47:109–200.

Rubinstein B, AI Stern (1986): Relationship of transplasmalemma redox activity to proton and solute transport by roots of Zea mays. Plant Physiol 80:805–811.

Sandelius AS, DJ Morré (1987): Characteristics of phosphatidylinositol exchange activity of soybean microsomes. Plant Physiol 84:1022–1027.

Sandelius AS, M Sommarin (1986): Phosphorylation of phosphatidylinositols in isolated plant membranes. Fed Eur Biochim Soc Lett 201:282–286.

Schacht, J (1978): Purification of polyphosphoinositides by chromatography on immobilized neomycin. J Lipid Res 19:1063–1067.

Schäfer A, F Bygrave, S Matzenauer, D Marmé (1985): Identification of a calcium-and phospholipid-dependent protein kinase in plant tissue. Fed Eur Biochem Soc Lett 187:25–28.

Scherer GFE, DJ Morré (1978): In vitro stimulation by 2,4-dichlorophenoxyacetic acid of an ATPase and inhibition of phosphatidate phosphatase of plant membranes. Biochem Biophys Res Commun 84:238–247.

Schilling EE, H Goldenberg, DJ Morré (1979): Distribution of insulin receptors among mouse liver endomembranes. Biochim Biophys Acta 555:504–511.

Schumaker KS, H Sze (1987): Inositol 1,4,5-triphosphate releases Ca^{2+} from vacuolar membrane vesicles of oat roots. J Biol Chem 262:3944–3946.

Sexton JC, TS Moore (1981): Phosphatidylinositol synthesis by a Mn^{2+}-dependent exchange enzyme in castor bean endosperm. Plant Physiol 68:18–22.

Sijmons PC, FC Langermeijer, AJ deBoer, HBA Prins, HF Bienfait (1984a): Depolarization of cell membrane potential during transplasma membrane electron transfer to extracellular electron acceptors in iron-deficient roots of *Phaseolus vulgaris* L. Plant Physiol 76:943–946.

Sijmons PC, W van den Briel, HF Beinfait (1984b): Cytosolic NADPH is the electron donor for extracellular Fe^{3+} reduction in iron-deficient bean roots. Plant Physiol 75:219–221.

Sommarin, AS Sandelius (1988): Phosphatidylinositol and phosphatidylinositolphosphate kinases in plant plasma membranes. Biochim. Biophys Acta 958:268–278.

Sun IL, FL Crane, C Grebing, H Löw (1984): Inhibition of plasma membrane NADH dehydrogenase by Adriamycin and related anthracycline antibiotics. J Bioenerg Biomemb 16:209–221.

Sun IL, R Garcia-Canero, W Liu, W Toole-Simms, FL Crane, DJ Morré, H Löw (1987): Diferric transferrin reduction stimulates the Na^+/H^+ antiport of HeLa cells. Biochem Biophys Res Commun 145:465–473.

Torruella M, LM Casano, RH Vallejos (1986): Evidence of the activity of tyrosine kinase(s) and of the presence of phosphotyrosine proteins in pea plantlets. J Biol Chem 261:6651–6653.

Van der Linde PCG, H Bonman, AM Mennes, KR Libbenga (1984): A soluble auxin binding protein from cultured tobacco tissues stimulates RNA synthesis in vitro. Planta 160:102–108.

Vanderwoude WJ, CA Lembi, DJ Morré (1972): Auxin (2,4-D) stimulation (in vivo and in vitro) of polysaccharide synthesis in plasma membrane fragments isolated from onion stem. Biochem Biophys Res Commun 46:245–253.

Varnold RL, DJ Morré (1985): Phosphorylation of membrane-located proteins of soybean hypocotyl: Inhibition by calcium in the presence of 2,4-dichlorophenoxyacetic acid. Bot Gaz 146:315–319.

Veluthambi K, BW Poovaiah (1986): In vitro and in vivo protein phosphorylation in *Avena sativa* L. coleoptiles: Effect of Ca^{2+}, calmodulin antagonists and auxin. Plant Physiol 81:836–841.

Venis MA (1980): Purification and properties of membrane-bound auxin receptors in corn. In Skoog F (ed): "Plant Growth Substances, 1979." New York: Springer-Verlag, pp 61–70.

Venis MA (1986): Receptors for plant auxin action and auxin transport. In Conn PM (ed): "The Receptors," Vol IV. New York: Academic Press, pp 275–314.

Venis MA (1987): Hormone receptor sites and the study of plant development. In Lenten JR, MB Jackson, Atkin RK (eds): "Hormone Action in Plant Development." London: Butterworths, pp 53–61.

Vreugdenhil D, A Burgers, KR Libbenga (1979): A particle-bound auxin receptor from tobacco pith callus. Plant Sci Lett 16:115–121.

Wheeler JJ, WF Boss (1987): Polyphosphoinositides are present in plasma membranes isolated from fusogenic carrot cells. Plant Physiol 85:389–392.

Williamson FA, DJ Morré, K Hess (1977): Auxin binding activities of subcellular fractions from soybean hypocotyls. Cytobiologie 16:63–71.

Yunghans WN, DJ Morré (1977): Adenylate cyclase activity not found in soybean hypocotyl and onion meristem. Plant Physiol 60:144–149.

Zbell B, G Walter (1987): About the search for the molecular action of high affinity auxin-binding sites on membrane-localized rapid phosphoinositide metabolism in plant cells. In Klämbt D (ed): "Plant Hormone Receptors." Berlin: Springer-Verlag, pp. 141–153.

Second Messengers in Plant Growth
and Development, pages 115–143
© 1989 Alan R. Liss, Inc.

5. Plasma Membrane Redox Reactions Involved in Signal Transduction

Frederick L. Crane

Department of Biological Sciences, Purdue University, West Lafayette,
Indiana 47907

OXIDOREDUCTASES IN THE PLASMA MEMBRANE

Plasma membranes contain a remarkable diversity of oxidoreductase enzymes. Enzymes that can oxidize the major substrates found in the cytoplasm and in the exterior environment —including NAD(P)H and ascorbate— are present in plasma membranes. As study of these enzymes has intensified, evidence is being developed that the oxidoreductase enzymes can have a role in signal transduction. The role may be by direct formation of a signal or through the indirect control of other signals.

None of the dehydrogenases responsible for the oxidation of cytosolic substrates by plant plasma membranes have been isolated or well defined. There also is little clear information about the oxidation–reduction carriers in the membrane. The orientation of the oxidation–reduction systems in the membrane has been partially discerned. Oxygen, semidehydroascorbate, or nitrate have been identified as probable natural electron acceptors.

NADH– and NADPH–cytochrome c reductase activity have been reported in isolated plasma membranes from plants (Crane et. al., 1979; Ramirez et al., 1984; Sandelius et al., 1986; Kjellbom and Larsson, 1984; Barr et al., 1986). Cytochrome c reductase activity can be mostly attributed to the activity of dehydrogenases, which have a natural function to reduce cytochrome P450 or cytochrome b_5. The NADPH–P450 reductase is required for mixed function oxidation; the NADH–cytochrome b_5 reductase is part of the fatty acyl coenzyme A (CoA) desaturase. These enzymes are found on endomembranes and would be expected to be present on the

Address reprint requests to Dr. Frederick L. Crane, Department of Biological Sciences, Purdue, IN 49707.

cytosolic side of the plasma membrane (Crane et al., 1985a; Morré et al., 1974).

Enzymes involved in transplasma membrane electron transport are well known in mitochondria and chloroplasts but, only recently, have been carefully examined in plasma membranes (Crane et al., 1985b). These enzymes transfer electrons across the membrane from internal electron donors, such as NADH, through electron transport systems that cross the membrane one or more times (Fig. 1). At the outer surface, electrons may be transferred to impermeable oxidants, such as ferricyanide, which can act as artificial probes for the transmembrane electron transport (Morré et al., 1987a; Craig and Crane, 1985b). The natural oxidants for these enzymes may be oxygen or semidehydroascorbate. Other natural oxidants may be ions, which must be reduced before they enter the cell in the reduced state, for example, ferric, cupric, or nitrate ions (Bienfait, 1985; Jones et al., 1987; Castignetti and Smarrelli, 1986).

Transplasma Membrane Electron Transport

The most common transplasma membrane electron transport system carries out the reduction of external ferricyanide with electrons from cytosolic NADH or NADPH. This type of enzyme is found in both animal and plant cells (Crane et at., 1985b) and has been found in all plants that have been studied (Table I). This enzyme can also account for part of the NAD(P)H ferricyanide reductase activity, which is observed in isolated plasma membranes. Since it is a transmembrane oriented enzyme with a dehydrogenase site on one side and a reduction site on the other side, when the donor and acceptor are impermeable—as with NADH and ferricyanide, respectively—closed membrane vesicles will not reveal this activity. Suitable detergents can be used to reveal the maximum amount of this activity in vesicular membrane preparations. Selective detergent stimulation of the NADH–ferricyanide reductase can be observed in contrast with the small amount of the NADH–cytochrome c reductase, which is oriented exclusively to the interior of the membrane (Sandelius et al., 1986). NADH oxidase activity in plasma membrane vesicles can also be activated by detergent treatment (De Luca et al., 1984; Brightman et al, 1988).

In addition to the universal transplasma membrane electron transport enzyme that carries out ferricyanide reduction by cells, an enzyme that carries out reduction of external iron chelates has been identified. This activity is induced by iron deficiency in certain dicots and has been referred to by Bienfait (1985) as the turbo system (Table II). It is proposed that this enzyme is necessary to convert impermeable and insoluble ferric iron com-

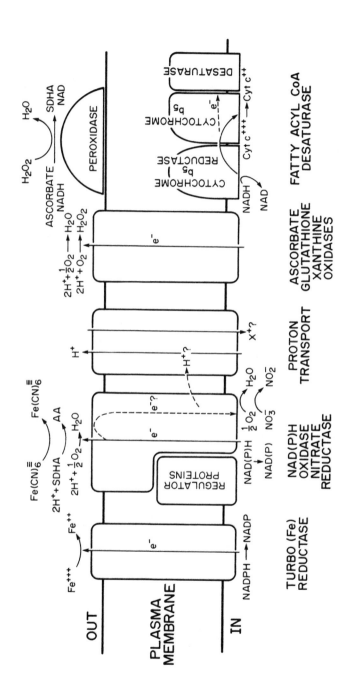

Fig. 1. Oxidoreductase enzymes intrinsic to the plasma membrane.

TABLE I. Ferricyanide Reduction by Plant Tissue or Cells

Species	Part	Condition	$Fe(CN)_6^{3-}$ reduction K_m	Rate	Reference
Corn	Root	0–5 mm	0.1 mM	64[a]	Qiu et al., 1985
Corn	Root	5–10 mm		13[a]	Qiu et al., 1985
Asparagus	Cells	Dark		1.5[b]	Neufeld and Bown, 1987
Asparagus	Cells	Light		22.9[b]	Neufeld and Bown, 1987
Oat	Leaves	Dark	0.15 mM	2.5[c]	Dharmawardhane et al., 1987
Oat	Leaves	Light	0.15 mM	4.8[c]	Dharmawaradhane et al., 1987
Cuscuta	Stem apex	Dark		3.8[d]	Revis and Misra, 1986
Cuscuta	Stem apex	Light		6.4[d]	Revis and Misra, 1986
Dendrophthoe	Leaf	Dark		10.5[d]	Revis and Misra, 1986
Dendrophthoe	Leaf	Light		11.5[d]	Revis and Misra, 1986
Orobanche	Shoot	Dark		1.3[d]	Revis and Misra, 1986
Orobanche	Shoot	Light		1.4[d]	Revis and Misra, 1986
Elodea	Leaves	Dark		55[a]	Revis and Misra, 1986
Elodea	Leaves	Dark		100[a]	Novak and Miklashevich, 1984
Maize	Roots	Aerobic		120[e]	Federico and Giartosio, 1983
Maize	Roots	Anaerobic		100[e]	Federico and Giartosio, 1983
Carrot	Cells	Dark, aerobic		3.5[d]	Craig and Crane, 1981
Carrot	Cells	Dark, anaerobic		2.0[d]	Craig and Crane, 1981
Tobacco callus	Cells	Dark		0.67[d]	Barr et al., 1984
Tobacco crown gall tumor	Cells	Dark		0.58[d]	Barr et al., 1984
Lemma gibba				2.2[c]	Lass et al., 1986
Acer pseudoplatinus	Cells			3.7[d]	Blein et al., 1986
Sugar cane	Protoplasts			1.33[f]	Thom and Maretzki, 1985

[a]nmole $min^{-1}g$ fresh wt^{-1}.
[b]nmole 10^6 $cells^{-1}$ min^{-1}.
[c]μmole g fresh wt^{-1} hr^{-1}.
[d]μmole g dry wt^{-1} min^{-1}.
[e]μmole g fresh wt^{-1}.
[f]μmole 10^4 $cells^{-1}$ min^{-1}.
The cells listed were grown in suspension culture. Sugar cane protoplasts were prepared from cultured cells.

TABLE II. Effects of Iron Nutrition on the Rate of Ferricyanide or Iron Complex
Reduction

Plant	Electron acceptor reduction rate[a]		Reference
	Ferricyanide	Ferric EDTA	
Soybean A2 (+ Fe)		0.06	Tipton and Towson, 1983
Soybean A2 (− Fe)		0.48	Tipton and Towson, 1983
Potato root (+ Fe)		0.1	Bienfait et al., 1987
Potato root (− Fe)		1.9	Bienfait et al., 1987
Bean root (+ Fe)	3.0	0.5	Lubberding et al., 1987
Bean root (− Fe)	7.0	5.0	Lubberding et al., 1987
Lemna gibba (+ Fe)	1.4	0.043	Lass et al., 1986
Lemna gibba (− Fe)	1.7	0.53	Lass et al., 1986
Barley root (+ Fe)	3.8	0.1	Lubberding et al., 1987
Barley root (− Fe)	3.8	0.1	Lubberding et al., 1987

[a] Rates are expressed as microequivalents (μequiv) of Fe reduced per hour per gram fresh
weight except for soybean, which represents an absorbance increase at 562 nm per 4½ min per
nine plants.
Barley is a monocot, all others are dicots.

pounds to soluble and permeable ferrous ions for uptake into the cell. Since
the activity of this enzyme is associated with a dramatic decrease in cytosolic
NADPH levels, this enzyme is apparently a specific transmembrane
NADPH dehydrogenase (Sijmons et al., 1984a; Qiu et al., 1985).

The turbo enzyme is only present in iron-deficient plants (Bienfait, 1985).
It is capable of reducing iron chelates, which have a much lower redox
potential than can be reduced by the standard transmembrane reductase.
The latter only reduces impermeable oxidants with a redox potential over 0
mV at pH 7.0 (Table III) (Crane et al.,1988). Allnutt and Bonner (1987)
have shown that iron deficiency in *Chlorella* induces a redox system that can
reduce ferric chelates of very low redox potential such as desferroxamine B
(− 454 at pH 8.0) (Cooper et al., 1978) and rhodotorulic acid; these compounds cannot be reduced by iron-sufficient cells. The reduction was measured by electron spin resonance, as well as by ferrous bathophenanthroline
sulfonate formation, so formation of the ferrous chelate to increase net redox
potential of the system is not the basis for the reaction. Reduction of ferric
desferroxamine B with NADPH (redox potential at pH 7.0 is − 340 mV) is
possible. Since ferricyanide inhibits the ferroxamine b reduction, the site of
iron reduction by the turbo enzyme can be at the cell surface. The ferricyanide is impermeable to the plasma membrane and can only compete for
electrons at the cell surface.

Monocots that secrete siderophores to increase iron-uptake capacity to
adapt to iron-deficiency conditions do not show an increase in the ability to
reduce iron chelates under iron deficiency (Table II).

TABLE III. Effect of Redox Potential on the Reduction of External Oxidants With Plant Cells

Compound reduced	Eo' pH 7.0 redox potential	Bean[a] depolarized membrane potential	Rate of reduction Diatoms[b] (femtomole cell^{-1} hr^{-1})	Carrot cells[c] (μmole min^{-1} g dry wt^{-1})
Cu BCDS	594		10.2	
Fe (CN)$_6^{3-}$	360	+	101	1.8
Cytochrome c	260		2	
DCPIP	210	+	3	
1,2 NQ-4-S	187			2.5
Cu phenanthroline	174		12	
Cu carbonate	167		19	
PMS	80			
Fe EDTA	94	+	1	0.2
Ru(NH$_3$)$_6$(Cl$_3$)	51			0.8
Fe oxalate	2			0.7
Nitroblue tetrazolium	− 30			0
Indigo tetrasulfonate	− 46			0
Cu EDTA	− 100		0	
Indigo disulfonate	− 125			0
Fe$_4$(P$_2$O$_7$)$_3$	− 140			0
AQ 1,5 disulfonate	− 174			0
FAD	− 220			0
FMN	− 220			0
Cu ethylenediamine	− 380		0	

Plus sign (+) indicates depolarization when incubated with cells.
[a] Sijmons et al., 1984b. Depolarization of bean root cells by oxidants. Cu BCDS, copper bathocuproindisulfonate; DCPIP, dichloroindophenol; PMS, phenazine methosulfate; 1,2 NQ-4-S, 1,2 naththoquinone-4-sulfonate; AQ 1,5 disulfonate, anthroquinone-1, 5-disulfonate; FAD, flavin adenine dinucleotide; FMN, flavin mononucleotide.
[b] Jones et al., 1987.
[c] Craig and Crane, 1985a.

Iron deficiency may also induce activity of redox systems on the exterior surface. Malate and citrate excretion is increased in iron-deficient, responsive dicots (Ric de Vos et al., 1986; Tipton and Thowsen, 1985), which could provide the substrate for malate dehydrogenase in the cell walls to form NADH. The latter could then reduce iron chelates through the external NADH dehydrogenase. Since malate inhibits ferric chelate reduction by soybean roots and the turbo system is 15 times more active than the malate–Fe reductase, Bienfait et al. (1987) concludes that malate dehydrogenases can contribute only a small part of high-efficiency iron reduction.

Blue Light Redox Changes

Other redox functions which have often been found to be associated with the plasma membrane are the blue-light-induced oxidation–reduction changes in flavin or cytochromes of the b type. These oxidoreductions are presumably connected with blue light control of cell growth. The site of the blue light reduction of flavin and cytochrome in relation to the NAD(P)H dehydrogenase activity, either internal or transmembrane, is not clear at this time. Presumably the light-induced reduction could affect the activity of the enzyme, which contains the photosensitive prosthetic group.

Light effects on the transplasma membrane electron transport also have been observed. In leaves the effect is maximum under light, which gives maximum photosynthesis, and the increase in activity can be based on a better supply of products of photosynthesis (Dharmawardhane et al., 1987; Novak and Miklashevich, 1984; Gepstein, 1982; Neufeld and Bown, 1987). In plants like *Cuscuta* (Revis and Misra, 1986), which are not photosynthetic, the stimulation may reflect an influence of the blue light receptor on the electron flow or peroxide production (Schmidt, 1984; Jesaitis et al., 1977; Leong et al., 1981; Kjellbom and Larsson, 1984; Widell and Larsson, 1983; Shinkle and Briggs, 1985).

The Oxidants for Plasma Membrane Electron Transport

In addition to ferric complexes, the extenal oxidant for the transplasma membrane electron transport can be oxygen, semidehydroascorbate, cupric complexes, or nitrate. Compounds with a redox potential as low as 0 mV are able to serve as electron acceptors at the membrane surface of carrot cells (Table III). With diatoms the redox potential of the reduction site on the exterior of the membrane was found to be between $+94$ and -100 mV (Jones et al., 1987). Thus the external site has a potential low enough to reduce oxygen (800mV) or semidehydroascorbate (53 mV). Isolated plant plasma membranes have NAD(P)H oxidase activity, but the enzyme responsible for the activity has not been completely defined. The NADH oxidase activity in soybean plasma membranes is stimulated by auxin (Morré et al., 1986). In the presence of ascorbic acid, the transmembrane NADH dehydrogenase may act as a semidehydroascorbate reductase to form ascorbate, which is oxidized by oxygen in a reaction catalyzed by ascorbate oxidase. Auxin stimulation of transmembrane ferricyanide reduction by tobacco cells also has been shown (Barr et al., 1984), so that dehydrogenase for the auxin-stimulated oxidase can be the transmembrane enzyme (Fig. 2). Further evidence for the relationship between the transmembrane ferricyanide reduction and the hormone-sensitive NADH oxidase developed

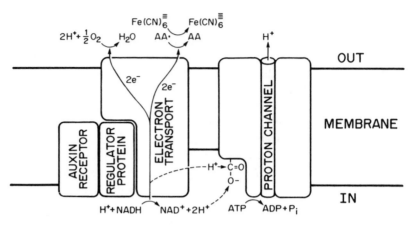

Fig. 2.   Possible components and arrangement of the transplasma membrane NADH oxidase, which is activated by auxin and can activate excretion of protons from the cell through an associated proton channel. Ascorbate radical (A·) in the cell wall may act as an alternate electron acceptor, which can be reoxidized by ascorbate oxidase in the wall. Ferricyanide can act as an alternate electron acceptor (cf. Morré et al., 1988). The auxin receptor may be facing inside or outside.

from studies on effects of ferricyanide on oxygen uptake by cells. Polevoy and Salamatova (1977) have shown that the auxin-stimulated oxygen uptake by maize coleoptile is inhibited by ferricyanide. Similar ferricyanide inhibition has been shown for oxygen uptake by carrot cells (Table IV). Treatment of carrot cells with the membrane-impermeable diazobenzene sufonate inhibits oxygen uptake in parallel with inhibition of ferricyanide reduction (Craig and Crane, 1985b). It has also been shown by Crane et al. (1985b) that auxin stimulates the cyanide and SHAM-insensitive (residual) oxygen uptake by carrot cells.

The nitroxide spin probe 3-doxyl-17-β-hydroxy-5-α-androstane places a nitroxide group at the mid-bilayer portion of the plasma membrane in a location not accessible to the aqueous environment. The nitroxide group is reduced to hydroxylamine by cells under anaerobiosis and the hydroxylamine is reoxidized by ferricyanide, Fe–EDTA, or oxygen. This is consistent with electron transfer from reductants in the cell to a mid-bilayer electron acceptor, which can then be reoxidized by external ferricyanide. Oxygen can also act as an electron acceptor (Dupont and Windle, 1987; Windle, 1985). Reduced nitroxide probe in pure lipid vesicles cannot be reoxidized by ferricyanide, so an electron carrier in the membrane is necessary for transfer of electrons from the interior of the membrane to external ferricyanide.

TABLE IV. Effect of Ferricyanide on Oxygen Uptake by Plant Cells

Cells	Addition	O_2 uptake (μl 15 section^{-1} hr^{-1})	Elongation (mm/section^{-1} 5 hr^{-1})
Corn coleoptile	None	113 ± 5	1.2 ± 0.09
Corn coleoptile	Fe(CN)$_6^{3-}$ (0.05 M)	119 ± 9	0.2 ± 0.13
Corn coleoptile	IAA (1 μM)	143 ± 7	2.9 ± 0.22
Corn coleoptile	IAA + Fe(CN)$_6^{3-}$ (0.05 M)	110 ± 12	0.5 ± 0.15
		O_2 uptake (nmole min^{-1} mg dry wt^{-1})	Growth (g dry wt/flask)
Carrot cells	None	4.4 ± 0.10	0.70
Carrot cells	0.2 mM Fe(CN)$_6^{3-}$ (FeIII)	3.5 ± 0.05	0.12
Carrot cells	0.2 mM Fe(CN)$_6^{4-}$ (FeII)	4.8 ± 0.06	
Carrot cells	Anti A + DSF	4.8 ± 0.08	
Carrot cells	Anti A + DSF + 0.2 mM Fe(CN)$_6^{3-}$	3.8 ± 0.02	

Corn coleoptile data are from Polevoy and Salamatova, 1977.
Carrot cells grown in suspension culture (Crane and Barr, in preparation, see also Crane et al., 1984).
The carrot cells are grown with 2,4-dichlorophenoxyacetic acid (2,4-D) in the culture media so the rate may be the auxin stimulated rate (Craig and Crane, 1982); 1% ethanol is added to give maximum electron transport by increasing NADH in the cytoplasm.
Anti A, antimycin A, 1 μM; DSF, disulfiram, 1 mM; IAA = indole-3-acetic acid.

The transplasma membrane elctron transport is highest during maximum growth of carrot cells (Crane et al., 1988; Barr and Crane, 1985) or in the most rapidly growing portion of roots (Qui et al., 1985) or shoots (Revis and Misra, 1986), which is the time when auxin effects would be most important. It is also most active during the most rapid phase of growth of yeast (Crane et al., 1982). Böttger and co-workers have shown that an oxidase in plant cells with a low affinity for oxygen is associated with auxin-induced proton movements; they have proposed that this oxidase is based on the trans-plasma membrane electron transport system (Böttger et al., 1985a and b; Böttger and Lüthen, 1986; Böttger, 1986) with NADH as the cytosolic substrate. Hormone-controlled NADH oxidase activity in plasma membrane is found in rapidly growing hypocotyl regions (Brightman et al., 1987a). Further evidence for a role of this oxidase in growth control derives from the fact that inhibitors of the plasma membrane NADH oxidase inhibit auxin-induced growth of soybean hypocotyls (Morré et al., 1987b). These inhibitors include actinomycin D, nitrophenyl acetate, Adriamycin, and cis-diamino dichloro platinum II. All of these inhibitors inhibit elongation of soybean hypocotyl (Morré et al., 1987b). Nitrophenyl acetate inhibits NADH–ferricyanide reductase activity in the membranes isolated by the free-flow procedure (100% at 200 µm). These membranes have the cytosolic side exposed (Barr et al., 1986); 40% of the membranes isolated by phase separation are partially right-side-out vesicles (Sandelius et al., 1986). Nitrophenyl acetate at 200 µM inhibits carrot cell proliferation by 90%.

Both NADH and NADPH may act as substrate for the transmembrane electron transport enzyme. Direct measurement of reduced pyridine nucleotide levels in plant tissue after addition of external oxidants has shown a decrease in NADPH, with a corresponding increase in NADP, whereas little change in NADH concentration is seen. For maize root segments Qiu et al. (1985) showed a 20% decrease in NADPH after 5 min incubation with ferricyanide and no significant change in NADH concentration. In bean roots ferricyanide addition caused a 33% decrease in NADPH for plants supplied nutrient iron and a 50% decrease for iron-deficient plants (Sijmons et al., 1984b). NADH levels were very low and showed no significant change. The stimulation of ferricyanide reduction by light in tissue capable of photosynthesis also has been attributed to an increased supply of NADPH inside the chloroplast, which, in turn, increases cytosolic substrates, since the increase in ferricyanide reduction can be inhibited by DCMU, which inhibits photosynthetic NADPH formation (Neufeld and Bown, 1987; Dharmawardhane et al., 1987). On the other hand, treatments that would increase cytosolic levels of NADH have also been shown to increase the rate of ferricyanide reduction.

Ethanol gives excellent stimulation of ferricyanide reduction by cells that contain alcohol dehydrogenase. The effect has been seen in yeast (Crane et al., 1982), carrot cells (Craig and Crane, 1981; Chalmers et al., 1984; Chalmers and Coleman, 1983), and corn roots (Böttger and Lüthen, 1986). The alcohol dehydrogenase inhibitor pyrazole inhibits the ethanol stimulation of ferricyanide reduction. Alcohols which do not act as substrate for alcohol dehydrogenase, such as methanol or 2-propanol, do not increase ferricyanide reduction. These results would be consistent with NADH acting as an electron donor for the transmembrane electron transport. The inhibition of glycolysis by iodoacetate and arsenite in carrot cells also causes a 50% decline in ferricyanide reduction, which would be consistent with an NADH role in the activity (Barr et al., 1985a). Proton release by carrot cells is stimulated by ethanol and inhibited by pyrazole, even when ferricyanide is not added as an artificial oxidant. This result is consistent with an activation of the proton release by an NADH oxidase in the membrane (Chalmers et al., 1984).

Since plasma membrane preparations from plants can use both NADH and NADPH as substrate for electron transfer to oxygen (Barr et al., 1985b) or to ferricyanide (Sandelius et al., 1986; Ramirez et al., 1984; Barr et al., 1986), both cytosolic NADH and NADPH pools may contribute to the transmembrane electron transport. A slower formation of NADPH, compared with NADH in the cytoplasm, would account for the observation of changes in NADPH but not NADH when ferricyanide is added to cells. Extraction of NADH and NADPH dehydrogenases from microsomes and plasma membranes reveals that more than one enzyme is present in the membranes and that the ability of these enzymes to react with artificial oxidants varies (Pupillo et al., 1986; Guerrini et al., 1987). The identification of the plasma membrane enzymes in extracts from a mixture of membranes is difficult.

Other Cytosolic Electron Donors

Attention has largely been focused on pyridine nucleotides as the primary electron donors for the plasma membrane electron transport. Other cytosolic substrates that should be considered are ascorbate, glutathione or thioredoxin, and xanthine (Crane et al., 1979). There are reports of xanthine oxidase (Somers et al., 1983; Fujihara and Yamaguchi, 1978) and thioredoxin reductase (Schallreuter and Wood, 1986), which can also function in plasma membranes. Ascorbate oxidase would provide for regeneration of ascorbate radical, which can act as electron acceptor for the plasma membrane NADH dehydrogenase (Morré et al., 1986). NADPH concentration in cytosol also will be influenced by the activity of the glutathione reductase and the redox state of glutathione. Nitrate reductase activity may also be controlled by ferricyanide (Tanaka and Asagami, 1986). Jones and Morel (1988) have shown that antibodies to nitrate reductase can inhibit transmembrane electron transport in diatoms. They propose that the dehydrogenase portion of nitrate reductase carries out transmembrane electron transport and proton release, while the molybdoprotein nitrate reductase subunit is oriented to the cytosol.

Redox Components Found in Plasma Membranes

Only a few studies have been made of possible redox carriers in plasma membranes, with the exception of cytochromes. Flavin has been reported in some membrane preparations (Table V). Cytochrome b type components also have been found. The identity of these cytochromes is defined only by spectra. In general, they have an α band at longer wavelengths than cytochrome b_5 (556 nm) (Table VI).

TABLE V. Redox Components in Plasma Membranes

Plant material	Flavin content (nmol mg prot^{-1})	Reference
Maize coleoptile	0.4	Yamamoto et al., 1985
Avena roots	0.2	Ramirez et al., 1984
Glycine hypocotyl	0.5	Barr et al., 1986
Saccharomyces	0.28	Ramirez et al., 1984

TABLE VI. Cytochromes in Plasma Membranes

Plant material	α peak λ	nmole/g protein	Reference
Neurospora	560	—	Munoz and Butler, 1975
Maize coleoptile	560	160	Widell et al., 1980
Saccharomyces	—	0	Ramirez et al., 1984
Saccharomyces	557	180	Schneider et al., 1979
Avena roots	560	217	Ramirez et al., 1984
Glycine hypocotyl	561	500	Sandelius et al., 1986
Barley roots	560	—	Kurkova and Verkhovskaya, 1984
Elodea shoots	560	—	Kurkova and Verkhovskaya, 1984

Major attention has been paid to redox components in the membrane that respond to blue light. This effect has been related to blue light growth responses in fungi and higher plants (Kjellbom and Larsson, 1984; Leong and Briggs, 1982; Widell and Larsson, 1983, Jesaitis et al., 1977).

Proton Movement

Transplasma membrane electron transport is associated with proton release from cells. The basis for the redox-associated proton movement is not clear, and the pathway by which the protons move out of the cell has not been defined. Ferricyanide as an external oxidant stimulates proton release from cells stoichiometrically with ferricyanide reduction (Table VII). This proton release can be seen immediately (Craig and Crane, 1985b; Böttger, 1986) or may take a minute or so to appear (Chalmers et al., 1984; Rubinstein and Stern, 1986) (Fig 3; Table VIII).

Selective inhibition of the proton movement in cells before and after addition of ferricyanide as an electron acceptor show that the channel for the redox-induced electron flow may be different from that for the ATP-dependent H^+ release (Table IX). On the other hand, it has been proposed that the electron transport may activate the transmembrane-proton-excreting ATPase by acidification of an internal control site on the ATPase (Morré et al., 1986).

TABLE VII. Changes in Membrane Potential Induced by Addition of External Oxidants to Cells and Tissues

Tissue or cells	Addition	Membrane Potential (mV)	Change from Control (ΔmV)
Bean root[a] (+ Fe)	Control	-130	
Bean root (+ Fe)	0.5 mM Fe(CN)$_6^{3-}$	-120	$+10$
Bean root (− Fe)	Control	-120	
Bean root (− Fe)	0.5 mM Fe(CN)$_6^{3-}$	-75	$+45$
Bean root (− Fe)	Control	-118	
Bean root (− Fe)	0.2 mM Fe Na EDTA	-90	$+28$
Lemna gibba[b] (+ Fe)	Control	-229	
Lemna gibba (+ Fe)	0.4 mM Fe(CN)$_6^{3-}$	-159	$+70$
Lemna gibba (+ Fe) in light	Control	-249	
Lemna gibba (+ Fe) in light	0.4 mM Fe(CN)$_6^{3-}$	-231	$+18$
Lemna gibba (+ Fe) anaer-obic	Control	-96	
Lemna gibba (+ Fe) anaer-obic	0.4 mM Fe(CN)$_6^{3-}$	-84	$+12$
Lemna gibba (+ Fe)	Control	-214	
Lemna gibba (+ Fe)	0.4 mM Fe(CN)$_6^{4-}$ (ferro)	-209	$+5$
Lemna gibba (+ Fe)	0.4 mM Fe Na EDTA	-222	$+9$
Lemna gibba (− Fe)	0.4 mM Fe Na EDTA	-172	$+70$
Lemna gibba (+ Fe)	0.2 mM DCPIP	-134	$+98$
Lemna gibba (+ Fe)	0.1 mM methylene blue	-140	$+77$
Lemna gibba[c]	1 mM CN$^-$, 1 mM SHAM	-166	$+83$
Elodea canadensis[d] (dark)	Control	-180	
Elodea canadensis (dark)	0.4 mM Fe(CN)$_6^{3-}$	-20	$+160$
Elodea canadensis (dark)	0.5 mM Fe(CN)$_6^{4-}$ (ferro)	-160	$+20$

(+ Fe), plants supplied iron in growth media; (− Fe), iron-deficient plants.
[a] Sijmons et al., 1984b.
[b] Lass et al., 1986.
[c] Loos and Lüttge, 1984.
[d] Ivankina et al., 1984.

Membrane Potential Changes

External electron acceptors induce a remarkable decrease in the plasma membrane potential of all cells tested. The change is consistent with a decrease of excess negative ions or an increase of cations in the cytoplasm (Table VII). Since reduction of external ferricyanide is accompanied by proton release on a one-for-one basis, according to most reports (Table X), the electron transport should be nonelectrogenic and cannot account for the decrease in membrane potential.

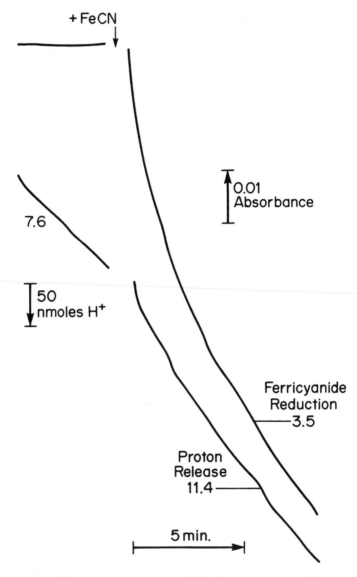

Fig. 3. Ferricyanide reduction and associated H^+ release by carrot cells by simultaneous recording in the same reaction vessel. Ferricyanide reduction measured by decrease in absorbance at 420–500 nm. Decrease in pH measured with a glass electrode in the spectrophotometer cuvette. Numbers on the tracings indicate μequiv min^{-1} (g dry weight)$^{-1}$. Ferricyanide induced proton release is calculated as 3.8 μequiv min^{-1} (g dry wt)$^{-1}$ by subtraction of the rate without ferricyanide from the rate with ferricyanide (Craig and Crane, 1985b).

TABLE VIII. Oxidation–Reduction-Induced Ion Movement

Cells	Oxidant or reductant addition	Effect on ion movement	Reference
Corn roots	$Fe(CN)_6^{3-}$	$K^+in \downarrow$ $H^+o \uparrow$	Kochian and Lucas, 1985
Sugar cane protoplast	$NADH\ K_i \downarrow, H^+o$ $\downarrow NADH +$ $Fe(CN)_6^{3-}$	$H^+i \uparrow \psi \uparrow$ $H^+o \uparrow$	Thom and Maretzki, 1985
Acer cells	$Fe(CN)_6^{3-}$	$H^+o \uparrow$	Blein et al., 1986
Maize coleoptile	$80\ \mu M\ O_2$	$H^+o \uparrow$ [a]	Böttger, 1986
(+ auxin)	$40\ \mu M\ O_2$	$H^+o \uparrow$ [b]	Böttger, 1986
	$10\ \mu M\ O_2$	$H^+o \uparrow$ [c]	Böttger, 1986
	$250\ \mu M\ O_2$	$H^+o \uparrow$ [d]	Böttger, 1986
Maize root	$Fe(CN)_6^{3-}$	$K^+i \uparrow$ $H^+o \uparrow$	Rubinstein and Stern, 1986
Maize root	NADH	$K_i \uparrow$ $H^+o \downarrow$	Lin, 1984
Carrot cells	NADH	$K_i \uparrow$	Misra et al., 1984
Carrot cells	$Fe(CN)_6^{3-}$	$H^+o \uparrow$	Chalmers et al., 1984
Carrot cells	$Fe(CN)_6^{3-}$	$H^+o \uparrow$	Barr et al., 1985b
Yeast	$Fe(CN)_6^{3-}$	$H^+o \uparrow$	Crane et al., 1982

All cells and protoplasts were from suspension culture.
[a] $4\ nmol\ sec^{-1}\ gww^{-1}$.
[b] $2\ nmol\ sec^{-1}\ gww^{-1}$.
[c] $0.3\ nmol\ sec^{-1}\ gww^{-1}$.
[d] $5.7\ nmol\ sec^{-1}\ gww^{-1}$.

Novak and co-workers (Novak and Ivankina, 1983; Ivankina et al., 1984; Novak and Miklashevich, 1984) and Böttger and co-workers (Böttger and Lüthen, 1986; Böttger et al., 1985a and b) have proposed that the ferricyanide-induced decrease in membrane potential can be based on a proton-pumping NADH oxidase in the plasma membrane, which transfers electrons to internal oxygen while releasing protons to the exterior, thus providing electrogenic proton movement to maintain part of the membrane potential. Ferricyanide would intervene to remove electrons from the redox system at the cell surface, thus eliminating the electrogenic effect.

The presence of an oxidase in the plasma membrane to maintain the proton movement would be necessary to support this idea. Some evidence has been reported for a relationship between oxygen uptake and proton release distinct from the role of oxygen in mitochondrial ATP production. The maintenance of membrane potential in Lemna requires a higher concentration of oxygen ($8\ \mu M$) than is required to maintain ATP levels ($1\ \mu M$) (Loppert, 1983). External ferricyanide does not cause a decrease in ATP in

TABLE IX. Inhibition of ATP-Dependent and Ferricyanide-Stimulated Proton Release by Plant Cells

Cells	Inhibitor	% of control activity Basal H^+ release	$Fe(CN)_6^{3-}$ - stim. H^+ release	References
Carrot	Diethylstilbesterol (100 μM; 20 min)	0	90	Barr and Crane, 1987
Carrot	Heptylhydroxyquinoline N oxide (60 μM)	7	70	Barr et al., 1987
Carrot	KCN (200 μM)	13	100	Barr et al., 1987
Carrot	Azide (100 μM)	26	64	Barr et al., 1987
Carrot	Argon (anaerobic)	8	46	Barr et al., 1987
Carrot	Gramicidin D (3 μM)	10	10	Barr et al., 1987
Carrot	1799 (0.3 μM)	20	18	Barr et al., 1987
Carrot	Oligomycin (100 μM)	10	160	Barr et al., 1987
Carrot	Antimycin A (4 μM)	16	88	Barr et al., 1987
Maize root	Fusicoccin	1,946	635	Federico and Giartosio, 1983
Aspara-gus	Diethylstilbesterol (100 μM)	0	100	Neufeld and Bown, 1987
Carrot	NaCl (50 mM)	80	163	
Carrot	DCCD (100 μM)	20	54	Barr and Crane, unpubl.
Carrot	Nitrophenylacetate (100 μM)	214	0	Barr and Crane, unpubl.
Carrot	Actinomycin D (10 μM)	129	0	Barr and Crane, unpubl.
Carrot	Chloroquine (60 μM)	186	0	Barr and Crane, unpubl.

Lemna (Lass et al., 1986), so a lack of ATP would not cause the membrane potential decrease. With young corn coleoptiles (1 cm), proton release is decreased at below 100 μM O_2; wheras in older coleoptiles (7 cm), high activity of proton release is maintained above 10 μM O_2. Fusicoccin stimulates the low affinity oxidase and proton secretion. Böttger et al. (1985a) propose that an oxidase in the plasma membrane with a low affinity for oxygen can account for the proton release dependent on high oxygen levels (Böttger et al., 1985b). The isolation of a hormone-stimulated NADH oxidase from elongating segments of soybean hypocotyl supports the presence of a plasma membrane oxidase (Brightman et al., 1987a and b).

The proposed intervention by ferricyanide to decrease membrane potential also should decrease oxygen uptake in young tissue. In several studies no inhibition of oxygen uptake by ferricyanide was observed (Lass et al., 1986; Novak and Miklashevich, 1984). This was explained by ferricyanide oxidation of cytosolic NADH, which would supply NAD to increase pyruvate

TABLE X.   Relation Between Transmembrane Ferricyanide Reduction and Proton Release by Plant Cells

Cell or tissue	$Fe(CN)_6^{3-}$ reduction	$Fe(CN)_6^{3-}$ stimulated proton release	Ratio H^+/e^-	Reference
Asparagus cultured cells (dark)	1.55	1.54	1.0	Neufeld and Bown, 1987
Asparagus cultured cells (light)	22.9	26.1	1.1	Neufeld and Bown, 1987
Acer cultured cells	3.7	3.4	0.93	Blein et al., 1986
Maize root	0.98	0.38	0.39	Rubinstein and Stern, 1986
Phaseolus root			0.5	Sijmons et al., 1984b
Elodea leaves (dark)	60	66	1.1	Ivankina et al., 1984
Maize root (aerobic) − FC	1.2	0.49	0.4	Federico and Giartosio, 1983
Maize root (anaerobic) − FC	1.0	0.03	0.03	Federico and Giartosio, 1983
Maize root (aerobic) + FC	3.1	1.5	0.48	Federico and Giartosio, 1983
Maize root (anaerobic) + FC	1.2	0.3	0.25	Federico and Giartosio, 1983
Carrot cultured cells	3.5	3.8	1.1	Craig and Crane, 1985b
Saccharomyces cerevisae	0.87	0.65	0.75	Crane et al., 1982

formation to increase mitochondrial oxygen uptake. On the other hand, ferricyanide inhibition of oxygen uptake by carrot cells can be demonstrated (Crane et al., 1987). This decrease of oxygen uptake by ferricyanide is favored both by ethanol, which increases cytosolic NADH, and by inhibition of mitochondrial oxidase activity with antimycin plus disulfiram (Table IV). This decrease in oxygen uptake in the presence of external ferricyanide is consistent with removal of electrons from a transmembrane electron transport chain before the site for oxygen reduction is reached.

It also is possible that the connection between the plasma membrane redox system and oxygen may be through oxidation and reduction of ascorbate (Morré et al., 1986). Hassidim et al. (1987) have shown that ferricyanide produces an inside positive membrane potential in vesicles formed with isolated plant plasma membranes only if ascorbate is inside the vesicle.

External Dehydrogenase and Peroxidase

Oxidation of external NADH or NADPH by plant cells has been widely observed (Lin, 1982; Møller and Lin, 1986; Askerlund et al., 1987). The reduction of external NAD may be accomplished with enzymes such as malate dehydrogenase in the cell wall. These enzymes use substrates such as malate excreted from the cell (Tipton and Thowsen, 1985). Two different oxidase systems for NADPH may then be detected on the cell surface.

At high concentrations of NAD(P)H there is a rapid oxidation, which is catalyzed by peroxidase. This activity is inhibited by cyanide, which inhibits peroxidase, and by catalase, which removes the H_2O_2 necessary for the peroxidation reaction (Mäder and Amberg-Fisher, 1982; Askerlund et al., 1987) and the activity is stimulated by SHAM (Møller and Berczi, 1985; Møller and Berczi, 1986).

The external NADH oxidation has been subject to extensive study because of reports that it stimulates transport activity in the plasma membrane (Lin, 1984; Misra et al., 1984). The peroxidase-catalyzed oxygen uptake is stimulated by SHAM, which may account for SHAM-stimulated respiration observed in older plant tissue (Brouwer et al., 1986; Sesay et al., 1986).

Redox and Messenger Function

Plasma membrane electron transport systems can be involved in second messenger function by direct action on cytosolic components or by modification of activity of other membrane enzymes, which are directly involved in messenger formation. Direct changes in cytosolic composition can include changes in pH, redox ratio (e.g., NAD:NADH, GSH:GSSH, or ascorbate:dehydroascorbate), membrane potential by electrogenic electron flow, or production of superoxide or hydrogen peroxide. Indirect effects can derive by modification of the activity of other membrane enzymes involved in second messenger production (Table XI).

Release of protons from growing plant cells under the influence of auxin has been related to stimulation of cell elongation in many studies (Cleland, 1971). This proton movement depends on the action of an ATPase in the plasma membrane (Sze, 1985; Marrè and Balarin-Denti, 1985). The transplasma membrane electron transport system also increases proton release from cells so that the ATPase or another proton channel can be activated. It has been proposed that the acidification of an internal site on the membrane may activate the proton pumping ATPase (Brummer and Parish, 1983). The transmembrane redox system can produce the internal proton for activation of this internal site (Morré et al., 1986). The effect would be analogous to the suggested redox activation of the Na^+/H^+ antiport in animal cells by release of an internal proton to activate the proton control site on the antiport (Sun et al., 1987). This type of activation involves a proton site on the membrane-bound protein and would not require acidification of the bulk cytoplasm. Direct transfer of the proton messenger from the redox site to the activation site could be shielded on the surface of the membrane (Williams, 1983; Dilley et al., 1987). Change in the bulk cytoplasmic pH tends to develop more slowly after auxin treatment, so it may depend on activation of the ATPase.

TABLE XI. Possible Relationships Between Oxidoreductase Activity in the Plasma Membrane and Transmembrane Signaling

Second messenger functions	Possible intervention in messenger production by plasma membrane oxidoreductase enzymes
Increase or decrease in cytosolic pH	1. $NADH^+$ oxidation releases internal H^+ 2. Proton carrier moves H^+ through membrane 3. Redox opens H^+ release channel or energizes H^+ pump 4. Change glycolytic enzyme activity
Increase in cytosolic Ca^{2+}	1. Redox H^+ release or conformation change opens Ca^{2+} channel in plasma membrane 2. Redox increases IP_3 release to open Ca^{2+} channels in internal membranes
Increase hydrolysis of phosphoinositides by phospholipase C Release IP_3 and DAG	1. H^+ activation of phospholipase at membrane surface 2. Control of inositol phosphate reductase by NAD^+ concentration
Phosphorylation of receptors by membrane-associated protein kinase C	1. Cytosolic pH change causes PKC association with membrane 2. Redox change in membrane electron carrier activates membrane-bound PKC
Change in membrane potential	1. Uncompensated H^+ movement across membrane with electron movement to internal site 2. Activation of channel with unequal ion exchange
Activation of guanylate cyclase	1. Redox activation of the cyclase or indirect by superoxide generation to modify other oxidant levels 2. Nitrate reductase to generate NO
Change in redox state of cytosolic pyridine nucleotides to change activity of pyridine-nucleotide-linked dehydrogenase	1. Ribonucleotide reductase 2. Poly ADP ribose synthesis 3. Inositolphosphate dehydrogenase 4. ADP ribosylation of G proteins
Change in thiol-disulfide in membrane proteins	1. Receptor or catalytic protein S-S site reduction 2. Thioredoxin or glutathione reductase

IP_3, inositol-1,4,5-trisphosphate; PKC, protein kinase C.

Electron transport across the plasma membrane can shift the oxidation reduction state of major redox carriers in the cytoplasm. Oxidation of NADPH by external oxidants has been demonstrated in plant cells (Qiu et al., 1985; Sijmons, 1984a), and NADH oxidation is seen in animal cells (Navas et al., 1986). Changes in ascorbate or glutathione (GSH) concentration may also be important in control of cellular function. For example, GSH has been shown to be related to carrot embryo development. Loss of GSH increases embryogenesis (Earnshaw and Johnson, 1985, 1987). Changes in NAD:NADH or NADP:NADPH ratio may affect the activity of enzymes, such as ribonucleotide reductase, which are important in DNA formation. NAD is necessary for ADP ribosylation of G proteins, which control signal transmission from agonist receptors in the plasma membrane (Khachatrain et al., 1987; Gilman, 1984). NAD-dependent ADP ribosylation is also necessary for poly ADP ribose synthesis, which, in turn, activates enzymes involved in DNA replication (Gaal et al., 1987).

Artificial external oxidants, which inhibit both growth of plant cells (Crane et al., 1984; Crane et al., 1988, 1987) and oxygen uptake through the plasma membrane oxidase (Böttger and Lüthen, 1986), also decrease membrane potential (Sijmons et al., 1984b; Novak and Miklashevich, 1984). This change in membrane potential indicates that the natural oxidase system in normal function induces a membrane potential that could be important for transport of ions and solutes into the cell (Novak and Ivankina, 1983; Crane et al., 1985b). Auxin, benzyladenine, and fusiccoccin increase membrane potential (Marré et al., 1974), but the basis for the redox-induced membrane potential is not well defined. Ion-gradient changes induced by the membrane potential may be important in transmembrane signaling (Lazlo et al., 1981).

The possible role of the transmembrane electron transport systems in control of thiol−disulfide transition in membrane proteins has not been explored. The presence of NADPH thioredoxin reductase as a transplasma membrane enzyme is indicative of this type of action (Schallreuter and Wood, 1986); NADH, NADPH, or GSH are potential reductants for disulfide reduction in the plasma membrane (Mukherjee and Mukherjee, 1981). Oxidation-reduction of thiols in membrane receptors, for example, insulin or transferrin receptors, can modify the receptor function or transport function (Sokol et al., 1986; Roth et al., 1983).

Oxidation reduction reactions have also been implicated in control of the formation of cAMP and cGMP in animal cells (Löw and Werner, 1976; Clark et al., 1977; Dohi and Murad, 1981; Makino et al., 1982). The control of adenylate cyclase may not occur in plant cells, but control of cyclic GMP may be important.

The oxidation–reduction state of membrane transport systems may control protein phosphorylation in the membrane. For example, the redox state of plastoquinone in chloroplasts controls phosphorylation of a protein in the membrane that regulates photosynthesis (Farchaus et al., 1982). This may be accomplished by increasing the binding of protein kinase C to the membrane (Malviya and Angland, 1986) or by activation of kinase by a conformation change in a redox carrier.

Proton release to a localized membrane site could activate phospholipase C. This process increases hydrolysis of phosphatidylinositol, with consequent activation of protein kinase C and calcium release, which are important in many proposed signal functions (Nishizuka, 1984). For example, Pfaffmann et al. (1987) have described a plasma membrane phosphatidylinositol-specific phospholipase C, which is most active at pH 6.6. In cell homogenates 85% of this enzyme is found in the supernatant and 10% in plasma membrane. Since cytosolic pH can control attachment of enzymes to the plasma membrane and in many cases a decrease in pH increases binding to the membrane (Crane et al., 1979), the protonation of the cytoplasm by the redox system may increase association of the phospholipase with the plasma membrane, to increase phospatidylinositol hydrolysis.

Ca^{2+} channels may be controlled by the membrane redox system. Opening a calcium channel may increase cytosolic Ca^{2+} concentration, which would activate protein kinases or other enzymes that depend on Ca^{2+}-calmodulin for activity (e.g., NADP kinase). The membrane oxidoreductase may directly control a calcium channel in the plasma membrane by a redox-induced confromational change or a protonation of a site on the channel protein (Houslay 1987). Alternatively, if the redox system activates phospholipse C to increase inositol trisphosphate release, then calcium release from internal membranes may be increased.

SUMMARY

It is clear from the evidence obtained in several laboratories with a variety of techniques that plasma membranes contain electron-transport enzymes. The recognition of these enzymes has been greatly improved by study of highly purified plasma membranes and the use of histochemical techniques. The use of external, impermeable oxidants has provided an approach to the study of these enzymes in intact cells and tissues. The redox proteins involved, their exact orientation in the membrane, and their electron-carrying prosthetic groups remain to be established.

Stimulation of transplasma-membrane electron transport has been shown to control cell elongation and proliferation. Growth hormones have been shown to control the electron transport. The nature of the interaction between the redox enzymes and hormones or growth factors is not known. The effect of the redox enzymes on signal transmission in control of cell growth and function remains to be defined. Further work needs to be done to define the natural electron donors and acceptors.

Electron flow across the plasma membranes to artificial external electron acceptors can control cell elongation and proliferation, increase proton excretion, and decrease membrane potential. How does this electron transport to artificial acceptors relate to electron transport to natural acceptors such as oxygen or semidehydroascorbate? How does electron transport and associated proton movement influence or induce signal transduction?

There are at least seven ways in which a redox system can generate a message or affect message generation, either through direct or indirect action, as follows:

A. Direct action
1. Oxidation of a substrate to generate a compound that can transmit the message, for example, formation of NAD from NADH to support ADP ribosylation reactions.
2. Reduction of an electron acceptor to generate a messenger, for example, reduction of oxygen to hydrogen peroxide.
3. Electrogenic movement of the electron to change membrane potential.
B. Indirect action
1. Oxidation of a protonated substrate to release protons or removal of protons by reduction of an acceptor to change local or bulk pH, which, in turn, can control the activity or location in the cell of enzymes that produce messages or can control transport channels.
2. Change in the redox state of an electron transport protein to change the conformational state of the protein, for example, reduction of a flavoprotein with conformation change transmitted to an adjacent G protein, or message generator.
3. Reduction of disulfide bonds in receptors or message generator proteins to activate or inactivate message production or oxidation of GSH to decrease reductant availability.
4. Reduction at the cell surface of nutrients essential for cell catalytic function for transport of these elements into the cell, for example, copper, iron, manganese, or nitrate.

The recognition of plasma membrane redox activity and development of methods to study the activity will allow examination of the redox effects on generation of recognized second messengers such as inositol phosphates or cytosolic calcium. In addition, the study of the redox process itself and the enzymes involved may allow introduction of new second messengers such as NAD. Clearly, the redox systems introduce a new parameter into control of cellular pH, membrane potential, and nutrient uptake. A complete picture of cellular control demands an evaluation of the participation of all messenger systems and their interactions.

REFERENCES

Allnutt FCT, WD Bonner Jr (1987): Evaluation of reductive release as a mechanism for iron uptake from ferrioxamine B by *Chlorella vulgaris*. Plant Physiol 85:751–756.

Askerlund P, C Larsson, S Widell, IM Møller (1987): NAD(P)H oxidase and peroxidase activities in purified plasma membranes from cauliflower influorescences. Physiol Plant 71:9–19.

Barr R, TA Craig, FL Crane (1985a): Transmembrane ferricyanide reduction in carrot cells. Biochim Biophys Acta 812:49–54.

Barr R, FL Crane (1985): Ca^{2+}-calmodulin antagonists affect plant growth regulator controlled plasma membrane redox in carrot cells. Biochem Biophys Res Commun 126:262–268.

Barr R, FL Crane (1987): Proton excretion by plant cells depends on the action of the H^+-ATPase and plasma membrane electron transport. J Cell Biol 105:187a.

Barr R, FL Crane, TA Craig (1984): Transmembrane ferricyanide reduction by tobacco callus cells. J Plant Growth Regul 2:243–249.

Barr R, O Martin Jr, FL Crane (1987): Redox induced proton excretion by cultured carrot cells is affected by protonophores and inhibitors of ATPase. Proc Indiana Acad Sci 96:139–144.

Barr R, AS Sandelius, FL Crane, DJ Morré (1985b): Oxidation of reduced pyridine nucleotides by plasma membranes of soybean hypocotyl. Biochem Biophys Res Commun 131:943–948.

Barr R, AS Sandelius, FL Crane, DJ Morré (1986): Redox reactions of tonoplast and plasma membranes isolated from soybean hypocotyls by free flow electrophoresis. Biochim Biophys Acta 852:254–261.

Bienfait HF (1985): Regulated redox processes at the plasmalemma of plant root cells and their function in iron uptake. J Bioenerg Biomemb 17:73–83.

Beinfait HF, LA de Weger, D Kramer (1987): Control of development of iron-efficiency reactions in potato as a response to iron deficiency is located in the roots. Plant Physiol 83:244–247.

Blein J-P, MC Canivec, X Decherade, M Bergon, JP Calmon, R Scalla (1986): Transplasma membrane ferricyanide reduction by sycamore cells. Plant Sci 46:77–85.

Böttger M (1986): Proton translocation systems at the plasmalemma and its possible regulation by auxin. Acta Horticul 179:83–93.

Böttger M, M Bigdon, H-J Soll (1985a): Proton translocation in corn coleoptiles: ATPase or redox chain? Planta 163:373–380.

Böttger M, H Lüthen (1986): Possible linkage between NADH oxidation and proton secretion in *Zea mays* L. roots. J Exp Bot 37:666–675.

Böttger M, HJ Soll, M Bigdon,(1985b): Influence of inhibitors od alternative respiration pathway and oxygen on growth and proton secretion. Biol Plant (Praha) 27:125–130.

Brightman AO, R Barr, FL Crane, DJ Morré (1987a): Purification of an auxin stimulated NADH oxidase from plasma membrane of soybean. J Cell Biochem (suppl) 11B:93.

Brightman AO, R Barr, FL Crane, DJ Morré (1987b): Auxin stimulated NADH oxidase purified from plasma membrane of soybean. Plant Physiol 86:1264–1269.

Brouwer KS, T van Valen, DA Day, H Lambers (1986): Hydroxamate stimulated O_2 uptake in roots of *Pisum sativum* and *Zea mays*, mediated by a peroxidase. Plant Physiol 82:236–240.

Brummer B, R Parish (1983): Mechanisms of auxin-induced plant cell elongation. Fed Eur Biochem Soc Lett 161:9–13.

Cakmak I, DAM van de Wetering, H Marschner, HF Bienfait (1987): Involvement of superoxide radical in extracellular ferric reduction by Fe deficient bean roots. Plant Physiol 85:310–316.

Castignetti D, J Smarrelli Jr (1986): Siderophores, the iron nutrition of plants and nitrate reductase. Fed Eur Biochem Soc Lett 209:147–151.

Chalmers JCD, JOD Coleman (1983): Ethanol stimulated extrusion by cells of carrot grown in suspension culture. Biochem Int 7:785–791.

Chalmers JCD, JOD Coleman, NJ Walton (1984): Use of an electrochemical technique to study plasmalemma redox reactions in cultured cells of *Daucus carota* L. Plant Cell Rep 3:243–246.

Clark MG, OH Filsell, IG Jarrett (1977): An effect of extracellular redox state on the glucagon-stimulated glucose release by rat hepatocytes and perfused liver. Hormone Metab Res 9:213–217.

Cleland R (1971): Cell wall extension. Annu Rev Plant Physiol 22:197–222.

Cooper SR, JV McArdle, KN Raymond (1978): Siderophore electrochemistry: Relation to intracellular release mechanism. Proc Natl Acad Sci USA 75:3551–3554.

Craig TA, FL Crane (1981): Evidence for a trans-plasma membrane electron transport system in plant cells. Proc Indiana Acad Sci 90:150–155.

Craig TA, FL Crane (1982): Hormonal control of a transplasma membrane electron transport system in plant cells. Proc Indiana Acad Sci 91:150–154.

Craig TA, FL Crane (1985a): Redox potential of the donor and H^+ release of the trans-plasmalemma redox system of carrot cells. Plant Physiol 77:suppl 145.

Craig TA, FL Crane (1985b): The trans-plasma membrane redox system in carrot cells: Inhibited by membrane-impermeable DABS and involved in H^+ release from cells. In Randall DD, DG Blevins, RL Larson (eds): "Current Topics in Plant Biochemistry and Physiology", Vol. 4. Columbia: University of Missouri Press, p 247.

Crane FL, R Barr, TA Craig, PC Misra (1984): Growth control by proton pumping transplasma membrane redox. Proc Plant Growth Regul Soc Am 11:87–95.

Crane FL, R Barr, TA Craig, DJ Morré (1988): Transplasma membrane electron transport in relation to cell growth and iron uptake. J Plant Nutr 11, (in press).

Crane FL, R Barr, DJ Morré, AO Brightman (1987): Plasmalemma redox systems and control of growth. Berlin: "Abstracts XIV International Botanical Congress." p 106.

Crane FL, H Goldenberg, DJ Morré, H Löw (1979): Dehydrogenases of the plasma membrane. In Roodyn DB (ed): "Subcellular Biochemistry", Vol. 6. New York: Plenum, pp 345–399.

Crane FL, H Löw, MG Clark (1985a): Plasma membrane redox enzymes. In Martonosi AN (ed): "The Enzymes of Biological Membranes," Vol. 4. New York: Plenum, pp 465–510.

Crane FL, H Roberts, AW Linnane H Löw, (1982): Transmembrane ferricyanide reduction by cells of the yeast *Saccaromyces cerevisiae*. J Bioenerg Biomemb 14:191–205.

Crane FL, Il Sun, MG Clark, C Grebing, H Löw (1985b): Transplasma membrane redox systems in growth and development. Biochim Biophys Acta 811:233–264.

De Luca L, U Bader, R Hertel, P Pupillo (1984): Detergent activation of NADH oxidase in vesicles derived from the plasma membrane of *Curcubita pepo* L. Plant Sci Lett 36:93–98.

Dharmawardhane S, AI Stern, B Rubenstein (1987): Light-stimulated transplasmalemma electron transport in oat mesophyll cells. Plant Sci 51:193–197.

Dilley RA, SM Theg, WA Beard (1987): Membrane proton interaction in chloroplast bio-energetics. Annu Rev Plant Physiol 38:347–389.

Dohi T, F Murad (1981): Effects of pyruvate and other metabolites on cyclic GMP levels in incubations of rat hepatocytes and kidney cortex. Biochem Biophys. Acta 673:14–25.

Dupont, FM, JJ Windel (1987): Do ESR spin probes detect a transplasma membrane electron transport system. In Ramirez, JM Redox. Functions of the Eukaryotic Plasma Membrane, Madrid, Consijo Superior Investigaciones Cientificas pp. 251–262.

Earnshaw BA, MA Johnson (1985): The effect of glutathione on development of wild carrot suspension culture. Biochem Biophys Res Commun 133:988–993.

Earnshaw BA, MA Johnson (1987): Control of wild carrot somatic embryo development by antioxidants. Plant Physiol 85:273–276.

Farchaus JW, WR Widger, WA Cramer, RA Dilley (1982): Kinase induced changes in electron transport rates of spinach chloroplasts. Arch Biochem Biophys 217:362–367.

Federico R, CE Giartosio (1983): A transplasma membrane electron transport system in maize root. Plant Physiol 73:182–184.

Fujihara S, M Yamaguchi (1978): Effects of allopurinol on metabolism of allantain in soybean plants. Plant Physiol 62:134–138.

Gaal JC, KR Smith, CK Pearson (1987): Cellular euthanasia mediated by a nuclear enzyme: A central role for nuclear ADP-ribosylation in cellular metabolism. Trends Biochem Sci 12:129-130.

Gepstein S (1982): Light induced proton secretion and its relation to senescence of oat leaves. Plant Physiol 70:1120–1124.

Gilman AG (1984): G-proteins and control of cyclase. Cell 36:577–579.

Guerrini F, V Valenti, P Pupillo (1987): Solubilization and purification of NAD(P)H dehy-drogenases of *Cucurbita* microsomes. Plant Physiol 85:828–834.

Hassidim M, B Rubinstein, H Lerner, L Reinhold (1987): Generation of a membrane potential by electron transport in plasmalemma-enriched vesicles of cotton and radish. Plant Physiol 85:872–875.

Houslay MD (1987): Ion channel controlled by guanine nucleotide regulatory proteins. Trends Biochen Sci 12:167–168.

Ivankina NG, VA Novak, AI Miklashevich (1984): Redox reactions with active H $^+$ transport in the plasmalemma of *Elodea* leaf cells. In Cram WJ, K Janácek, R Rybova, L Sigler (eds): "Membrane Transport in Plants." Chichester, Wiley: pp 404–405.

Jesaitis AJ, PR Heuers, R Hertel, WR Briggs (1977): Characterization of a membrane fraction containing *b*-type cyctochrome. Plant Physiol 59:941–947.

Jones GL, MM Morel (1987a): Plasmalemma redox enzymes and nitrogen assimilation in marine phytoplankton. Berlin: "Abstracts XIV International Botanical Congress." p. 21.

Jones GL, MM Morel (1988b): Plasmalemma redox activity in the diatom *Thalassiosira:* A possible role for nitrate reductase. Plant Physiol 87:143–147.

Jones GJ, BP Palenik, FMM Morel (1987): Trace metal reduction by phytoplankton: The role of plasmalemma redox enzymes. J Phycol 23:237–244.

Khachatrain L, C Klein, A Howlett (1987): Pertussis and cholera toxin ADP-ribosylation in *Dictyostelium discoidium* membrane. Biochem Biophys Res Commun 149:975–981.

Kjellbom P, C Larsson (1984): Preperation and polypeptide composition of chlorophyll-free plasma membranes from leaves of light-grown spinach and barley. Physiol Plant 62:501-509.

Kochian LV, WJ Lucas (1985): Potassium transport in corn roots. III. Perturbation by exogenous NADH and ferricyanide. Plant Physiol 77:429-436.

Kurkova EB, ML Verkhovskaya (1984): Redox components of plant cell plasmalemma. Soviet Plant Physiol 31:496-501.

Lass B, G Thiel, CI Ullrich-Eberius (1986): Electron transport across the plasmalemma of Lemna gibba G₁, Planta 169:251-259.

Lazlo PS, E Barros, P de la Pena, R Ramos (1981): Ion gradients as candidates for transmembrane signalling. Trends Biochem Sci 6:83-86.

Leong T-Y, WR Briggs (1982): Evidence from studies with acifluorifen for participation of a flavin-cytochrome complex in blue light photoreception for phototropism of oat coleophiles. Plant Physiol 70:875-881.

Leong T-Y, RD Vierstra, WR Briggs (1981): A blue light sensitive cytochrome-flavin complex from corn coleoptiles: Further characterization. Photochem Photobiol 34:697-703.

Lin W (1982): Response of corn roots to exogenous NADH: Oxygen consumption, ion uptake and membrane potential. Proc Natl Acad Sci USA 79:3773-3776.

Lin W (1984): Further characterization on the transport property of the plasmalemma NADH oxidation system in isolated corn protoplasts. Plant Physiol 74:219-222.

Loos S, U Lüttge (1984): Effects of mercuric chloride on membranes of Lemna gibba in the energized and non-energized state. Physiol Veg 22:171-179.

Löppert H (1983): Energy coupling for membrane hyperpolarization in Lemna: Respiration rate, ATP level and membrane potential at low oxygen concentration. Planta 159:329-335.

Löw H, S Werner (1976): Effects of reducing and oxidizing agents on the adenylate cyclase activity in adipocyte plasma membranes. Fed Eur Biochen Soc Lett 65:96-98.

Lubberding HJ, FHJM de Graaf, HF Bienfait (1988): Ferric reducing activity in roots of Fe-deficient phaseolus vulgaris: Source of reducing equivalents. Biochem Biophys Pflanzen (in press).

Mäder M, V Amberg-Fisher (1982): Role of peroxidase in lignification of tobacco cells. I. Oxidation of NADH and formation of H_2O_2 by cell wall peroxidases. Plant Physiol 70:1128-1131.

Makino H, A Kanatsuka, M Osegawa, A Kumazai (1982): Effects of dithiothreitol on insulin-sensitive phosphodisterase in fat cells. Biochem Biophys Acta 704:31-36.

Malviya AN, P Anglund (1986): Modulation of cytosolic protein kinase c activity by ferricyanide: Primary event seems transmembrane redox signaling. Fed Eur Biochem Soc Lett 200:265-270.

Marrè E, A Ballarin-Denti (1985): Proton pumps of plasmalemma and tonoplasts. J Bioenerg Biomemb 17:1-21.

Marrè R, P Lado, A Ferroni, A Ballarin-Denti (1974): Transmembrane potential increase induced by auxin, benzyladenine and fusicoccin: Correlation with proton extrusion and cell enlargement. Plant Sci Lett 2:257-265.

Misra PC, TA Craig, FL Crane (1984): A link between transport and plasma membrane redox systems in carrot cells. J Bioenerg Biomemb 16:143-152.

Møller IM, A Berczi (1985): Oxygen consumption by purified plasmalemma vesicles from wheat roots. Fed Eur Biochem Soc Lett 193:180-184.

Møller IM, A Berczi (1986): Salicylhydroxamic acid stimulated NADH oxidase by purified plasmalemma vesicles from wheat roots. Physiol Plant 68:67-74.

Møller IM, W Lin (1986): Membrane bound NAD(P)H dehydrogenases in higher plant cells. Annu Rev Plant Physiol 37:309–334.

Morré DJ, G Auderset, G Penel, H Canut (1987a): Cytochemical localization of NADH–ferricyanide oxido-reductase in hypocotyl segments and isolated membrane vesicles of soybean. Protoplasma 140:133–140.

Morré DJ, AO Brightman, L-Y Wu, R Barr, B Leak, FL Crane (1988): An essential role of plasma membrane redox in elongation growth in plants. Physiol Plant 73:187–193.

Morré DJ, FL Crane, R Barr, C Penel, LY Wu (1987b): Inhibition of plasma membrane redox and of elongation growth of soybean. Physiol Plant 72:236–240.

Morré DJ, TW Keenan, CN Huang (1974): Membrane flow and differentiation: Origin of Golgi apparatus membranes from endoplasmic reticulum. In Ceccarelli B, F Clementi, J Meldolosi (eds): "Advances in Cytopharmacology," Vol. 2, New York: Raven, pp 107–125.

Morré DJ, P Navas, C Penel, FJ Castillo (1986): Auxin stimulated NADH oxidase (semidehydroascorbate reductase) of soybean plasma membrane: Role in acification of the cytoplasma. Protoplasma 133:195–197.

Mukherjee SP, C Mukerjee (1981): Role of sulfhydryl oxidation in adipocyte plasma membrane surface in the response of adenylate cyclase to isoprotenol and glucagon. Biochem Biophys Acta 677:339–349.

Munoz V, W Butler (1975): Photoreceptor pigment for blue light in Neurospora crassa. Plant Physiol 55:421-426.

Navas P, IL Sun, DJ Morré, Fl Crane (1986): Decrease in NADH in HeLa cells in the prescence of transferrin or ferricyanide. Biochem Biophys Res Commun 135:110–115.

Neufeld E, AW Bown (1987): A plasma membrane redox system and proton transport in isolated mesophyll cells. Plant Physiol 83:895–899.

Nishizuka Y (1984): Phosphatidylinositol, diacylglycerol phorbol myristate acetate and protein kinase c in cell surface signal transmission and tumor promotion. Nature 308:693–697.

Novak VA, NG Ivankina (1983): Influence of nitroblue tetrazolium on membrane potential and ion transport in water thyme. Soviet Plant Physiol 30:845–853.

Novak VA, AI Miklashevich (1984): Ferricyanide reducing activity in Elodea leaves and its relation to energy metabolism. Soviet Plant Physiol 31:380–386.

Pfaffman H, E Hartmann, AO Brightman, DJ Morré (1987): Phosphatidylinositol specific phospholipase c of plant stems: Membrane associated activity concentrated in plasma membrane. Plant Physiol 85:1151–1155.

Polevoy VV, T Salamatova (1977): Auxin, proton pump and cell trophics. In Marré E, Ciferri O (eds): "Regulation of Cell Membrane Activities in Plants." Amsterdam: Elsevier, pp 209–216.

Pupillo P, V Valenti, L de Luca, R Hertel (1986): Kinetic characterization of reduced pyridine nucleotide dehydrogenases (duroquinone dependent) in Curcubita microsomes. Plant Physiol 80:384–389.

Qiu Z-S, B Rubinstein, AI Stern (1985): Evidence for electron transport across the plasma membrane of Zea mays root cells. Planta 165:383–391.

Ramirez JM, G Gimenez-Gallego, R Serrano (1984): Electron transfer constituents in plasma membrane fractions of Avena sativa and Saccharomyces cerevisiae. Plant Sci Lett 34:103–110.

Revis S, PC Misra (1986): Transplasma membrane electron transport in angiospermic parasites. J Plant Physiol 122:337–345.

Ric de Vos C, HJ Lubberding, HF Bienfait (1986): Rhizosphere acidification as a response to iron deficiency in bean plants. Plant Physiol 81:842–846.

Roth Z, N Chayen, S Dikstein (1983): The involvement of the intracellular redox state and pH in the metabolic control of stimulus response coupling. Int Rev Cytol 85:39–61.

Rubinstein B, AI Stern (1986): Relation of transplasmalemma redox activity to proton and solute transport by roots of Zea mays. Plant Physiol 80:805–811.

Sandelius AS, R Barr, FL Crane, DJ Morré (1986): Redox reactions of plasma membranes isolated from soybean hypocotyls by phase partition. Plant Sci 48:1–10.

Schallreuter KV, JM Wood (1986): The role of thioredoxin reductase in the reduction of free radicals of the surface of the epidermis. Biochem Biophys Res Commun 136:630–637.

Schmidt W (1984): Blue light physiology. Bioscience 34:698-704.

Schneider H, GF Fuhrmann, A Fiechter (1979): Plasma membrane from Candida tropicalis grown on glucose or hexadecane. Biochem Biophys Acta 554:309–322.

Sesay A, CR Stewart, RM Shibles (1986): Effects of KCN and salicylhydroxamic acid on respiration of soybean leaves at different ages. Plant Physiol 82:443–447.

Shinkle JR, WR Briggs (1985): Physiological mechanism of the auxin induced increase in light-sensitivity of phytochrome mediated growth responses in aveno coleoptile sections. Plant Physiol 79:349–356.

Sijmons PC, W van den Briel, HF Bienfait (1984a): Cytosolic NADPH is the electron donor for extracellular Fe^{+++} reduction in iron-deficient bean roots. Plant Physiol 75:219–221.

Sijmons PC, FC Lanfermeijer, AH de Boer, HBA Prins, HF Bienfait (1984b): Depolarization of cell membrane potential during transplasma membrane electron transfer to extracellular electron acceptors in iron deficient roots of Phaseolus vulgaris L. Plant Physiol 76:943–946.

Sokol PP, PD Holohan, CR Ross (1986): Essential disulfide and sulfhydryl groups for organic cation transport in renal brush border membranes. J Biol Chem 261:3282–3287.

Somers DA, T-M Kuo, A Kleinhofs, RL Warner (1983): Nitrate reductase deficient mutants in barley. Plant Physiol 71:145–149.

Sun IL, R Garcia-Cañero, W Liu, W Toole-Simms, FL Crane, DJ Morré, H Löw (1987): Diferric transferrin reduction stimulates the Na^+/H^+ antiport of HeLa cells. Biochem Biophys Res Commun 145:467–473.

Sze H (1985): H^+ translocating ATPases: Advances using membrane vesicles. Annu Rev Plant Physiol 36:175–208.

Tanaka O, K Asagami (1986): Ferricyanide induces flowering by supression of nitrate assimilation in Lemna panacostate 6746. Plant Cell Physiol 27:1013–1068.

Thom M, A Maretzki (1985): Evidence for a plasmalemma redox system in sugar cane. Plant Physiol 77:873–876.

Tipton CL, J Thowson (1983): Reduction of iron by soybean roots: Correlation with iron efficiency on calcareous soils. Iowa State J Res 57:409–422.

Tipton CL, J Thowsen (1985): Fe^{III} reduction by cell walls of soybean roots. Plant Physiol 79:432–435.

Widell S, C Larsson (1983): Distribution of cytochrome b photoreductions mediated by endogenous photosensitizer or methylene blue in fractions from corn and cauliflower. Physiol Plant 57:196–202.

Widell S, JS Britz, WR Briggs (1980): Characterization of the red light induced reduction of particle associated with b-type cyctochrome from corn in the presence of methylene blue. Photochem Photobiol 32:669–677.

Williams RJP (1983): Mitochondrial compartments and chemiosmosis. Trends Biochem Sci 8:48.

Windle JJ (1985): An in vivo spin probe study of plant cell plasma membranes. Plant Physiol 77s:87.

Yamamoto Y, E Niki, J Eguchi, Y Kamiya, H Shinasaki (1985): Oxidation of biological membranes and its inhibition: Free radical chain oxidation of erythrocyte ghost membranes by oxygen. Biochim Biophys Acta 819:29–36.

Second Messengers in Plant Growth
and Development, pages 145–166
© 1989 Alan R. Liss, Inc.

6. pH as a Second Messenger in Plants

Hubert Felle

Botanisches Institut I der Justus Liebig-Universität, D-6300 Giessen, Federal Republic of Germany

INTRODUCTION

Although the concept of signal transduction and amplification has been accepted for animal cells for quite some time, plant physiologists found it difficult to transfer these principles to their systems until recently. Certainly, the importance of calcium, for instance, as an activator and regulator of numerous intracellular processes is undisputed. Yet an equivalent role of H^+ (or pH) as second messenger is much less widely accepted.

Since all life is based on aqueous chemistry and because water spontaneously ionizes, protons cannot be excluded from the intracellular milieu. Their activity must be regulated. Besides being both the substrate and product of metabolic pathways, protons have an apparent potential to communicate information about the cellular energy balance to enzymes and structures that may share no other common effector. Since small internal pH changes cause large effects on the activity of enzymes (membrane-bound transport proteins included), the potential impact on intracellular processes are manifold and, in principle, need not be mediated by specialized receptors such as those required by cytosolic free calcium and cyclic AMP. Thus, as Busa and Nuccitelli (1984) point out, the principal regulatory role of cytosolic pH may be to coordinate the activities of diverse enzyme-catalyzed pathways, membrane transport, and regulatory agents (Ca^{2+}, inositol 1,4,5-trisphosphate [IP_3], and so on) in a manner appropriate to fulfill the required task.

Address reprint requests to Dr. Hubert Felle, Botanisches Institut I der Justus Liebig-Universität, Senckenbergstr. 17-21, D-6300 Giessen, Federal Republic of Germany.

For several reasons, including technical ones, much more is known about regulatory effects of cytosolic pH (pH_c) in animal systems than in plant cells. To give an insight into how shifts in pH_c might activate, regulate, or merely accompany biological events, Table I and a well-studied example from the animal kingdom, the activation of the sea urchin egg, are presented. For a full review the reader is referred to the excellent article of Busa and Nucci-telli (1984).

TABLE I. Examples of Cytosolic pH Changes Associated With Changes in Cellular Metabolism and Development

Organism	Event	pH change	Reference
Sea urchin			
Lytechinus pictus	Fertilization of egg	Alkalinization (0.4 units)	Shen and Steinhardt (1978)
Lytechinus pictus	Mobility initiation of spermatozoon	Alkalinization (0.5 units)	Lee et al. (1983)
Lytechinus pictus	Acrosome reaction	Initial alkalinization followed by permanent acidification	Lee et al. (1983)
Shrimp			
Artemia salina	Arousal from dormancy	Transient alkalinization ± 1.6 units	Busa et al. (1982)
Frog			
Xenopus laevis	Fertilization of egg	Alkalinization (0.3 units)	Webb and Nuccitelli (1982)
Rana pipiens	Insulin stimulation of glycolysis	Alkalinization (0.1–0.2 units)	Moore (1979)
Ciliate			
Tetrahymena pyriformis	Cell cycle	pH oscillation starting with acidification (period 200 min)	Gillies and Deamer (1979)
Yeast			
Pichia pastoris	Spore germination	Alkalinization (1 unit)	Barton and Den Hollander (1980)
Saccharomyces cerevisiae	Starvation/refeeding	Transient alkalinization (0.4 units)	Den Hollander et al. (1981)
Slime mold			
Physarum polycephalum	Starvation/refeeding	Alkalinization (0.4 units)	Morizawa and Steinhardt (1982)
Physarum polycephalum	Cell cycle	Alkalinization by 0.4 units followed by acidification by 0.25 units	Morizawa and Steinhardt (1982)

Fertilization terminates the quiescent state (low respiratory rate, reduced synthetic activity) of sea urchin eggs. Whereas plasma membrane depolarization and transient release of Ca^{2+} are referred to as "early events," hyperpolarization, formation of microvilli, increase in protein synthesis and mRNA, activation of DNA synthesis, and finally cell division occur 2 min to 1.5 hr after fertilization and are the so-called "late events." In the sea urchin egg these late events are accompanied by an increase in internal pH (pH_i).

The first indirect evidence that pH_i alkalinization is involved in egg activation comes from work of Steinhardt and Mazia (1973), who demonstrated that NH_3 (added as NH_4OH at pH 9) activated the appearance of new K^+ conductance, leading to hyperpolarization in unfertilized sea urchin eggs. This finding was supported by later experiments showing that protein synthesis (Epel et al., 1974), chromosome condensation (Mazia, 1974), and DNA synthesis (Mazia and Ruby, 1974; Vaquier and Brandriff, 1975) were likewise elicited by increasing pH_i with ammonia. Direct evidence was brought forward by Shen and Steinhardt (1978) who, using Thomas recessed-tip pH microelectrodes, recorded a pH_i increase from 6.8 to 7.3 within 5 min after fertilization of the *Lyntechinus pictus* egg. Although it is still a topic of debate as to how pH_i is alkalinized after fertilization, there is evidence that the Na^+/H^+ antiporter is involved. This follows from experiments showing that fertilization is accompanied by a release of protons, requires external Na^+, and is inhibited by amiloride.

As to the targets of pH_i mediation during egg activation, two important experiments and approaches should be mentioned. First, when the pH_i of fertilized eggs is lowered by introducing acetic acid, protein synthesis is inhibited. Second, protein synthesis in a cell-free system, prepared from *Lyntechinus pictus* eggs, is strongly pH dependent (10–20-fold greater at pH 7.4 than at pH 6.9), and the optimum pH (7.4) equals the pH_i of fertilized eggs. Despite this compelling evidence, one important objection to the pH_i hypothesis exists. The NH_3 activation of the sea urchin egg is accompanied by a transient intracellular calcium pulse, raising the possibility that this pulse, not intracellular alkalinization, is responsible for the activating effect of NH_3.

This example, selected from many, indicates that pH_c is involved in cellular processes, either as an apparent prerequisite or as a necessary precondition for the initiation of important cellular events. In some of the examples cited, a role of cytosolic pH or changes thereof as (second) messenger seems experimentally already established or can at least be tacitly assumed. Also with plants pH changes of the cytosol are probably necessary to trigger signal response cascades or to activate certain enzymes. These can now be measured reliably and may be elicited through various external stimuli perceived either at the plasma membrane or at internal membranes.

REGULATION OF CYTOSOLIC pH

To elicit the target reactions, protons might act as intracellular messengers by changing pH_c. Other possibilities involve cell wall acidification, intramolecular or local proton shifts within membranes, and last, but not least, pH shifts localized only to certain cytoplasmic regions. Regardless of whether cytosolic changes occur toward the alkaline side or toward the acidic side, cellular pH homeostasis will at once attempt to restore the control pH, if the pH shift is sensed. The sensing of pH changes could be accomplished by basically two groups of compounds: those with dissociable side chains and H^+ transporters.

The question remains: Is there some sort of a superior center that sets the control pH and thus coordinates the different reactions in a cell to pH shifts? Although there is no conclusive answer to that question at present, we will have to assume that a cell must have the ability to distinguish between pH shifts induced or triggered by a membrane-bound receptor, and pH shifts coming from metabolism or, in rarer instances, from the outside. Therefore, a cell must receive some kind of additional information that prevents the countering of, for instance, a receptor-induced pH change.

Metabolic pH shifts are mostly slow and always present. Thus they involve long-term pH homeostasis. Sudden cytosolic pH shifts require an altogether different, short-term pH regulation. In these instances, metabolism runs differently. It is conceivable that feedback loops, consisting of proton-consuming and proton-producing elements, are altered in a way to set a new control pH or even to retard active pH regulatory processes. The setting of a new control pH may not be too simple, since most cellular processes, as well as general metabolism, are pH dependent and hence function optimally only if the cytosolic pH is maintained within narrow limits. On the other hand, it must be the aim of, for example, a phytohormone to alter cellular metabolism and thereby influence membrane transport, the activity of intracellular enzymes, calcium activity and (binding) properties of calcium-binding proteins, and a variety of other cellular components.

Membrane Transport

While cell physiologists are armed with a battery of techniques to monitor cytosolic pH directly and continuously, we are still far from understanding cytosolic pH regulation, although this refers more to intracellular cytosolic events than to proton transport across membranes. The plasma-membrane-located H^+ transporters, for instance, the H^+-export pump (Gradmann et al., 1978; Poole, 1978; Felle, 1981; Spanswick, 1982; Tazawa and Shimmen, 1982), and H^+ cotransporters (Komor and Tanner, 1974; Jung and Lüttge, 1980; Kinraide and Etherton, 1980; Felle, 1983; Johannes and Felle, 1985, 1987) are reasonably well studied, and their immediate effects on cytoplasmic pH are known.

Externally induced shifts in cytosolic pH are countered by plant cells as demonstrated in Figures 1–3. In Figure 1 the relative insensitivity of the pH_c to large pH gradients across the plasma membrane is shown. For example, a change in external pH of 1 unit generally shifts pH_c by about 0.1 unit or even less. Such responses of the cells, which have been found in all investigated systems, animal, bacteria, fungi, and plants, implicate either a low proton permeability at the plasmalemma, or a good pH_c regulation, or both. In Figure 2, the response to H^+ import via a H^+/Cl^- symport is shown. The pH recovery is much faster than observed after the addition of acetic acid to the cells (Fig. 3). These two observations are not controversial, because only a limited amount of H^+ can be transported by the H^+/Cl^- symporter at a time, whereas the protonated acetic acid (HAc) diffuses through the membrane(s) in a concentration dependent manner. Moreover, the acidification of the cytosol due to externally applied weak acids is of a different kind, because it originates from the dissociation of HAc which will depend on its pK_a and the pH_c. Since the external weak acid concentration is kept con-

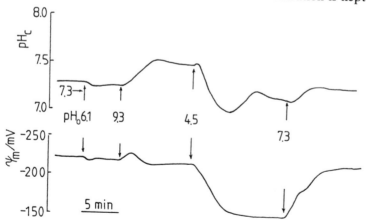

Fig. 1. Cytosolic pH (pH_c) and membrane potential (ψ_m) measured in rhizoid cells of the aquatic liverwort *Riccia fluitans* in the presence of different external pH (pH_o), as indicated. Method: Double-barreled pH-sensitive microelectrodes, as described in Felle (1987).

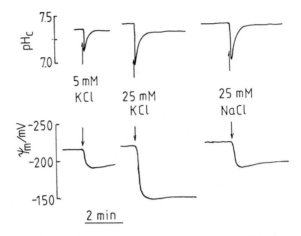

Fig. 2. Rapid response of cytoplasmic pH of *Sinapis* root hairs to sudden increases in external chloride from 0.01 mM to the indicated concentrations. The initial fast acidification is interpreted as a result of a H^+/Cl^--symport in the plasma membrane, the recovery of cytosolic pH (pH_c) as the reaction of cellular pH regulation. The method was as described in (Felle, 1987; Felle and Bertl, 1986a).

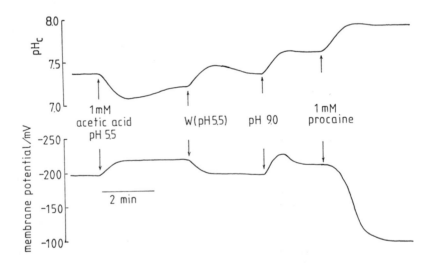

Fig. 3. Effect of acetic acid and procaine on cytosolic pH (pH_c) and membrane potential (ψ_m) of *Riccia fluitans* rhizoid cells. The method was as described (Felle, 1987).

stant, cytosolic protons removed by cellular pH regulation will be replaced by further uptake and dissociation of HAc. Nevertheless, as Figure 3 demonstrates, a partial recovery of pH_c is observable, indicating pH regulatory processes to be faster and more effective than HAc import and subsequent dissociation. Apart from the obvious H^+ export by the H^+ ATPases at the plasma membrane and the tonoplast, such processes may include Na^+/H^+ exchange by a Na^+/H^+ antiport, as known from animal cells and certain bacteria (Guern et al., 1986) and K^+/H^+ exchange by a K^+/H^+ antiport (Marrè et al., 1987).

Apart from the fact that all transport proteins, as well as all known enzymes, are pH sensitive in their activity, the H^+-transporting proteins are complicated in their mode of action. We can distinguish two major groups of H^+ transporters: 1) Primary active systems, which build up an electrochemical proton gradient ($\Delta\mu_H^+/F$) across the pertinent membrane or maintain it ATPases, redox systems, pyrophosphatase); 2) secondary active systems such as so-called "H^+ cotransporters," which are intrinsically passive in that they use H^+ as driver ions and thereby dissipate cellular energy. While as a consequence of primary active electrogenic H^+ transport the membrane is hyperpolarized, H^+ cotransport is electrophoretic and usually leads to (transient) depolarization. It is obvious that all these systems have some influence on cytosolic pH, either by shifting pH_c or by restoring and maintaining it. Additionally, we have to realize that any pH_c change will alter the activity of these transporters, either through H^+ being a substrate or through direct interaction with the protein itself.

The pH dependence of the proton pump. Acidification (e.g., with weak acids) of the cytosol stimulates primary active H^+ export, and, as such, leads to membrane hyperpolarization and cell well acidification. Alkalinization (with weak bases) deactivates the H^+ pump and leads to depolarization (Fig. 3). An interesting question with regard to pH regulation is: What will happen with pH_c, when the H^+ pump gets deactivated by means of inhibitors? Since all acid produced internally will have to leave the cell after the vacuole and cellular buffers are exhausted, one would expect an acidification of the cytosol as soon as the H^+ export is inhibited. This, however, is not so trivial. In Figure 4 several apparently contradictory results are given. 1) Vanadate, a potent direct inhibitor of the plasma membrane proton pump only slightly causes cytosolic acidification. 2) Cyanide, by inhibiting oxidative phosphorylation and therefore ATP synthesis, clearly and substantially acidifies pH_c. 3) CCCP, an uncoupler of oxidative phosphorylation, does not lead to significant cytosolic acidification, although cellular ATP levels are low, and the pump is undoubtedly deactivated. 4) In the presence of CCCP, cyanide acidifies pH_c.

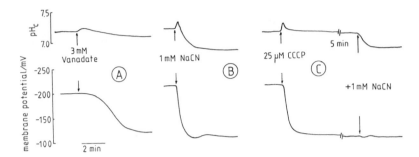

Fig. 4.   Cytosolic pH (pH_c) and membrane potential (ψ_m) of *Riccia fluitans* rhizoid cells, before and after the addition of **A:** 3 mM vanadate, **B:** 1 mM NaCN, **C:** 25 μM CCCP, and 25 μM CCCP followed by 1 mM NaCN. External conditions: 1 mM KCl, 0.1 mM NaCl, 0.1 mM $CaCl_2$, Mes/Tris-buffer 5 mM, external pH (pH_o) = 7.3. The methods were as described (Felle, 1987).

These findings can be interpreted in several ways. 1) The cyanide-induced acidification is seemingly not a result of the ceased proton pumping because CCCP had shut it down already. The observed acidification must have metabolic reasons and is presumably due to an accumulation of glycolytic end products. 2) A deactivation of the proton pump seemingly does not lead to rapid acidification of the cytosol. 3) In spite of this, the observation of the H^+ pump reacting to pH shifts with altered transport activity, leads to the suggestion that the pump is an important link in long-term pH regulation (Morré, Chapter 4).

Interaction of H^+ pump and H^+ cotransport. ΔpH is one signal and $\Delta\psi_m$ is another signal for the H^+ export pump to react on. A depolarization activates the pump. It is widely accepted that, in plants and fungi, amino acids and sugars (mainly hexoses) are cotransported together with protons. It could therefore be expected that as a result of this proton uptake, the cytosolic pH would acidify. However, this is not true. As the data of Figure 5 demonstrate, the cell is transiently depolarized, as predicted, but the cytosolic pH is increased by 0.2–0.3 units. Since no such alkalinization is observed when the pump is deactivated, as for instance in the presence of cyanide, it is very likely that H^+ export has been stimulated as a consequence of the depolarization. Similar results and interpretation are given by Rodriguez-Navarro et al. (1986) for a H^+/K^+-symport in the fungus *Neurospora*.

These examples indicate that with membrane transport the cell has several ways to regulate and maintain cytosolic pH. In reacting to both ΔpH and $\Delta\psi_m$ as described above, the cell may compensate for a pH change by activating or deactivating not only active H^+ export, but also by regulating

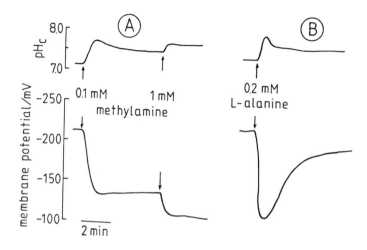

Fig. 5. Alkalinization of cytosolic pH (pH_c) of *Riccia fluitans* rhizoid cells after the addition of the depolarizing agents **A:** methylamine and **B:** L-alanine at the indicated concentrations. The methods were as described (Felle, 1987).

so-called secondary active H^+ (co)transporters. Considerable experimental support for this view comes from the activation and regulation of the Na^+/H^+ exchanger in animal cells (Moolenaar, 1986). When pH_i falls below a certain "threshold," the activity of the exchanger is increasingly stimulated. Aronson et al. (1982) have demonstrated that cytosolic H^+ acts as an allosteric activator of this Na^+/H^+ exchanger by binding to an inward-facing regulatory site. Thus, by changing the H^+ affinity of the regulatory site, extracellular stimuli could control the physiological state of the exchanger and thereby affect the value of pH_i.

In the liverwort, *Riccia fluitans*, Johannes and Felle (1987) showed that cytosolic acidification (acetic acid) by 0.2–0.5 units stimulated the amino acid/H^+ transporter to a much greater extent than would have been expected from the concomitant change in electrochemical gradient for protons (the driving force). This is indicative of a very pH-sensitive internal binding (or regulatory) site for amino acids but could also point to allosteric binding of H^+ to the transporter, as has been demonstrated for the Na^+/H^+-exchanger in animal cells.

pH Dependency of Intracellular Events

An obvious but experimentally challenging starting point for the targets of ΔpH-mediated metabolic regulation is the control of intracellular enzyme activities (Davies, 1973, 1986; Smith and Raven, 1979). In crassulacean acid metabolism (CAM) plants, the importance of pH_c and vacuolar pH (pH_v) and especially the diurnal changes thereof for the control of malate production (malate dehydrogenase) and translocation across the tonoplast have been demonstrated (Marigo et al., 1983). In isolated plantlets of the succulent plant *Kalanchoë tubiflora*, the experimentally increased cytosolic pH (induced by weak acids) caused inhibition of nocturnal malic acid accumulation. This change is accompanied by an inhibition of diurnal remobilization of malic acid, associated with decreased pH_c. If we accept that a ΔpH_c is the prerequisite for such changes in malic acid metabolism and the second messenger, what is the nature of the first messenger? Since these pH changes occur with a diurnal fluctuation, light may be the inducer. All green plants tested so far change their pH_c under a light/dark regime (Felle and Bertl, 1986a,b; Steigner et al., 1988).

Figure 6 gives a few examples of the pH transients accompanied by changes in membrane potential. Switching from light to dark transiently acidifies pH_c by about 0.3 units, whereas light-on causes the opposite effect. Presumably, these pH shifts are due to light/dark-dependent H^+ translocation across the thylakoid membranes. In the green alga *Eremosphera viridis*, turning off the light activates a potassium channel to generate an action potential (Köhler et al., 1985; Steigner et al., 1988). This may be due to the above demonstrated light/dark-dependent acidification of the cytosol, because other treatments leading likewise to cytosolic acidification (e.g., weak acids) also triggered an action potential (Fig. 7). These results, together with the recently reported light-induced depletion in cytosolic free calcium in *Nitellopsis* (Miller and Sanders, 1987), open interesting perspectives as to light-triggered stimulus–response cascades in plants for future studies.

Since the activity of intracellular enzymes depends not only on direct effects of pH_c on activity but also indirectly from effects due to the pH response of other enzymes in the pathway, we always deal with a complicated network of reactions. Thus it becomes difficult to judge "real" effects of pH on activity solely from in vitro experiments. A single enzyme may react directly to small pH shifts and reveal, for instance, quite a different pH optimum than physiologically meaningful in the complete pathway or cascade in the cytosol.

Interaction of Cytosolic Free Calcium and pH

Since a change in pH_c can be regarded as a very common signal, it must not be the only effector or mediator necessary for the proper functioning of certain intracellular events. The important role of calcium as a second

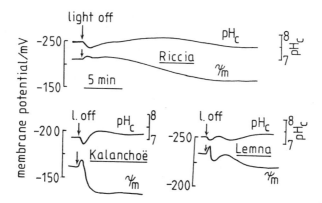

Fig. 6.   Effect of light/dark regimes on cytosolic pH (pH_c) and membrane potential (ψ_m) of *Riccia fluitans* green thallus cells, leaf cells of *Kalanchoë daigremontiana*, and green cells of *Lemna gibba*.

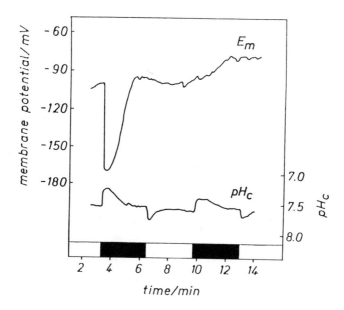

Fig. 7.   Recording of membrane potential and cytosolic pH (pH_c) of the green alga *Eremosphera viridis* during light-off/-on regimes. External pH ($pH_o = 5.5$). Shaded portion of abscissa describes dark period. (Reproduced from Steigner et al., 1988 [Fig. 2], with permission of the publisher).

messenger is well established; therefore, a functional interrelation of pH_c and calcium was predicted. Using various means to manipulate either intracellular calcium or pH_c direct or inverse relationships have been observed in such diverse cell types as vertebrate muscle fibers, snail and squid neurons, insect salivary gland, frog blastomeres, and mammalian myocardium (see review: Busa and Nuccitelli, 1984). However, the physiological implications of these relationships often remain incompletely understood because 1) direct and continuous measurements of free calcium and pH is still rather difficult—simultaneous monitoring of both almost impossible—and 2) some of the observed effects may have little biological meaning because the experimental interferences in cellular metabolism were too drastic (e.g., application of calmodulin inhibitors or large internal pH shifts).

With plants, until recently, intracellular pH was monitored with indirect methods, for example, the distribution of weak acids (DMO), or by ^{31}P nuclear magnetic resonance (NMR) spectroscopy. NMR spectroscopy, while being a very promising and noninvasive method, has shortcomings. For example, it is difficult to measure rapid changes in pH_c in response to a stimulus. More recently, direct and quantitative methods have been introduced for the continuous measurement of intracellular pH (Sanders and Slayman, 1982; Felle and Bertl, 1986a) or of free calcium (Miller and Sanders, 1987; Felle, 1988a,b) using (double-barreled) ion-sensitive microelectrodes. In the cytosol of *Zea mays* and rhizoid cells of the liverwort *Riccia fluitans,* clear and direct relationships between pH_c and Ca^{2+} were observed. When pH_c decreased, either after changing external pH (pH_o) or as a result of weak acid action, a simultaneous increase in cytosolic calcium ensued (Fig. 8, Felle, 1988b). Alkalinization of the vacuole by neutral red (Fig. 9) also increased cytosolic calcium. This result suggests the additional possibility of a direct and functional relationship of H^+ and Ca^{2+} in plants across the tonoplast, perhaps involving among other possibilities the H^+/Ca^{2+} exchanger (Schumaker and Sze, 1987).

CYTOSOLIC pH AND PHYTOHORMONES

Among the outcomes of the development of reliable pH microelectrodes for use in plant cells has been the ability to detect pH_c directly in single cells (Bertl et al., 1984) as well as in tissues (Felle, 1987, 1988a,b) in response to phytohormones. These studies have the advantage that here the primary signal, the hormone, may be administered precisely and controlled to elicit the ΔpH.

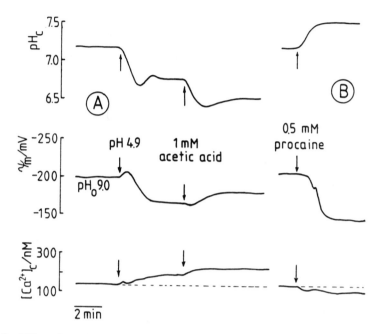

Fig. 8.   Effect of cytosolic acidification (**A:** acetic acid) and alkalinization (**B:** procaine) on cytosolic free calcium and pH_c of *Sinapis alba* root hairs, method: Ion-sensitive double-barreled microelectrodes as described (Felle, 1987; 1988a,b).

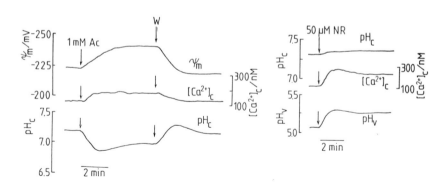

Fig. 9.   Interaction of cytosolic pH and free calcium of *Riccia fluitans* rhizoid cells measured in the presence of 1 mM acetic acid (Ac, left side) and 50 µM neutral red (NR, right side). Note, vacuolar pH (pH_v) became more basic when neutral red was added.

Auxin

Lately, the primary action of auxin in elongation growth is discussed in terms of auxin binding to a plasma-membrane-bound receptor/transducer system, which actuates a signal–response cascade appropriate to growth control (Morré, Chapter 4). A well-established signal–response cascade in animal cells is the formation of IP_3 and of diacylglycerol (DAG) (Berridge, 1984). After the binding of an agonist to a membrane-bound receptor, cytosolic free calcium and pH_c are changed. There is evidence that similar pH and calcium changes occur also in plant cells as a consequence of lipid breakdown in response to auxins or other phytohormones (Morré et al., 1984).

Beside the onset of elongation growth, hyperpolarization, after a lag of 8–10 min following the addition of indole-3-acetic acid (IAA), was one of the most prominent effects to be monitored from coleoptiles of *Zea* and *Avena* (e.g., Cleland et al., 1977; Bates and Goldsmith, 1983; Brummer et al., 1985). Stahlberg and Polevoy (1979). Felle et al. (1986) showed that this hyperpolarization was part of an oscillation with a period of 25–35 min, which, once initiated, continued for several hours, even after the removal of the IAA (Fig. 10). Determination of membrane potential and cytosolic pH in epidermal cells of *Zea* coleoptiles with double-barreled microelectrodes gave exciting results. Within 4–5 min after the addition of 1 μM IAA (external pH 6.0), the cytosolic pH began to fall slowly and then to oscillate with the same period as found for the membrane potential, but roughly displaced by λ/2 (see also Fig. 13, Morré, Chapter 4). From where does this ΔpH arise?

1. Because of the low IAA concentration added (1 μM) at an external pH of 6, the ΔpH of about 1.3 across the plasma membrane is insufficient to acidify the cytosol by an ion trap mechanism.

2. Basically the same argument holds for an assumed IAA^-/H^+ symport, because it is electrically silent and the same ΔpH of 1.3 units is effective as the only driving force.

3. A putative $IAA^-/2H^+$ symport (Hertel, 1983; Goldsmith and Goldsmith, 1981; Sabater and Rubery, 1987), however, could account for the acidification observed, because a driving force of -195 to -200 mV (ΔpH = 1.3, $\Delta\psi_m = -120$ to -125mV) could well accumulate IAA^- from 1 μM outside to about 2 mM in the cytosol. However, the acidification arising therefrom should start right after the addition of IAA and not after a lag of several minutes.

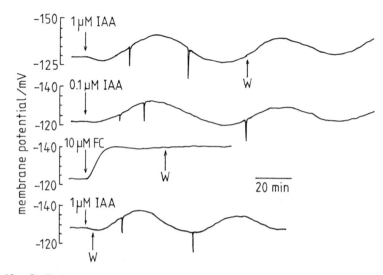

Fig. 10. Oscillations of membrane potential of maize coleoptile cells following the addition of 0.1 and 1 μM indole-3-acetic acid (IAA), or 10 μM fusicoccin (FC) to the external medium. W = removal of stimulus. External conditions: 1 mM KCl, 0.1 mM NaCl, 0.1 mM $CaCl_2$, 5 mM Mes/Tris, external pH (pH_o) = 6.0.

4. Another possibility follows from the above demonstrated interaction of calcium and protons (Figs. 8 and 9). Formation of DAG and IP_3 as a response to auxin binding to a receptor (Löbler and Klämbt, 1985) may change cytosolic pH as well as increase free calcium, respectively. What has been found is that 1 μM IAA, added to maize coleoptiles at an external pH of 6.0, causes cytosolic-free calcium to increase first and then, most interestingly, also to oscillate with a period that compares well with the pH oscillations (Fig. 11). A close interrelation of H^+ and Ca^{2+} is thus demonstrated, although unfortunately technical difficulties do not yet permit measurements of H^+ and Ca^{2+} simultaneously. In at least one respect, the observed intracellular events in plant cells differ from the ones described for animal systems. In animal cells formation of DAG, etc. leads to cytosolic alkalinization. The reason for this inconsistency may be found in the different way in which animal cells transport H^+ across the plasma membrane, in their lack of a proton pump and possibly in the vacuole.

5. A most intriguing possibility of a ΔpH generation by auxin is given by Morré et al. (1986). According to this model, IAA activates a plasma membrane redox system, specifically, NADH oxidation, which leads to a

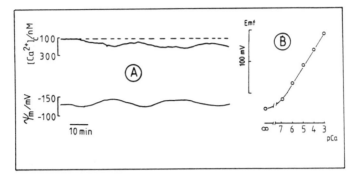

Fig. 11. **A:** Oscillations in cytosolic calcium of maize coleoptile cells following addition of 1 μM indole-3-acetic acid (IAA) to the external medium. **B:** Calibration of the calcium electrode. External conditions as described in Figure 10. Emf, electromotive force; relative units, mV.

stimulation of electron export and presumably to a pH_c change. More detailed information is in the article of Morré (Chapter 4).

Physiological Implications of IAA-Induced ΔpH

A primary target of an acidification of the cytosol is certainly the H^+ ATPase (H^+ export pump), since H^+ is substrate and will lead therefore to an immediate and direct activation. This has been demonstrated repeatedly before (Sanders and Slayman, 1982; Marré et al., 1986; Felle and Bertl, 1986a; Frachisse et al., 1988) and need not be discussed in more detail. An activation of the H^+ export will result in the well-known hyperpolarization (or oscillation) of the membrane potential and to cell wall acidification.

The idea that cell wall acidification is a prerequisite for elongation growth induced by IAA (Cleland, 1976; Hager et al., 1971) has been challenged lately by Kutschera and Schopfer (1985). They demonstrate that the time courses of elongation growth and cell wall acidification differ. Additionally, elongation growth takes place also in the presence of alkaline buffers. This questions whether the H^+ pump is involved in auxin action. On the other hand, it is difficult to measure cell wall pH and thus to judge critically auxin-dependent ΔpH there. Yet there are arguments that still implicate proton pumping in auxin action (acid growth theory) as follows. 1) Weak acids, externally applied, stimulate elongation growth. This is not because of external acidification of the cell wall space (pH_o was kept constant), but it seems a direct consequence of cytosolic acidification after dissociation of the protonated acid (Fig. 12). 2) Alkalinization of the cytosol and 3) deactivation of the H^+-pump inhibit elongation growth (Parish et al., 1986). In general, inactivation of proton pumping will lead immediately to severe inhibition of

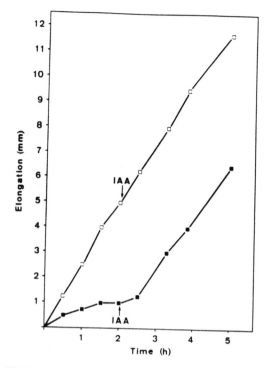

Fig. 12. Effect of IAA on elongation growth of maize coleoptile cells (□) 20 mM Na acetate-HCl, pH 4.5; (■) 20 mM Na acetate-HCl, pH 7.0. IAA (20 μM) was added after 122 min to both treatments. (Reproduced from Brummer et al., 1985 [Fig. 1], with permission of the publisher.)

all H^+-driven transport. This is not because the membrane potential is reduced but mainly because the proton circulation in and out of the cell, driven by the H^+ pump, is interrupted. Transport of many substrates and probably also that of cell wall material or precursors may be connected to proton circulation. With regard to the auxin-induced elongation growth, a stimulation of this pump and hence increased circulation of protons, seems mandatory.

Cytosolic pH, a Messenger in Fusicoccin Action?

There is a wide consensus that the phytotoxin fusicoccin (FC) stimulates active proton export (Marrè, 1979), which is most commonly measured as H^+ extrusion or rapid hyperpolarization. The mechanism of FC stimulation, however, is as yet unknown. Craig and Crane (1982) demonstrated that FC stimulates electron transfer across the plasmalemma, measured as stimulation of ferricyanide reduction, an event that is accompanied by external acidification. According to their data, FC primarily acts on a reductase and not on a plasma membrane H^+ ATPase. This is an interesting parallel to the auxin stimulated redox system reported by Morré et al. (1986).

Since proton translocation was measured in accordance with the stimulation of electron transport, cytosolic pH shifts may have occurred. Such changes have indeed been monitored by several groups. Unfortunately, however, they differ in the observed direction. Specifically, whereas Bertl and Felle (1985), Brummer et al. (1985), and Hager and Moser (1985) report cytosolic acidification immediately following FC application, Reid et al. (1985) and Marrè et al. (1986, 1987) measured an increase in pH_c, 15–30 min after the FC had been added. These findings need not necessarily be contradictory, however, because they were carried out at different times after the FC application and could well reflect a pH_c oscillation similar to that observed in the presence of auxin. Recently, evidence for a FC binding protein has been brought forward by deBoers and Cleland (International Symposium: Physiology and Biochemistry of Auxins in Plants, Libliče, CSSR, 1987). It will be interesting to find out whether this protein is in any way functionally connected with the redox system or is even part of it.

As to the pH changes, the role of pH_c in FC action still remains unclear. It could be that the pH_c changes occur as a transient concomitant phenomenon, not directly involved in the primary conspicuous FC effects but possibly in metabolic aftereffects.

CONCLUSIONS

There are a number of criteria mandatory for a molecule to be a messenger in cellular metabolism. 1) The respective molecule must be present in small quantities. This is important for its regulation (transport, binding molecules), and the sensing of its changes. 2) The changes should be transient. 3) The changes must initiate, mediate, or regulate cellular enzyme activities. 4) The changes should lead to signal amplification.

In asking whether protons are to be regarded as second messengers in plants, no conclusive answer based on experimental evidence can be offered at present. Compared with other ions, protons are different in that they are multifunctional, but they clearly show the characteristics of second messengers. For example, pH changes are known to be induced by extracellular stimuli transduced after the perturbation of stimulus-perceiving structures. On the other hand, regardless of whether it is induced by phytohormones, light, or by regular metabolic cell function, a ΔpH will act on a wide range of cellular enzymes and will provoke at least the counter action of cellular pH regulation. This may not always hold for localized pH changes that occur close to the affected target, since they may not be sensed. It is conceivable, therefore, that as a direct consequence of a stimulus, a ΔpH is triggered and is restored in the bulk cytosol, but not in membrane or protein pockets. Can a cell distinguish a ΔpH originating from metabolism or from other sources? Presumably yes, but certainly only if additional information is available, for example, when ΔpH oscillates or when other messengers like Ca^{2+} are involved. Therefore, it may be meaningful to refer to pH or ΔpH simply as a *messenger,* or as Busa and Nuccitelli (1984) suggest, as *synergistic messenger,* providing a metabolic correlation through which the actions of other effectors are integrated. In this context, light perception in plants, whether through chlorophyll or phytochrome, may be similarly mediated. As demonstrated in Figures 6 and 7, changes in light intensity lead to changes in pH_c (and Ca^{2+}), with the potential to trigger or elicit other events. Future experiments will decide whether regulation of cellular events through phytochrome involves not only Ca^{2+} but also pH_c.

REFERENCES

Aronson PS, J Nee, MA Suhm (1982): Modifier role of internal H^+ in activating the Na–H exchanger in renal microvillis membrane vesicles. Nature 299:161–163.

Barton JK, JA Den Hollander, TM Lee, A MacLaughlin, RG Shulman (1980): Measurement of the internal pH of yeast spores by ^{31}P nuclear magnetic resonance. Proc Natl Acad Sci USA 77:2470–2473.

Bates GW, MHM Goldsmith (1983): Rapid response of the plasma-membrane potential in oat coleoptiles to auxin and other weak acids. Planta 159:231–237.

Berridge MJ (1987): Inositol trisphosphate and diacylglycerol: Two interacting second messengers. Annu Rev Biochem 56:156–193.

Bertl A, H Felle, FW Bentrup (1984): Amine transport in *Riccia fluitans.* Cytoplasmic and vacuolar pH recorded by a pH-sensitive microelectrode. Plant Physiol 76:75–78.

Bertl A, H Felle (1985): Cytoplasmic pH of root hair cells of *Sinapis alba* recorded by a pH-sensitive microelectrode. Does fusicoccin stimulate the proton pump by cytoplasmic acidification? J Exp Bot 36:1142–1149.

Brummer B, A Bertl, I Potrykus, H Felle, RW Parish (1985): Evidence that fusicoccin and indole-3-acetic acid induce cytosolic acidification of *Zea mays* cells. Fed Eur Biochem Soc 189:109–114.

Busa WB, JH Crown, GB Matson (1982): Intracellular pH and the metabolic status of dormant and developing *Artemia* embryos. Arch Biochem Biophys 216:711–718.

Busa WB, R Nuccitelli (1984): Metabolic regulation via intracellular pH. Am J Physiol 246:409–438.

Cleland RE (1976): Kinetics of hormone-induced H^+-excretion. Plant Physiol 58:210–213.

Cleland RE, HBA Prins, JR Harper, N Higinbotham (1977): Rapid hormone-induced hyperpolarization of the oat coleoptile potential. Plant Physiol 59:395–397.

Craig TA, FL Crane (1982): Hormonal control of a transplasma membrane electron transport system in plant cells. Proc Indiana Acad Sci 91:150–154.

Davies DD (1973): Metabolic control in higher plants. In Milborow (ed): "Biosynthesis and its Control in Plants." London: Academic Press, pp 1–20.

Davies DD (1986): The fine control of cytosolic pH. Physiol Plant 67:702–706.

Den Hollander JA, K Ugurbil, TR Brown, RG Shulman (1981): Phosphorus-31 nuclear magnetic resonance studies of the effect of oxygen upon glycolysis in yeast. Biochemistry 20: 5871–5880.

Epel D, RA Steinhardt, T Humphries, D Mazia (1974): Analysis of the partial metabolic depression of sea urchin eggs by ammonia: The existence of independent pathways. Dev Biol 40:245–255.

Felle H (1981): A study of the current-voltage relationships of electrogenic active and passive membrane elements in *Riccia fluitans*. Biochim Biophys Acta 646:151–160.

Felle H (1983): Driving forces and current-voltage characteristics of amino acid transport in *Riccia fluitans*. Biochim Biophys Acta 730:342–350.

Felle H (1987): Proton transport and pH control in *Sinapis alba* root hairs: A study carried out with double-barrelled pH microelectrodes. J Exp Bot 38:340–354.

Felle H (1988a): Auxin causes oscillations of cytosolic free calcium and pH in *Zea mays* coleoptiles. Planta 174:495–499.

Felle H (1988b): Cytoplasmic free calcium in *Riccia fluitans* and *Zea mays*. Interaction of Ca^{2+} and pH? Planta (in press).

Felle H, A Bertl (1986a): The fabrication of H^+-selective liquid-membrane microelectrodes for use in plant cells. J Exp Bot 37:1416–1428.

Felle H, A Bertl (1986b): Light-induced cytoplasmic pH changes and their interrelation to the activity of the electrogenic proton pump in *Riccia fluitans*. Biochim Biophys Acta 848:176–182.

Felle H, B Brummer, A Bertl, RW Parish (1986): Indole-3-acetic acid and fusicoccin cause cytosolic acidification of corn coleoptile cells. Proc Natl Acad Sci USA 83:8992–8995.

Frachisse JM, E Johannes, H Felle (1988): The use of weak acids as physiological tools: A study of the effects of fatty acids on intracellular pH and electrical plasmalemma properties of *Riccia fluitans* rhizoid cells. Biochim Biophys Acta 938:199–210.

Gillies RJ, DW Deamer (1979): Intracellular pH changes during the cell cycle in *Tetrahymena*. J Cell Physiol 100:25–33

Goldsmith MHM, TH Goldsmith (1981): Quantitative predictions for the chemiosmotic uptake of auxin. Planta 153:25–33.

Gradmann D, UP Hansen, WS Long, CL Slayman, J Warnke 1978: Current-voltage relationships for the plasma membrane and its principal electrogenic pump in *Neurospora crassa*. I. Steady state conditions. J Membrane Biol 39:333–367.

Guern J, Y Mathieu, M Pean, C Pasquier, JC Beloeil, JY Lallemand (1986): Cytoplasmic pH regulation in *Acer pseudoplatanus* cells. I. A ^{31}P NMR description of acid-load effects. Plant Physiol 82:840–845.

Hager A, H Menzel, A Krauss (1971): Versuche und Hypothese zur Primärwirkung des Auxins beim Streckungswachstum. Planta 100:47–75.

Hager A, I Moser (1985): Acetic acid esters and permeable weak acids induce active proton extrusion and extension growth of coleoptile segments by lowering the cytoplasmic pH. Planta 163:391–400.

Hertel R (1983): The mechanism of auxin transport as a model for auxin action. Z Pflanzenphysiol 112:53–67.

Johannes E, H Felle (1985): Transport of basic amino acids in *Riccia fluitans:* Evidence for a second binding site. Planta 166:244–251.

Johannes E, H Felle (1987): Implications for cytoplasmic pH, protonmotive force and amino acid transport across the plasmalemma of *Riccia fluitans*. Planta 172:53–59.

Jung KD, U Lüttge (1980): Amino acid uptake by *Lemna gibba* by a mechanism with affinity to neutral L- and D-amino acids. Planta 150:230–235.

Kinraide TB, B Etherton (1980): Electrical evidence for different mechanisms of uptake for basic, neutral, and acidic amino acids in oat coleoptiles. Plant Physiol 65:1085–1089.

Köhler K, W Steigner, W Simonis, W Urbach (1985): Potassium channels in *Eremosphera viridis*. I. Influence of cations and pH on resting membrane potential and on an action potential-like response. Planta 166:490–499.

Komor E, W Tanner (1974): The hexose–proton cotransport system of *Chlorella*: pH-dependent change in K_m-values and translocation constants of the uptake system. J Gen Physiol 64:568–581.

Kutschera U, P Schopfer (1985): Evidence against the acid growth theory of auxin action. Planta 163:483–493.

Lee HC, C Johnson, D Epel (1983): Changes in internal pH associated with initiation of motility and acrosome reaction of sea urchin sperm. Dev Biol 95:31–45.

Löbler M, D Klämbt (1985): Auxin-binding protein from coleoptile membranes of corn. I. Purification by immunological methods. J Biol Chem 260:9849–9853.

Marigo G, U Lüttge, JAC Smith (1983): Cytoplasmic pH and the control of crassulacean acid metabolism. Z Pflanzenphysiol 109:405–413.

Marrè E (1979): Fusicoccin: A tool in plant physiology. Annu Rev Plant Physiol 30:273–288.

Marrè E, N Beffagna, G Romani (1987): Potassium transport and regulation of intracellular pH in *Elodea densa* leaves. Bot Acta 0:17–23.

Marrè MT, G Romani, M Bellando, E Marré (1986): Stimulation of weak acid uptake and increase in cell sap pH as an evidence for FC- and K^+-induced cytosol alkalinization. Plant Physiol 82:316–323.

Mazia D (1974): Chromosome cycles turned on in unfertilized sea urchin eggs exposed to NH₄OH. Proc Natl Acad Sci USA 71:690–693.

Mazia D, A Ruby (1974): DNA synthesis turned on in unfertilized sea urchin eggs by treatment with NH₄OH. Exp Cell Res 85:167–172.

Miller T, D Sanders (1987): Depletion of cytosolic free calcium induced by photosynthesis. Nature 326:397–376.

Moolenaar WH (1986): Effects of growth factors on intracellular pH regulation. Annu Rev Physiol 48:363–376.

Moore RD (1979): Elevation of intracellular pH by insulin in frog skeletal muscle. Biochim Biophys Res Commun 91:900–904.

Morisawa M, RA Steinhardt (1982): Changes in intracellular pH of *Physarum plasmodium* during the cell cycle and in response to starvation. Exp Cell Res 140:341–351.

Morré DJ, B Gripshover, A Monroe, JT Morré (1984): Phosphatidylinositol turnover in isolated soybean membranes stimulated by the synthetic growth hormone 2,4-dichlorophenoxyacetic acid. J Biol Chem 259:15364–15368.

Morré DJ, P Navas, C Penel, CJ Castillo (1986): Auxin-stimulated NADH oxidase (semidehydroascorbate reductase) of soybean plasma membrane: Role in acidification of cytoplasm. Protoplasma 133:195–197.

Parish RW, H Felle, B Brummer (1986): Evidence for a mechanism by which auxins and fusicoccin may induce elongation growth. In Trewavas AJ (ed): "Molecular and Cellular Aspects of Calcium in Plant Development." New York: Plenum, pp 301–308.

Poole R (1978): Energy coupling for membrane transport. Annu Rev Plant Physiol 29:437–460.

Reid RJ, LD Field, MG Pitman (1985): Effects of external pH, fusicoccin, and butyrate on the cytoplasmic pH in barley root tips measured by ³¹P-nuclear magnetic resonance spectroscopy. Planta 166:341–347.

Rodriguez-Navarro A, MR Blatt, CL Slayman (1986): A potassium–proton symport in *Neurospora crassa*. J Gen Physiol 87:649–674.

Sabater M, PH Rubery (1987): Auxin carriers in *Cucurbita* vesicles. II. Evidence that carrier-mediated routes of both indole-3-acetic acid influx and efflux are electro-impelled. Planta 171:507–513.

Sanders D, CL Slayman (1982): Control of intracellular pH: Predominant role of oxidative metabolism, not proton transport, in the eucaryotic microorganism *Neurospora*. J Gen Physiol 80:377–402.

Schumaker KS, H Sze (1987): Inositol 1,4,5-trisphosphate releases Ca^{2+} from vacuolar membrane vesicles of oat roots. J Biol Chem 262:3944–3946.

Shen SS, RA Steinhardt (1978): Direct measurements of intracellular pH during metabolic derepression of the sea urchin egg. Nature 272:253–254.

Smith FA, JA Raven (1979): Intracellular pH and its regulation. Annu Rev Plant Physiol 30:289–311.

Spanswick RM (1982): The electrogenic pump in the plasma membrane of *Nitella*. In Kleinzeller A, F Bronner, CL Slayman (eds): "Current Topics in Membranes and Transport. Electrogenic Pumps." New York: Academic Press, pp 35–47.

Stahlberg R, VV Polevoy (1979): Nature of rhythmic oscillations of the membrane potential in corn coleoptile cells. Dokl Acad Nauk USSR 247:1022–1024.

Steigner W, K Köhler, W Simonis, W Urbach (1988): Transient cytoplasmic pH changes in correlation with opening of potassium channels in *Eremosphera*. J Exp Bot 39:23–36.

Steinhardt RA, D Mazia (1973): Development of K-conductance and membrane potentials in unfertilized sea urchin eggs after exposure to NH₄OH. Nature 241:400–401.

Tazawa M, T Shimmen (1982): Control of electrogenesis by ATP, Mg^{2+}, H^+, and light in perfused cells of *Chara*. In Kleinzeller A, F Bronner, CL Slayman (eds): "Current Topics in Membranes and Transport: Electrogenic Pumps." New York: Academic Press, pp 49–67.

Vaquier VD, B Brandriff (1975): DNA synthesis in unfertilized sea urchin eggs can be turned on and off by the addition and removal of procaine hydrochloride. Dev Biol 47:12–31.

Webb DJ, R Nuccitelli (1982): Intracellular pH changes accompany the activation of development in frog eggs: Comparison of microelectrode and ^{31}P NMR measurements. In Nuccitelli R, DW Deamer (eds): "Intracellular pH: Its Measurement, Regulation, and Utilization in Cellular Functions." New York: Alan R. Liss, pp 293–324.

Second Messengers in Plant Growth
and Development, pages 167–179
© 1989 Alan R. Liss, Inc.

7. Ether Phospholipid Platelet-Activating Factor (PAF) and a Proton-Transport-Activating Phospholipid (PAP): Potential New Signal Transduction Constituents for Plants

Günther F.E. Scherer

Botanisches Institut, Universität Bonn, D-5300 Bonn 1, Federal Republic of Germany

IMPORTANCE OF PROTON TRANSPORT IN PLANTS

ATPases are the primary ion pumps on the plasma membrane and the tonoplast or vacuolar membrane of plant cells. At least one plasma membrane-associated ATPase is Ca^{2+} activated (Dieter and Marmé, 1981). The function of the ion-translocating ATPases is to distribute the energy stored in an ion gradient over a membrane surface. This can be used to drive other energized transport processes. A membrane potential is created by an electrogenic ion pump, and, with appropriate charge compensation, a pH gradient also is formed. The energy from the electrochemical proton gradient can be used to energize secondary transport processes, which are driven by the flow of protons back into the cytoplasm. Such secondary transport processes are known, for example, for anions, cations, sugars, amino acids, organic acids, and alkaloids. Depending on the charge of the molecule to be transported, the secondary transport can be constructed either as a cotransport or as an antiport, with a variable number of protons participating. This general scheme of changing one energy currency, ATP in the cyto-

Address reprint requests to Günther F.E. Scherer, Botanisches Institut, Universität Bonn, Venusbergweg 22, D-5300 Bonn 1, Federal Republic of Germany.

plasm, into another one, a proton gradient across the membrane surface to the extracellular space, allows the cell to employ basically one type of ion-translocating ATPase to drive several different transport processes. The scheme highlights the importance of the primary pumps in the plant cell, that is, proton-translocating ATPases in both the plasma membrane and in the tonoplast. Proton transport by H^+-ATPases also is integrated into physiological reactions or responses of the plant to various environmental stimuli. Methods of proton transport in plants have been reviewed (Poole, 1978; Spanswick, 1981; Sze, 1985).

PHYSIOLOGICAL REACTIONS

Proton transport is necessary to drive sugar transport via the phloem (*phloem loading*), since the sugar crosses the cell walls on its way from leaf mesophyll to the phloem. Another important physiological reaction that involves the proton-translocating ATPases as the driving force is the opening and closing of the stomata. The rapid folding of *Mimosa* leaves on touch is a unique physiological response. In a much slower form, a similar movement is the day–night rhythm of leaf folding (nyctinasty) of *Samanea*. Both responses are known to involve a massive ion translocation in the pulvinary nodes of the leaves.

In roots, gravitropic bending requires signal transduction at the tissue level. The stimulus of gravity must be transformed into a growth response that makes the root always grow downward or, more precisely, in the direction of the vector of gravity. This physiological reaction must involve very rapid electric responses mediated somehow by the plasma membrane H^+-ATPase. The corresponding reaction, the gravitropic bending of the shoot, has not yet been investigated at the same level of sophistication, but a similar sequence of events might be anticipated.

The last important physiological process to be mentioned here is the crassulacean acid metabolism (CAM), which involves the storage of CO_2 in the form of malic acid in the vacuole during the night. CAM plants such as *Kalanchoe* release the malic acid during the day and transform it back into CO_2 and pyruvate, so that the CO_2 can be used for photosynthetic carbon fixation without the necessity of opening their stomata during the day. This transport of malic acid in and out of the vacuole allows such plants to close their stomata during the day so that they can survive in arid habitats with much less water loss than plants without such an adaptation. These short illustrations highlight the importance of H^+-ATPases in plant physiology.

STIMULATION OF PROTON TRANSPORT BY THE PLATELET-ACTIVATING FACTOR

Platelet-activating factor (PAF, 1-O-alkyl-2-acetyl-*sn*-glycero-3-phosphocholine), when added to zucchini microsomes, was capable of stimulating proton transport as shown in Figure 1 (Scherer, 1985). Phosphate was required as a permeant anion, instead of the usual chloride.

The rationale for trying PAF as an effector in plant microsomal proton transport was its superficial similarity to oleylacetyldiglyceride (OAG), a known artificial stimulator of protein kinase C in animal cells (Nishizuka, 1984). At the time, PAF was tested in the proton transport experiments, its

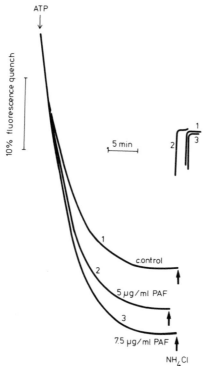

Fig. 1. Kinetics of proton transport in zucchini microsomes. A decrease in fluorescence means an increase in the pH gradient across the vesicle membrane. Ammonium chloride added at the arrows collapses the pH gradient. In the presence of platelet-activating factor (PAF) or the plant proton-transport-activating phospholipid (PAP) (not shown here, but compare Scherer and Stoffel [1987]), a steeper pH gradient is obtained at the end of a proton-transport test. (Reproduced from Scherer, 1985, courtesy of Biochem Biophys Res Commun, Academic Press.)

chemical structure had been determined (Demopoulos et al., 1979; Benveniste et al., 1979) and is presented in Figure 2 (for a review, see Hanahan, 1986). Among its structural characteristics are the acetyl group at the C2 position of the glycerol backbone and an ether bond to a long-chain aliphatic alcohol at the C1 position. In animals both these features are of utmost importance for biological activity (Demopoulos et al., 1979; Satouchi et al., 1981; Hillmar et al., 1984; Hanahan, 1986). This is understandable because the PAF binds to a protein receptor in the plasma membrane of sensitive cells with extremely high affinity (Hwang et al., 1986). Additionally, PAF in animal cells stimulates both an influx of Ca^{2+} and a protein kinase C activity.

The biological activity of this hormone-like phospholipid in animal cells extends down to 10^{-12} M. One obvious difference in the effect of PAF on plant and animal membranes is that micromolar concentrations of PAF are

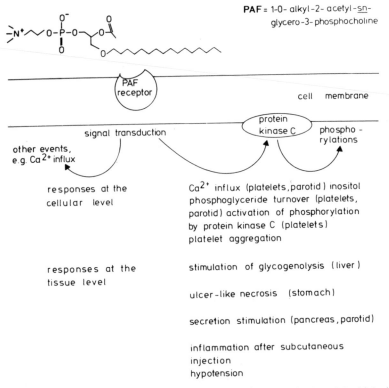

Fig. 2. The chemical structure of the platelet-activating factor (PAF) and its biological activities in mammals. The information contained in this figure is taken from Hanahan (1986). Active concentrations of PAF are in the range of 0.1–10 nM.

required in order to observe its stimulatory activity on proton transport on plant microsomes. When PAF was added to microsomes from soybean, lupin root, and red beet, usually about 20–40 μM of the PAF was optimal. The effect of PAF decreased with increasing concentrations greater than 40 μM. With these microsomal fractions about 80% or more of the transport activity was derived from the tonoplast (Scherer and Fischer, 1985).

As shown in Figure 1, the effect of the PAF on H^+ transport was on the steady-state ΔpH rather than on the initial rate of transport; therefore, the effect was quantified on the basis of an increase of the steady-state ΔpH. The steady-state ΔpH in leaky vesicles with constant backflow of protons was used as the measure of proton transport activity (Gogarten-Boekels et al., 1985; Scherer and Fischer, 1985). The effect of PAF on proton transport exhibited a clear lipid specificity for the stimulatory action (Table I). Quite remarkably, no other phospholipid and none of the other glycerides tested stimulated proton transport appreciably. Since we wanted to know whether lipids known to stimulate animal protein kinase C (Nishizuka, 1984) also

TABLE I. Stimulation of Proton Transport by Lipids in Zucchini Microsomes Containing Cytosolic Proteins.**

Comparisons	Concentrations (μg/ml, unless otherwise given)	Control (%)
Comparison with other glycerides		
Platelet-activating factor	10 μg/ml (= 19 μM)	191 ± 22
1-O-Hexadecylglycerol	50μM	111 ± 14
1-Monoolein	10	114 ± 20
1,2-Diolein	10	112 ± 10
Triolein	10	111 ± 5
Phorbol-12-myristate-13-acetate	10	98 ± 17
Comparison with phospholipids		
Platelet-activating factor	10	197 ± 20
Phosphatidylserine	10	108 ± 9
Phosphatidylinositol	10	108 ± 12
Phosphatidic acid	10	103 ± 9
Phosphatidylglycerol	10	107 ± 11
Phosphatidylcholine	10	105 ± 7
Phosphatidylethanolamine	10	107 ± 13
Comparison with lysolipids		
Platelet-activating factor	10	162
Platelet-activating factor	20	216
Lysophosphatidylcholine	10	106
Lysophosphatidylcholine	20	126

**From Scherer et al. 1988a, Courtesy of Planta, Spinger Verlag.

might stimulate plant proton transport, we tested phorbol ester. Both phorbol ester and diacylglyceride, another activator of protein kinase C, had no effect on the proton transport of plant microsomes. The possibility that PAF acted as a detergent was excluded by the fact that lysophosphatidylcholine, a known detergent and very similar in its chemical structure to PAF, had little effect on proton transport in comparison with PAF.

The stimulation of the steady-state ΔpH in a vesicle with ATP-dependent proton transport could, in principle, be accomplished by making the membranes less permeable for proton diffusion outward, by alleviating the anion cotransport inward, or by stimulating the velocity of the proton-translocating ATPase without other changes in the membrane. The first two possibilities were checked by an experiment where the apparent passive proton permeability was tested in the presence of different anions and of a PAF concentration high enough to increase the steady-state ΔpH. When the passive flow of protons was recorded in the presence of either chloride or phosphate, PAF had no influence on the apparent passive proton permeability, demonstrating that neither the diffusion of protons nor of chloride or phosphate was changed by the PAF. When ATP hydrolysis was measured, the observed effect of PAF on the Km of ATP hydrolysis was minimal, whereas the maximal velocity of the reaction was stimulated in agreement with the transport measurements (Scherer and Martiny-Baron, 1987). Taken in toto, these data indicated that the effect of PAF was to increase the velocity of the proton-translocating ATPase.

PLANTS CONTAIN A LIPID WHICH IS A PROTON-TRANSPORT-ACTIVATING PHOSPHOLIPID (PAP) AND WHICH AGGREGATES BLOOD PLATELETS

For PAF activation of proton translocation to be physiologically significant, a similar lipid would have to occur naturally in plants. A plant lipid corresponding to PAF in its biological activity and, presumably with a very similar chemical structure, was found in plant lipid extracts (Fig 3, taken from Scherer and Stoffel, 1987). This lipid has been called the *proton-transport-activating phospholid* (PAP). The plant lipid was found when plant lipid extracts were compared with authentic PAF, using two-dimensional thin-layer chromatography (TLC). One minor spot on a two-dimensionally developed plate cochromatographed with PAF. The plant lipid also contained lipid-bound phosphorus. A quantitative determination of the amount of the PAP in plant membranes showed that it represented about 1% of the total phospholipids in zucchini microsomes, assuming that it contains

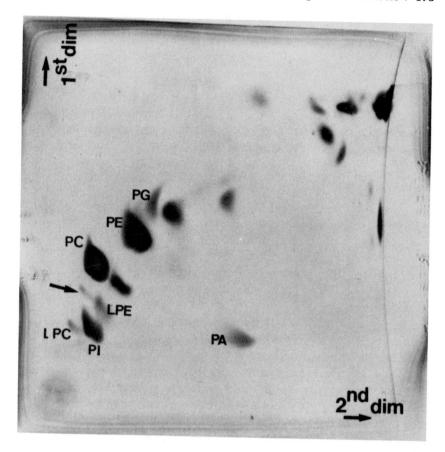

Fig. 3. Two-dimensional thin-layer chromatogram of a total lipid extract from crude zucchini microsomes. The position of the PAP is indicated by an arrow. PA, phosphatidic acid; PAP, proton-transport-activating phospholipid (arrow); PC, phosphatidylcholine; PE, phosphatidylethanolamine; PG, phosphatidylglycerol; PI, phosphatidylinositol; LPC, lysophosphatidylcholine; LPE, lysophosphatidylethanolamine. (Reproduced from Scherer and Stoffel, 1987, courtesy of Planta, Springer-Verlag.)

also only 1 mole phosphorus per molecule as does PAF. When the lipid was isolated by TLC and by high-pressure liquid chromatography (HPLC), the lipid-stimulated proton transport in a manner similar to PAF. PAP also had biological activity similar to PAF in animal cells in that it aggregated human blood platelets (Scherer et al., 1988b). When the plant lipid was compared with authentic PAF, it had about a 30–300-fold lower specific activity in the platelet bioassay. Since the platelet aggregation assay is very sensitive to even small changes in the structure of PAF, the fact that PAP causes aggregation and the fact that PAP and PAF have similar chromatographic properties suggest that PAP and PAF have a very similar chemical structure (Demopoulos et al., 1979; Satouchi et al., 1981; Hillmar et al., 1984; Hanahan, 1986).

PAF INTERACTS WITH A SOLUBLE PROTEIN KINASE AND PROTON-TRANSLOCATING ATPASE

Critical to increasing our understanding of how PAF affects plant ATPases was the observation that the stimulation of microsomal proton transport could be removed by washing the membranes. Moreover, the stimulation could be restored by re-adding the supernatant proteins. This was not a nonspecific protective effect of an increased protein concentration, since added bovine serum albumin did not restore the PAF effect. That a protein was a likely candidate for the mediation of the stimulatory effect of the PAF was shown by the fact that the addition of boiled supernatant was ineffective.

The protein fraction responsible for the stimulatory effect was partially purified by DEAE-Sephacel column chromatography. Protein eluting with about 0.3 M salt stimulated proton transport in a PAF-dependent manner. Since we suspected that a protein kinase could mediate the PAF stimulation, protein kinase activity profile was measured in a parallel experiment. Indeed, at the same relative position in the protein elution profile, a peak of protein kinase activity was found. The comparison of the protein profiles of many of such experiments showed that the elution profile of the DEAE-Sephacel chromatography was reproducible. To test the hypothesis that a protein kinase mediated the PAF stimulation of proton transport, the specific stimulation of protein kinase by PAF and PAP was measured. The lipid specificity of the activation of protein kinase activity clearly showed both PAF and PAP stimulated protein kinase activity appreciably to values above the average for other lipids. The relatively low level of stimulation may have been due to the high background of unspecific protein kinase in microsomes. Importantly, PS, diacylglycerol (DAG), and phorbol esters, the known activators for animal protein kinase C, did not stimulate protein phosphorylation when compared with other lipids. Thus PAF and PAP specifically stimulated protein kinase activity and activated proton transport. The PAF-stimulated protein kinase activity copurifies with the protein that mediates proton-transport stimulation. Therefore, it is postulated that the PAP, a

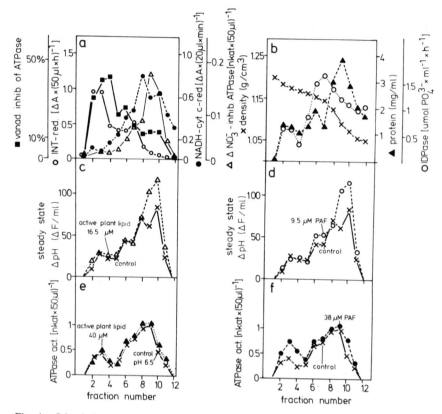

Fig. 4. Stimulation of ATPase activity and proton transport activity by PAF and proton-transport-activating factor (PAP), the active plant lipid. Comparison of marker enzymes after centrifugation of the membranes (zucchini hypocotyl) on an isopycnic linear sucrose gradient. **a:** ■——■ orthovanadate-inhibited ATPase; Δ——Δ nitrate-inhibited ATPase; ○——○ INT-reductase; ●···● cytochrome c-reductase nicotinamide adenine dinucleotide, reduced (NADH). **b:** ×——× density; ▲ . . . ▲ protein; ○ . . . ○ IDPase. **c:** ×——× proton transport activity, control; Δ . . . Δ proton transport activity in the presence of 16.5 μM lipid-bound phosphorus of the PAP. **d:** ×——× proton transport activity, control; ○ . . . ○ proton transport activity in the presence of 9.5 μM PAF. **e:** ATPase activity without ×——× and in the presence of ▲ . . . ▲ 40 μM lipid-bound phosphorus of the PAP. **f:** ATPase activity without ×——× and with ●···● 38 μM PAF (Stoffel and Scherer, unpublished results).

protein kinase, and one or more H^+-ATPases are (part of) a system in plant cells that regulates proton transport (Scherer et al., 1988a).

Since plants possess a H^+-ATPase in the plasma membrane and in the tonoplast (Sze, 1985), we investigated the possibility that one or both H^+-ATPases or a partially purified protein kinase are stimulated by either PAF or PAP (Scherer et al., 1988a). The proton transport activation by the PAF is most evident in the tonoplast-containing fractions and less so in the plasma-membrane-containing fractions from linear sucrose gradients (Fig. 4, from Stoffel, 1987). The stimulation of ATP hydrolysis is more evident in the plasma-membrane-containing than in the tonoplast-containing fractions.

The phosphorylation of specific proteins is stimulated by the PAF, and these proteins are currently being identified. An example is given in Figure 5. It can be seen that the phosphorylation of two proteins was stimulated by the PAF; one of the proteins had an apparent molecular weight (MW) of 55 kDa and one had an MW of 35 kDa. Phosphorylation of a protein with a molecular weight of 100 kDa was not observed. Evidence for regulatory phosphorylation of the plasma membrane ATPase (having an apparent MW of 100 kDa) has been obtained in other systems (Bidway and Takemoto, 1987; Schaller and Sussman, 1988) where phosphorylation increased ATPase activity. We could find no PAF-dependent phosphorylation of a 100-kDa protein of the plasma membrane H^+-ATPase in our experiments. Tognoli and Basso (1987) observed that fusicoccin stimulated the phosphorylation of a 33-kDa protein, even though fusicoccin is known to activate the plasma membrane H^+-ATPase specifically. It is conceivable that the 33 kDa protein that Tognoli and Basso observed in *Acer pseudoplatanus* cells is similar to the 35 kDa protein that we observed in zucchini, and that the 55-kDa protein observed by us might be a subunit of the tonoplast H^+-ATPase, which has two large subunits of ~ 55–65 kDa and 65–75 kDa (Sze, 1985). The only evidence for regulation of H^+-ATPase in the tonoplast is the observation by Struve and Lüttge (1987), where they found that ATPase activity increases during the transition from C3 metabolism to CAM metabolism of *Mesembranthemum chrystallinum,* but a mechanism for this regulatory event has not yet been suggested.

In the model presented in Figure 6, we suggest that the PAP is a constituent of the membrane; thus the soluble protein kinase is bound to the membrane in close proximity to the H^+-ATPase, so that regulatory phosphorylation can occur. In the absence of counteracting stimuli, the PAP would regulate the proton pump, the ATPase, or both, but the same processes also might be activated by cytosolic Ca^{2+}.

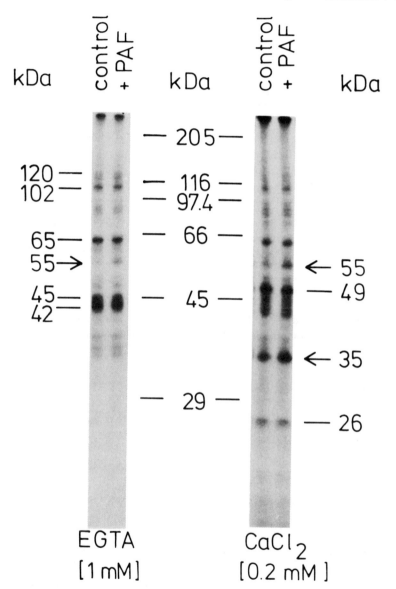

Fig. 5. Phosphorylation of membrane proteins in zucchini microsomes by endogenous protein kinase activity in the presence or absence of platelet-activating factor (PAF). Arrows indicate the position of the two proteins whose phosphorylation is stimulated by PAF (Martiny-Baron and Scherer, unpublished results).

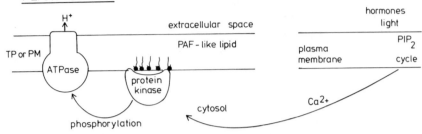

Fig. 6. Hypothetical mechanism of action of proton-transport-activating factor (PAP) and platelet-activating factor (PAF) in plants. The PAF-stimulated protein kinase is thought to phosphorylate proteins in the tonoplast and the plasma membrane. Since the subunit structure of the two H^+ ATPases in the tonoplast and the plasma membrane are completely different, presumably different proteins are phosphorylated in these two membranes, which may be either subunits of these enzymes or only associated regulatory protein(s). About 10 μM PAF is required in plants for activity. Ca^{2+} released by activation of the phosphatidylinositol 4,5-bisphosphate (PIP2) cycle could also stimulate this system.

ACKNOWLEDGMENTS

This work was supported by grants from the Bundesministerium für Forschung und Technologie and from the Deutsche Forschungsgemeinschaft.

REFERENCES

Benveniste J, M Tence, R Varenne, J Bidault, C Boullet, J Polonsky (1979): Semi-synthèse et structure proposée due facteur activant les plaquetts (P.A.F.): PAF-acether, un alkyl ether analogue de la lysophosphatidylcholine. C R Acad Sci Ser D 289:1037–1040.

Bidway AB, JY Takemoto (1987): Bacterial phytotoxin, syringomycin, induces a protein kinase-mediated phosphorylation of red beet plasma membrane polypeptides. Proc Natl Acad Sci USA 84:6755–6759.

Demopoulos CA, RN Pickard, DJ Hanahan (1979): Platelet activating factor, evidence for 1-0-alkyl-2-acetyl-sn-phosphorylcholine as the active component (a new class of lipid chemical mediators). J Biol Chem 254:9355–9358.

Dieter P, D Marmé (1981): A calmodulin-dependent, microsomal ATPase from corn (Zea mays L.). Fed Eur Biochem Soc Lett 125:245–248.

Gogarten-Boekels M, JP Gogarten, FW Bentrup (1985): Kinetics and specificity of ATP-dependent proton translocation measured with acridine orange in microsomal fractions from green suspension cells of Chenopodium rubrum L. J Plant Physiol 118:309–329.

Hanahan DJ (1986): Platelet activating factor: A biologically active phosphoglyceride. Annu Rev Biochem 55:483–509.

Hillmar J, T Muramatsu, N Zollner (1984): Effects of a thio analog of platelet activating factor on platelet aggregation and adenosine 3′, 5′-monophosphate concentration in hepatocyte suspensions and platelets: A comparison with the naturally occurring compound. Hoppe-Seyler's Z Physiol Chem 365:33–41.

Hwang SB, MH Lam, HN Chang (1986): Specific binding of (3) dihydrokadsurenone to rabbit platelet membranes and its inhibition the receptor agonists and antagonists of platelet-activating factor. J Biol Chem 261:13720–13726.

Nishizuka Y (1984): The role of protein kinase C in cell surface signal transduction and tumour production. Nature 208:693–698.

Poole RJ (1978): Energy coupling for membrane transport. Annu Rev Plant Physiol 29:437–460.

Satouchi K, RN Pincard, LM McManus, DJ Hanahan (1981): Modification of the polar head group of acetyl glyceryl ether phosphorylcholine and subsequent effects on platelet activation. J Biol Chem 256:25–32.

Schaller GE, MR Sussman (1988): Phosphorylation of the plasma membrane H^+-ATPase of oat roots by a calcium-stimulated protein kinase. Planta 73:509–518.

Scherer GFE (1985): 1-0-Alkyl-2-acetyl-sn-glycero-3-phosphocholine (platelet activating factor) stimulates plant H^+ transport in vitro and growth. Biochem Biophys Res Commun 133:1160–1167.

Scherer GFE, G Fischer (1985): Separation of tonoplast and plasma membrane H^+-ATPase from zucchini hypocotyls by consecutive sucrose and glycerol gradient centrifugation. Protoplasma 129:109–119.

Scherer GFE, G Martiny-Baron (1987): Stimulation of in vitro H^+ transport in zucchini microsomes by the ether lipid platelet activating factor and a soluble protein. In Klämbt D (ed): "Plant Hormone Receptors. NATO ASI Series H: Cell Biology," Vol 10, pp 163–175.

Scherer GFE, G Martiny-Baron, B Stoffel (1988a): A new set of regulatory molecules in plants: Platelet activating factor stimulates protein kinase and proton transport in plants. Planta (in press).

Scherer GFE, B Stoffel (1987): A plant phospholipid and platelet activating factor stimulate H^+ transport in isolated membrane vesicles. Planta 172:127–130.

Scherer GFE, B Stoffel, G Martiny-Baron (1988b): Platelet-activating factor (= PAF) and a PAF-like plant lipid are biologically active in a cross-wise fashion: PAF activates ATP-dependent H^+ transport and the PAF-like lipid aggregates platelets and vice versa. Biol Chem Hoppe-Syeler 369:7.

Spanswick RM (1981): Electrogenic pumps. Annu Rev Plant Physiol 32:267–289.

Stoffel B (1987): Platelet-Activating-Factor Analoge Lipide in Planzenmembranen. Diploma Thesis. Bonn University.

Struve I, U Luttge (1987): Characteristics of Mg^{2+} ATP-dependent electrogenic proton transport in tonoplast vesicles of the facultative crassulacean-acid-metabolism plant Mesembryanthemum crystallinum L. Planta 170:11–120.

Sze H (1985): H^+-translocating ATPases: Advances using membrane vesicles. Annu Rev Plant Physiol 36:175–208.

Tognoli L, B Basso (1987): The fusicoccin-stimulated phosphorylation of a 33kDA polypeptide in cells of Acer pseudoplatanus as influenced by extracellular and intracellular pH. Plant Cell Env 10:233–240.

Second Messengers in Plant Growth
and Development, pages 181–212
© 1989 Alan R. Liss, Inc.

8. Membrane-Derived Fatty Acids as Precursors to Second Messengers

James Michael Anderson

Agricultural Research Service, United States Department of Agriculture, and Departments of Crop Science and Botany, North Carolina State University, Raleigh, North Carolina 27695–7631

INTRODUCTION

The generation of second messengers often correlates with stimulus reception at cellular membranes, and therefore, the phospholipids in membranes should be an ideal source for the generation of second messenger compounds. In animal cells, several types of second messengers are derived directly from phospholipids, including the inositolphosphates, diacylglycerol (DAG), and platelet-activating factor. For each of these classes of animal second messengers, a counterpart has been identified in plant tissues. In fact, a significant proportion of the present volume is devoted to phospholipid-derived second messengers. This chapter will explore the possibility that plants utilize second messengers that are derived from two classes of phospholipid hydrolysis products: DAG and free fatty acids.

The fatty acids released during hydrolysis of lipid esters in animal tissues are the precursors for the synthesis of a large number of physiologically active compounds. In particular, arachidonic acid is processed into the eicosanoids, a diverse group of chemical messengers that includes the prostaglandins, prostacyclin, leukotrienes, lipoxins, and thromboxanes (Needleman et al., 1986). Although most eicosanoids appear to be used primarily in intercellular communication, some may also act as second messengers. The biochemical pathways utilized by animals to produce eicosanoids have direct corollaries in plants. In fact, enzyme preparations

Address reprint requests to James M. Anderson, USDA-ARS, Plant Science Research, 3127 Ligon Street, Raleigh, NC 27607.

isolated from plant tissues can catalyze the formation of both prostaglandins and leukotrienes (Bild et al., 1978; Shimizu et al., 1984).

Two fatty acid-derived compounds found in plants, traumatic acid (TA) and jasmonic acid (JA), have been implicated as plant growth regulators (Vick and Zimmerman, 1987). However, there are several other compounds that have been isolated from plants that might also play a messenger role. The possibility that some of these compounds are fatty acid-derived second messengers will be discussed. The possible involvement of fatty acid-derived chemical messengers in stress signal transduction also will be considered. The association of membrane disruption with stress and the implication of a possible role for TA in wounding makes stress a logical signal for transduction by fatty acid-derived messengers.

Elucidation of the mechanisms involved in the transduction of hormonal signals in animals has seen many significant advances in the last few years. These advances have prompted plant physiologists to look for similar transduction mechanisms in plants. The discovery of calmodulin (Marmé, Chapter 3; Anderson and Cormier, 1978; Anderson et al., 1980; Muto and Miyachi 1977), polyphosphoinositides (Boss, Chapter 2), and phospholipid activated protein kinases in plants (Schafer et. al., 1985) has added weight to the assumption that animals and plants use similar transduction mechanisms. It is this assumption of similarity on which the discussions in this chapter are based. Therefore, the approach will be to present a brief and highly selective overview of the animal literature pertaining to fatty acid-derived chemical messengers. Using models from the animal literature, arguments will be presented for the possibility that similar chemical messengers exist in plant tissues.

It should be noted at the outset that the presence of an animal second messenger in plant tissues does not necessarily mean that the compound has a similar function in plants. Cyclic nucleotides and acetylcholine, for example, have been identified in plants (Cyong et al., 1982; Hoshino, 1983; Smallman and Maneckjee, 1981; Hartmann and Gupta, Chapter 11), but no physiological function has been identified. It is possible that messenger compounds found in plants that are related to those found in animals may play a role in the interaction between plants and animals. Plants synthesize analogues to animal messengers, and these analogues may alter the growth and development of predators or disease organisms. Plants provide a source of many pharmacological agents, and many of these compounds are probably utilized for defensive purposes. The synthesis of compounds similar to prostaglandins by onions may be an example of this tactic (Attrep et al., 1973; Claeys et al., 1986). Some plants apparently have evolved sensitivity to an animal hormone in order to react to the presence of that animal. Dyer (1980) found that mouse submaxillary gland epidermal growth factor, a peptide hormone, stimulated the growth of young sorghum seedlings at micromolar concentrations. The suggestion was made that the epidermal growth factor was deposited on the leaves by grazing herbivores and thus stimulated the growth of the plants. Components of animal cells not found in plants may be used as a signal of predation or disease invasion. Elicitors made from fungal membrane lipids, such as arachidonic acid, are probably examples of this strategy (Bloch et al., 1984).

Finally, plants may utilize a chemical regulator that is similar to one found in animals because both compounds were made by similar biosynthetic pathways in both groups of organisms. This situation probably exists for the fatty acid-derived regulators discussed in this chapter. What makes these compounds even more interesting is the fact that their synthesis in both animals and plants appears to be the result of the initiation of similar signal transduction pathways.

DIACYLGLYCEROL AND PROTEIN KINASE C
Diacylglycerol as a Second Messenger

The transduction of extracellular signals has often been linked to the phosphorylation of specific polypeptides by protein kinases. PKC is a phospholipid- and Ca^{2+}-dependent protein kinase, which can be activated by DAG (Nishizuka, 1984). This activation is a consequence of a DAG-dependent increase in PKC's affinity for Ca^{2+} and thereby a reduction in the amount of Ca^{2+} needed for activity (Donnelly and Jensen, 1983; Niedel and Blackshear, 1986; O'Brian et al., 1984). The in vivo enzyme is probably completely dependent on DAG for activation because cellular levels of Ca^{2+} alone would not be sufficient to activate the enzyme. Tumor-promoting phorbol esters can substitute for DAG by possibly binding to the DAG binding site on the enzyme (Ashendel, 1985; Berridge, 1984; Niedel and Blackshear, 1986). Using phorbol esters as extracellular activators of PKC, activation of PKC has been shown to be required for transduction of a large number of chemical messengers, including growth factors, hormones, and neurotransmitters (Nishizuka, 1984). Response to most of these chemical messengers has also been linked to phospholipase C-dependent turnover of phosphoinositide in the plasma membrane and the release of the second messenger, inositol trisphosphate (IP_3) (Berridge, 1984; Majerus et al., 1986). The role of inositol phosphates as second messengers is reviewed by Boss, Chapter 2, but two parts of the phosphoinositol cascade (Fig. 1) are important to the present discussion.

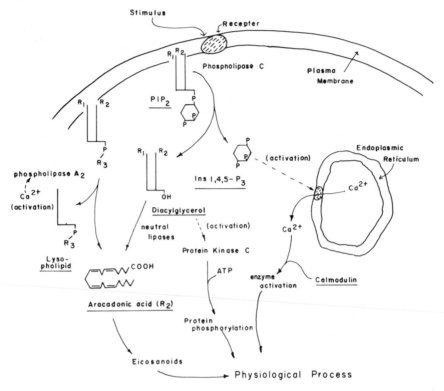

Fig. 1. Model of the phosphoinositol cascade found in animals and the possible coupling of this cascade to the production of eicosanoids.

The IP$_3$ produced during signal transduction causes the release of Ca^{2+} from endoplasmic reticulum stores (Berridge, 1984; Majerus et al., 1986) and possibly from other sites following phosphorylation of IP$_3$ to inositol tetrakisphosphate (IP$_4$) (Michell, 1986; Irvine and Moor, 1986). The increase in cytoplasmic free Ca^{2+} can activate several Ca^{2+} and calmodulin-dependent enzymes, including protein kinases other than PKC (Niedel and Blackshear, 1986) and phospholipases (Van den Bosch, 1982). The release of IP$_3$ from phosphatidylinositol bisphosphate (PIP$_2$) produces, transiently, DAG, which in animals is composed mostly of 1-stearoyl-2-arachidonyl diglyceride (Holub et al., 1970). Therefore, PKC activation can be linked to the release of DAG from PIP$_2$ by phospholipase C, implicating DAG as a second messenger.

In addition to the DAG activator, PKC requires phospholipid for activity. The phospholipid requirement is probably met in vivo by the binding of the enzyme to membranes. In vitro, however, the requirement can be met by various phospholipids, with phosphatidylserine providing the largest stimulation (Ashendel, 1985). Other phospholipids, including the phosphoinositides, also will work as the cofactor (O'Brian et al., 1987).

PKC phosphorylates proteins at serine and threonine residues and will phosphorylate a large number of proteins in vitro (O'Brian et al., 1984; Niedel and Blackshear, 1986). The in vivo substrates of PKC are not fully understood and often are polypeptides that are identified in polyacrylamide gels (Nishizuka, 1984; Takeda et al., 1987). One probable substrate of PKC is inositol 1,4,5-trisphosphate 5'-phosphomonoesterase, the enzyme that degrades IP_3. Phosphorylation of the esterase stimulates enzyme activity that causes relaxation of IP_3-dependent Ca^{2+} release (Connolly et al., 1986). Recently, PKC was shown to be activated by Ca^{2+}-activated proteases that often copurify with the PKC (Kishimoto et al., 1983; Savart et al., 1987). Although Ca^{2+}-dependent proteases are common in animal tissues, they have not been isolated from plants.

Phorbol esters have recently been implicated in the stimulation in vivo of myristylation of an 82 kDa polypeptide in cultured cells (Malvoisin et al., 1987). Acylation of proteins with fatty acids may be one mechanism that allows proteins to associate with membranes. Protein acylation has been implicated in carcinogenesis in animals. The mechanism of the stimulation has not been determined and may be a secondary consequence of phosphorylation of an enzyme by PKC.

Protein Kinase C in Plants

The first evidence for a Ca^{2+}- and DAG-stimulated protein kinase activity in plants came from preliminary work by Morré et al. (1984). Subsequently, Schafer et al. (1985) partially purified such an enzyme from zucchini. The zucchini enzyme was phospholipid- and Ca^{2+}-dependent, and the phospholipid requirement could be met with phosphatidylserine (PS) and phosphatidylethanolamine but not by phosphatidylcholine or PI. Also, a putative PKC has been isolated from wheat by Oláh and Kiss (1986). Although the wheat enzyme was stimulated by a mixture of PS and either phorbol ester or 1,2-diolein, a requirement for DAG or phorbol ester to activate the enzyme was not documented. A similar protein kinase was isolated from *Amaranthus tricolor* by Elliott and Skinner (1986); again, they were not able to show a phorbol ester activation. The properties of the plant enzymes are similar to a putative PKC isolated from *Neurospora*, inasmuch as a phorbol ester-dependent activation of the fungal enzyme could not be demonstrated (Favre and Turian, 1987). No physiological activity has been connected to the extracellular addition of phorbol esters to higher plant tissues.

The absence of activation of plant PKC by phorbol esters may only be a technical problem encountered with isolation of the plant enzymes so that eventually a DAG activation will be demonstrated, but plant PKC may be regulated in a different manner than the animal enzyme. One possibility might be that in plants, phorbol esters cannot substitute for DAG. The DAG in plants would be expected to have predominately linoleic and linolenic acids esterified in the C_2 position of glycerol, while DAG in animals has mostly arachidonic acid in this position (Holub et al., 1970). Thus plant PKC might require a different type of DAG from that utilized in animals.

In endothelial cells, DAG has recently been shown to be released from phosphatidylcholine by a novel phospholipase C that does not require Ca^{2+} (Martin et al., 1987), and a phospholipase C has been described that has high specificity for ether lipids such as platelet-activating factor (Nishihira and Ishibashi, 1986). Furthermore, membrane glycerides can be obtained with other phospholipases besides phospholipase C (Wolfe, 1982). Phospho-lipases A_1 and A_2 catalyze the release of sn-1 and sn-2 acyl side chains leaving a lysophospholipid (Van den Bosch, 1982). Lysophospholipids have been found in the plasma membrane fraction of plant tissues (Wheeler and Boss, 1987), and lysophospholipids could possibly stimulate the plant PKC. Phos-pholipase A_2 appears to be regulated by Ca^{2+} (Fig. 1) both in plant and animal tissues (Berridge, 1984; Morré et al., 1984), so that activation of PKC by a product of phospholipase A_2 might be dependent on the phospho-inositide-dependent release of Ca^{2+}. Therefore, PKC stimulation might still be coupled to phosphoinositide turnover.

FORMATION OF OXYGENATED FATTY ACIDS
Lipoxygenases

Lipoxygenase (E.C. 1.13.11.12) catalyzes the introduction of molecular oxygen into the cis,cis-1,4-pentadiene structure of polyunsaturated fatty acids, resulting in a 1-hydroperoxy-$trans,cis$-2,4-pentadiene derivative (Schewe et al., 1986; Kuhn et al., 1986).

The most common substrate in animals is the C_{20} unsaturated fatty acid, arachidonic acid. In plants, the normal substrates are linoleic and linolenic acids. However, the animal enzymes can oxidize linoleic acid (Engels et al., 1986), and plant lipoxygenases will oxidize other unsaturated fatty acids, including arachidonic (Bild et al., 1978; Vick and Zimmerman, 1987).

Animal Lipoxygenases

Animal lipoxygenases and the types of chemical messengers derived from each lipoxygenase reaction will be reviewed only briefly. For reviews of chemical messengers derived from lipoxygenase activity, see Needleman et al. (1986) and Wolfe (1982). Four distinct lipoxygenases have been isolated from mammalian tissues, and each enzyme catalyzes the introduction of one or more molecules of oxygen into arachidonic acid to yield various hydroperoxy fatty acid products (Fig. 2). The products of the various lipoxygenases are each starting points in the synthesis of a specific class of chemical messengers. Mammalian lipoxygenases are designated as 5-, 12-, and 15-lipoxygenase, with the number referring to the double bond that is oxygenated in arachidonic acid. Cyclooxygenase is a specialized lipoxygenase that inserts two molecules of oxygen into arachidonic acid to yield the 15-hydroperoxy-9,11-endoperoxide product. The enzyme also catalyzes ring closure and formation of the cyclopentane ring characteristic of the prostanoids. A second activity of the enzyme is a peroxidase activity that produces the 15-hydroxy derivative. The prostanoic acid product of cyclooxygenase provides the starting point for the synthesis of prostaglandins, prostacyclin, and thromboxane (Fig. 2). The activity of cyclooxygenase is inhibited by nonsteroidal, anti-inflammatory agents such as aspirin and acetaminophen, and this inhibition is the basis of the pharmacological activity of these compounds (Needleman et al., 1986).

5-Lipoxygenase is a Ca^{2+}-dependent enzyme that catalyzes the introduction of hydroperoxide into the C_5 position of arachidonic acid to yield 5-hydroperoxy-6,8,11,14-eicosatetraenoic acid (5-HPETE; Parker, 1985). The enzyme has been identified in several mammalian tissues, and 5-HPETE is the precursor to the leukotrienes (Fig. 2). Platelet 12-lipoxygenase converts arachidonic acid to 12-S-hydroperoxy-5,8,10,14-eicosatetraenoic acid (Needleman et al., 1986). The hydroperoxy derivative is reduced to the hydroxide by a glutathione-dependent peroxidase. The 15-lipoxygenase forms a 15-hydroperoxy derivative, which is further oxidized by 5-lipoxygenase. The dihydroperoxy derivative leads to the formation of the lipoxins (Fig. 2).

Plant Lipoxygenases

Plant lipoxygenases insert oxygen at either the 9- or 13-position of linoleic or linolenic acids. For a comprehensive review of the subject, the reader is

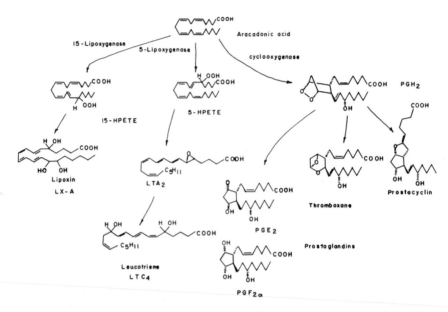

Fig. 2. Examples of how eicosanoids can be produced by the lipoxygenase-dependent oxidation of arachidonic acid in animals.

referred to Vick and Zimmerman (1987). The individual enzymes isolated from various species have much lower specificity than the corresponding animal enzymes (Needleman et al., 1986), so a mixture of 9- and 13-hydroperoxy products is usually obtained (Fig. 3; Vick and Zimmerman, 1987). The enzymes are generally classified by the pH optimum of the reaction and the relative proportion of 9- and 13-hydroperoxy derivatives made. The ratio of products produced can often be varied by changing the reaction pH and divalent ion concentration. It should be noted that at least one lipoxygenase isolated from plants has been shown to be inhibited by nonsteroidal anti-inflammatory drugs. This inhibition is similar to the action of these drugs on cyclooxygenase (Sircar et al., 1983).

The best characterized lipoxygenases are from soybean seed. The type-1 (LOX-1) enzyme from soybean has maximum activity at pH 9 and produces mainly 13-hydroperoxy derivatives. The LOX-1 behaves as a 15-lipoxygenase with arachidonic acid as the substrate and at higher concentrations produces the 8,15-dihydroperoxy derivative (Bild et al., 1977). The type-2 enzyme (LOX-2) has a pH maximum at 6.5 and produces a 50:50 mixture of 9-, 13-derivatives. The activity of LOX-2 is calcium stimulated, which is reminiscent of the calcium regulation of 5-lipoxygenase from mammals.

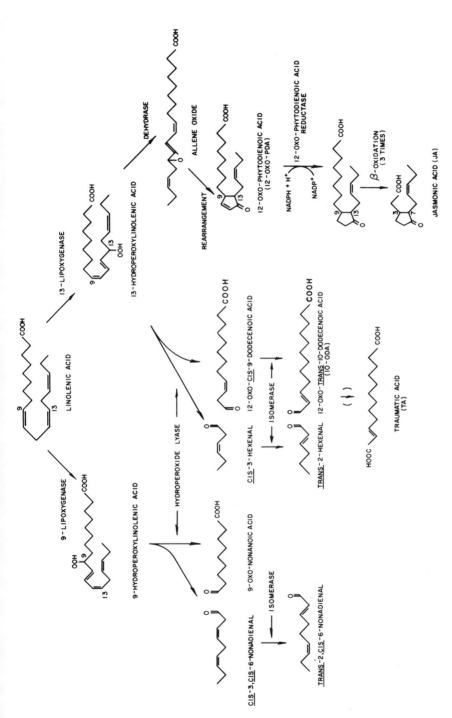

Fig. 3. Production of lipoxygenase products from linolenic acid in plants.

LOX-2, however, acts as a 15-lipoxygenase with arachidonic acid. Under certain in vitro conditions LOX-2 also has cyclooxygenase activity with arachidonic acid (Bild et al., 1978).

Metabolites of C_{18} Hydroperoxides

There are two known pathways for metabolizing the hydroperoxides of linoleic and linolenic acid in plants (Fig. 3; Vick and Zimmerman, 1987). The first pathway involves the enzyme, hydroperoxide lyase, which cleaves hydroperoxy fatty acids to aldehydes and oxoacids, so that cleavage of 13-hydroperoxylinolenic acid yields 12-oxo-cis-9-dodecanoic acid and cis-3-hexenal (Fig. 3). In most plants, the cis-3-enal structure is enzymatically isomerized to the trans-2-enal form so that trans-2-hexenal and 12-oxo-trans-10-dodecenoic acid (10-ODA) are the final products in vivo (Zimmerman and Coudron, 1979; Hatanaka et al., 1977). The plant growth regulator, traumatic acid, is derived from 10-ODA. The hydroperoxide lyase isolated from most species shows a preference for the 13-hydroperoxy derivative. However, examples of enzymes with a 9-hydroperoxy preference have been found (Galliand, 1979). The aldehyde products of hydroperoxide lyase can be reduced to the corresponding alcohols in some plants. The products hexanal, nonenal, and hexenol are flavor and aroma components of leaves and fruit (Berger et al., 1987; Hamilton-Kemp and Andersen, 1984; 1986; Kumar and Motto, 1986). These compounds also have been shown to have antifungal activity (Urbash, 1984). Recently, Hamilton-Kemp et al. (1987), using a specific inhibitor of lipoxygenase, demonstrated that synthesis of C_6 volatiles required lipoxygenase in wheat extracts.

Fatty acid hydroperoxides are also metabolized to form α- and γ-ketols (Vick and Zimmerman, 1987). Originally, this conversion was attributed to a hydroperoxide isomerase enzyme (Vick and Zimmerman, 1987), but it is now known that the reaction is catalyzed by a hydroperoxide dehydrase (Hamberg, 1987). The unstable allene oxide product of the dehydrase reaction is quickly hydrolyzed to α- and γ-ketols. Specificity for either the 13- or 9-hydroperoxy derivatives again varies with species. Recently, Claeys et al. (1986) presented evidence that hydroperoxides can also undergo rearrangement to trihydroxy fatty acids so that 13-hydroperoxy linoleic acid would form 9,12,13-trihydroxy-10-octadecenoic acid. Trihydroxy fatty acids have been shown to have physiological effects in both plants and animals.

Zimmerman and Feng (1978) reported the synthesis of a cyclic fatty acid, 12-oxo-phytodienoic acid (12-oxo-PDA), from linolenic acid using an extract from flaxseed (Fig. 3). The enzyme responsible for this synthesis was originally identified as a hydroperoxide cyclase, which, unlike the cyclooxygenase-dependent synthesis of prostanoids, required the 13-hydroperoxy derivative of linolenic acid as a substrate (Vick and Zimmerman, 1979).

Recent reports now indicate that the actual cyclization mechanism involves the formation of an unstable allene oxide intermediate catalyzed by hydroperoxide dehydrase (Baertschi et al., 1988), followed, by a nonenzymatic rearrangement to form 12-oxo-PDA (Fig. 3). Vick and Zimmerman (1983) subsequently showed that 12-oxo-PDA was further metabolized to jasmonic acid (JA) by a series of enzymatic steps, including reduction by 12-oxo-PDA reductase, and β-oxidation of the product to JA (Vick and Zimmerman, 1983, 1986). The fact that several species were able to synthesize JA from added 12-oxo-PDA (Vick and Zimmerman, 1984) and the wide occurrence of JA in plants both suggest that the synthetic pathway is general in higher plants.

Hydroxylation of Fatty Acids

Fatty acids are hydroxylated in both animals and plants by cytochrome P450 type monooxygenases (Hendry, 1986). In plants, these monooxygenases are localized in the microsomal fraction of cell extracts and are found in tissues that do not contain chlorophyll. Although P450 monooxygenases appear to function primarily in detoxification reactions in both plants and animals (Hendry, 1986), arachidonic acid can be oxidized in mammalian tissues to epoxides and vic-diols that have physiological activity (Needleman et al., 1986). The P450 monooxygenases in plants can also produce hydroxylated derivatives of fatty acids, and several of the resultant compounds are used in the synthesis of suberin and cutin (Kolattukudy, 1980). It is possible that some chemical messengers in plants are the result of hydroxylation reactions, but at present there is no indication that P450 type oxidations are involved in stimulus–response coupling. Therefore, these pathways will not be discussed further in this chapter.

EICOSANOIDS
Eicosanoids in Animals

The eicosanoids are a diverse class of potent metabolic regulators in mammals. The main classes of eicosanoids are the prostaglandins, prostacyclin, leukotrienes, thromboxanes, and lipoxins (Fig. 2; Andersen et al., 1986; Borgeat and Samuelsson, 1979; Samuelsson, 1983, 1985; Samuelsson et al., 1985; Wolfe, 1982). No attempt will be made in this chapter to review the pharmacological activity of such a diverse group as the eicosanoids, but selected traits of these animal hormones will be given in order to draw comparisons to similar compounds synthesized in plants.

Eicosanoids are synthesized from 20-carbon unsaturated fatty acids, primarily arachidonic, as the result of tissue-specific, hormonal and nonhormonal, stimuli (Wolfe, 1982). The rate-limiting step in the synthesis is thought to be the release of arachidonic acid from membrane esters because enzymes that metabolize arachidonic acid into eicosanoids are apparently constitutive. Usually, synthesis of eicosanoids is coupled with phosphoinositide turnover so that the major source of arachidonate is probably DAG (Majerus et al., 1985). In animals, some DAG is degraded to free fatty acids and glycerol by DAG and monoacylglycerol lipases (Fig. 1; Majerus et al., 1986; Wolfe, 1982). Phospholipase A_2 has also been shown to be an important source of free arachidonate in some animal tissues (Majerus et al., 1986). In fact, the anti-inflammatory effects of a group of proteins, the lipocortins, has been attributed to their ability to inhibit phospholipase A_2 (Fauvel et al., 1987).

Eicosanoids are not required for growth and development of animal tissues and therefore should be considered as secondary hormones (Wolfe, 1982). In general, eicosanoids have local hormonal effects that either reinforce or antagonize the action of other chemical mediators. One of the main classes of stimuli that elicit eicosanoid synthesis are tissue stresses such as trauma, disease, and allergy; eicosanoids often are involved in potentiating tissue reactions to such stresses (Musch and Siegel, 1985; Samuelsson, 1983, 1985; Hedqvist et al., 1985; Ford-Hutchinson, 1985; Ohuchi et al., 1985; Tomioka et al., 1985; Wolfe, 1982).

Eicosanoids in Plants

In animals, the eicosanoids are synthesized exclusively from arachidonic acid and related C_{20} fatty acids. The presence of eicosanoids in plants would, therefore, appear to require that arachidonic acid be a constituent of plants. Arachidonic acid makes up a significant proportion of the unsaturated fatty acid in mammalian membranes (Samuelsson, 1985). Arachidonic acid is also a significant lipid in mosses, ferns (Gellerman et al., 1975; Haigh et al., 1969; Jamieson and Reid, 1975; Lytle et al., 1976), and certain algae (Jamieson and Reid, 1972, Nichols and Appleby, 1969). In fact, in some species of mosses, arachidonic and eicosapentaenoic acid represent more than 30% of the total extractable lipid (Gellerman et al., 1975). Similar fatty acids have been identified in some gymnosperms, including arachidonic acid in *Ginkgo bilboa* (Hitchcock and Nichols, 1971) and 5,11,14-eicosatrienoic acid in *Podocarpus nagi* (Tagagi, 1964).

In contrast with lower plants and gymnosperms, there is no conclusive evidence that arachidonic acid is present as a natural component of angiosperm membranes (Hitchcock and Nichols, 1971). However, arachidonic acid is much more labile to autooxidation than linolenic acid and, therefore, minute amounts of arachidonic acid may have escaped detection. Only limited amounts of arachidonic acid would presumably be required to produce second messengers. Janistyn (1982b) did present evidence for arachidonic acid in wheat germ oil, using gas chromatographic–mass spectroscopic techniques, but the data presented do not appear to rule out other possibilities. Although the presence of arachidonic acid in higher plant membranes appears to be a remote possibility, this possibility should be investigated further using tissues that do not normally store oils.

Higher plants may not need arachidonic acid to produce chemical messengers because similar compounds can be made from 18-carbon fatty acids. The ability to synthesize arachidonic acid may have been lost during evolution. The absence of arachidonic acid could provide a selective advantage to plants because this fatty acid could serve as a specific signal of the presence of foreign organisms such as fungi or insects. Arachidonic acid has, in fact, been shown to be an elicitor of phytoalexin synthesis in some species of the *Solanaceae* (Bloch et al., 1984). Both arachidonic acid and eicosapentaenoic acid can specifically elicit the accumulation of phytoalexins in potato tubers (Bloch et al., 1984; Bostock et al., 1982), and both fatty acids are natural components of the fungus *Phytophthora infestans,* which also elicits the accumulation of the phytoalexins in potato. The actual signal compound is probably not arachidonic acid but a metabolite produced by lipoxygenase (Zook et al., 1987).

Plant tissues are capable of converting arachidonic acid into compounds similar to eicosanoids produced by mammals. Shimizu et al. (1984) have shown that potato tuber homogenates can convert arachidonic acid into 6-*trans*-leukotriene A_4 (LTA_4). Although the purified lipoxygenase from potato tubers normally converts linolenic acid to the 9-hydroperoxy derivative, it also will convert arachidonic acid to (5S)-hydroperoxy-6-*trans*-8,11,14-*cis*-eicosatetraenoic acid. The potato enzyme also converted the 5-hydroperoxy derivative into LTA_4 with good yield. It should be noted that leukotrienes are highly unstable compounds and could easily escape detection in plant extracts (Samuelsson et al., 1985). LOX-2 isolated from soybeans also can catalyze the in vitro cyclooxidation of arachidonic acid, yielding significant amounts of prostaglandin $F_{2\alpha}$ (Bild et al., 1978). Both these results should, however, be viewed with caution because the reaction conditions used in vitro may not represent the in vivo conditions, and neither prostaglandins or leukotrienes have been isolated from soybeans.

Compounds with chromatographic properties similar to prostaglandins and which have pharmacological activity in mammalian tissues have been isolated from *Allium cepa* (Attrep et al., 1973). These compounds subsequently have been identified as trihydroxy derivatives of linoleic acid, which apparently have cochromatographed with prostaglandins (Claeys et al., 1986). Janistyn (1982a) reported the identification by gas chromatographic–

mass spectroscopy of prostaglandin $F_2\alpha$ in flowering *Kalanchoe blossfeldiana*, but this result has not been confirmed by other researchers. The physiological activity most often associated with adding prostaglandins to plant tissues are changes in membrane permeability or ion transport (Christov and Vaklinova, 1987; Larqué-Saavedra, 1979; Roblin and Bonmort, 1984). Some examples are lowering of stomatal aperture of *Commeliana communis* (Larqué-Saavedra, 1979), altering the leaflet movements in *Cassia fasciculata* (Roblin and Bonmort, 1984), and lowering the threshold voltage necessary for electrofusion of pea protoplasts (Christov and Vaklinova, 1987). In the electrofusion experiments, 10 μM PGE_2 was shown to change membrane fluidity both in the presence and absence of Ca^{2+}. Prostaglandins have also been shown to alter ion transport in animal tissues (Wolfe, 1982). Other effects of prostaglandins are more difficult to explain. These effects include inhibition of crown-gall tumor formation in potato discs (Favus et al., 1977), induction of barley endosperm acid phosphatase (Curry and Galsky, 1975), and shortening of the time to flowering of *Pharbitis nil* (Groenewald and Visser, 1978). Several of the physiological effects of prostaglandins can be mimicked by addition of arachidonic acid to the tissues (Groenewald and Visser, 1978; Roblin and Bonmort, 1984) and can be inhibited by nonsteroidal, anti-inflammatory drugs that specifically inhibit cyclooxygenase (Groenewald and Visser, 1974; Roblin and Bonmort, 1984).

The question as to whether plants utilize products of arachidonic or other polyunsaturated C_{20} acids as chemical messengers remains unanswered. More data are needed to establish the natural presence of either arachidonic acid or eicosanoids in higher plant tissues. The data are quite convincing that plants can synthesize compounds similar to prostaglandins and that these compounds have physiological activity in both plants and animals (Saniewski, 1979). One plausible explanation of the data might be that eicosanoids are synthesized from extracellular arachidonic acid supplied by invading organisms and used by the plant to sense the invasion. Another possibility is that metabolites of linolenic and linoleic acid serve the role of eicosanoids in plant tissues.

OCTADECANOIDS AND RELATED COMPOUNDS
Octadecanoids

The term octadecanoids was used recently by Vick and Zimmerman (1987) to refer to the oxygenated 18-carbon fatty acids that are the plant equivalent to the 20-carbon eicosanoids found in animals. Much less is known about the octadecanoids and their possible physiological activities than about the eicosanoids. Compounding the problem of determining whether any of the octadecanoids are chemical regulators in plants is the fact that oxygenated lipids are used in the synthesis of cutin, suberin, and waxes (Kolattukudy, 1980). Dicarboxylic acids and hydroxy substituted acids of both 16- and 18-carbon fatty acids are common constituents of both cutin and suberin.

The precursor to jasmonic acid, 12-oxo-PDA, has a structure quite similar to the prostaglandins (Figs. 2, 3), and it is possible that this compound has a regulatory role distinct from jasmonic acid (Vick and Zimmerman, 1987). However, most tissues would probably metabolize the 12-oxo-PDA rapidly to JA (Vick and Zimmerman, 1984). Derivatives of 12-oxo-PDA have also been isolated from *Chromolaena morii*, a composite. The chromolones are 13-hydroxyderivatives of 12-oxo-PDA (Blumenfeld and Gazit, 1969) with the general structure shown in Figure 4.

Plants synthesize several polyhydroxy-octadecanoids. From extracts of the roots of *Bryonia alba*, Panossian et al. (1983) isolated a fraction consisting of a mixture of trihydroxyoctadecenoic acid derivatives that had pharmacological activity similar to prostaglandins in experiments on isolated strips of rat colon (Fig. 4). However, the plant isolate was only 1/1,000 as active as $PGF_2\alpha$. Similar compounds have recently been isolated from onions, and these compounds were shown to be derived from lipoxygenase oxidation of fatty acids (Claeys et al., 1986). Albro and Fishbein (1971) isolated 5,8,12-trihydroxy-*trans*-9-octadecenoic acid from wheat bran. This compound co-chromatographed with $PGF_2\alpha$ in more than one system; however, it had no activity as a prostaglandin. Finally, the structure of the trihydroxyoctadecenoic acids is reminiscent of the lipoxins (Fig. 2).

At present, there is no report of a physiological activity in plant tissues associated with the trihydroxyoctadecanoids. However, an unusual acetyl ester, 1-acetoxy-2,4-dihydroxy-*n*-heptadeca-16-ene, was isolated from avocado fruits (Bittner et al., 1971; Blumenfeld and Gazit, 1969). This compound inhibited IAA-induced growth of wheat coleoptiles and kinetin-supported growth of soybean callus at concentrations above 50 μM. These compounds have also been isolated from other plant tissues (Kashman et al., 1969). These results should be viewed with caution because relatively high concentrations of added compound were required to inhibit tissue growth.

Dodecanoids

Oxygenated 12-carbon fatty acids have been shown to have physiological activity in plants. Redemann et al. (1968) isolated a trihydroxy, α-keto-dodecanoic acid, phaseolic acid, from bean seeds. This acid acted as a

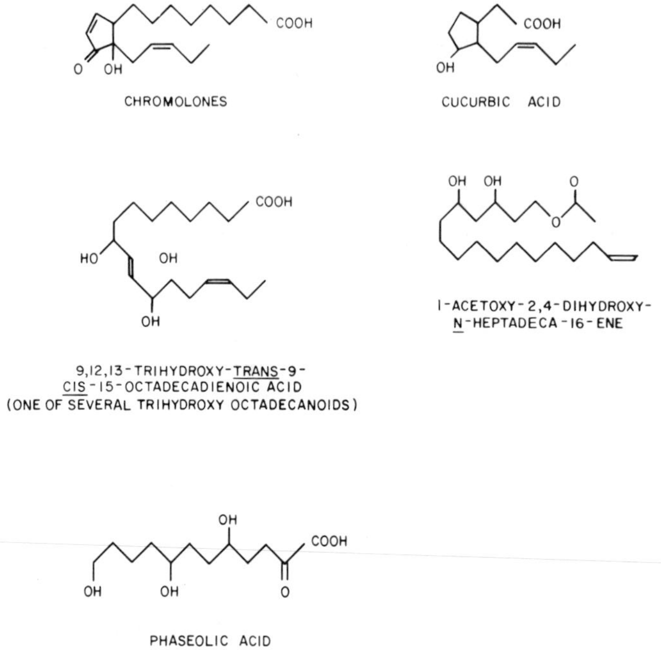
COOH
O
OH
CHROMOLONES
COOH
OH
CUCURBIC ACID
COOH
HO
OH
OH
9,12,13-TRIHYDROXY-TRANS-9-
CIS-15-OCTADECADIENOIC ACID
(ONE OF SEVERAL TRIHYDROXY OCTADECANOIDS)
OH OH
O
O
1-ACETOXY-2,4-DIHYDROXY-
N-HEPTADECA-16-ENE
OH
COOH
OH
OH
O
PHASEOLIC ACID

Fig. 4. Some examples of compounds with growth-regulating properties that are probable derivatives of fatty acids.

gibberellin synergist (Fig. 4). Phaseolic acid was less than 1/1,000 as active as GA_3 in stimulating growth of dwarf peas and maize. At similar concentrations phaseolic acid substituted for GA_3 in inducing α-amylase in barley endosperm, and it substituted for kinetin in inhibiting senescence of barley leaves at the same concentration as cytokinin.

JASMONIC AND TRAUMATIC ACIDS

Jasmonic Acid

Jasmonic acid (JA) (Fig. 3) and its methyl ester are probably the best characterized of the potential chemical messengers derived from the lipoxygenase-dependent oxidation of fatty acids. Methyl jasmonate (Me-JA) was first isolated as an odoriferous constituent of the essential oil of *Jasminum grandiflorum* L. (Demole et al., 1962) and has been found in the essential oils of other species as well (Nishida and Acree, 1984; Crabalona, 1967). A physiological activity for JA was first demonstrated by Aldridge et al. (1971), with the isolation of JA from the culture filtrates of the fungus, *Lasiodiplodia theobromae* and the demonstration that JA inhibited plant growth. Both JA and Me-JA have been isolated from numerous species in at least nine families of higher plants (Anderson, 1985a, 1986; Dathe et al., 1981; Meyer et al., 1984; Ueda and Kato, 1980, 1982; Vick and Zimmerman, 1987; Yamane et al., 1981, 1982). This wide distribution of JA provides further support for a physiological role for JA in plant tissues.

JA shares both chemical and biological similarities to the plant growth regulator abscisic acid (ABA) (Ueda and Kato, 1981; Vick and Zimmerman, 1987). Both ABA and JA are keto-acids with similar molecular weights, solubility properties, and pK's. Both compounds inhibit plant growth and promote the senescence of detached leaves (Dathe et al., 1981; Miersch et al., 1986; Ueda et al., 1981; Ueda and Kato, 1982; Yamane et al., 1981). ABA and JA also have distinct physiological activities. In general, ABA is a better inhibitor of seedling growth than JA (Yamane et al., 1981; Dathe et al., 1981; Fletcher et al., 1983; Ueda and Kato, 1982). The inverse is true with the promotion of senescence in detached leaves where Me-JA is better than ABA (Ueda and Kato, 1981; Ueda et al., 1981). JA can also elicit physiological responses that are not apparent with ABA addition such as the inhibition of *Camellia* pollen germination (Yamane et al., 1981, 1982). Other examples of the physiological activity of JA include the inhibition of the auxin- and light-induced opening of pulvinules of *Mimosa pudica* L. (Tsurumi and Asahi, 1985) and the promotion of rooting of mung bean seedlings (Zimmerman and Vick, 1983).

Up to this point in the discussion, both JA and Me-JA have been considered to be the same growth regulator, but this may not be correct. Me-JA is a lipophilic compound that is volatile at room temperature. This volatility introduces the possibility that Me-JA may be similar to ethylene and act as a gaseous hormone in plants. Also, Me-JA has been shown to promote the senescence of excised green leaves at much lower concentrations than JA or ABA (Ueda and Kato, 1981; Ueda et al., 1981). Recently, Me-JA has been shown to inhibit the formation of lycopene in ripening green tomatoes and stimulate β-carotene accumulation (Saniewski and Czapski, 1983; Saniewski et al., 1987). The application of Me-JA in a lanolin paste also stimulated the loss of chlorophyll and increased the level of ethylene. However, the concentration of Me-JA used in the lanolin was >10 mM, so the amount of Me-JA that diffused into the tissue could have been extremely high.

The stimulation of senescence of excised leaves has given JA and Me-JA the reputation of being senescence hormones, but it must be noted that young tissues contain the highest levels of the JA-synthetic enzymes (Groenewald and Visser, 1978). Recent work in our laboratory has shown that JA > 50 μM is required to cause growth inhibition and loss of chlorophyll in soybean photomixotrophic suspension cultures (Anderson, 1988). JA < 30 μM does not affect growth but specifically stimulated the synthesis of several glycoproteins and can actually increase chlorophyll levels slightly. The increase in the level of specific polypeptides elicited by JA has been noted in tissues other than soybean. Weidhase et al. (1987a, 1987b) demonstrated that treatment with Me-JA or JA stimulates the synthesis of at least three new abundant polypeptides in excised barley leaf segments, and although treatment with cytokinins can inhibit Me-JA-stimulated senescence, cytokinins did not inhibit the increase in polypeptides. Therefore, senescence promotion and growth inhibition by JA may be the result of toxic levels of the compound and not necessarily relate to the in vivo activity of JA. It is interesting that in soybean tissue cultures (Anderson, 1988), barley leaf segments (Weidhase et al., 1987b), and sunflower seedlings (Anderson, 1987), JA increased the level of three similar polypeptides with apparent molecular weights near 25,000, 35,000, and 65,000 kDa.

The naturally occurring form of JA was originally thought to be (−)-jasmonic acid (Dathe et al., 1981), but recently (+)-7-iso-jasmonic acid (7-iso-JA) also has been found in plants (Miersch et al., 1986). Improved extraction procedures have shown that the two isomers exist at a ratio of 65:35 JA:7-iso-JA in *Vicia faba* L. fruits and at 90:10 in commercial preparations. Only a small difference in biological activity (growth inhibition and senescence promotion) between the two compounds was found.

The nature of the physiological role of JA remains obscure but probably does not entail a second messenger role. JA appears to be transported in the phloem stream because we were able to detect JA from exudates of the cut tips of soybean pods using a fluorescence labeling technique (Anderson, 1985a, 1985b, 1986). It is interesting to note that in contrast to ABA, which was also measured in the exudates, the concentration of JA did not increase with drought stress. However, JA may be involved in signaling a stress different from water stress. This concept will be developed further in the last section.

A growth inhibitory compound, cucurbic acid (Fig. 4; (+)-2-(*cis*-pent-2′-enyl)-3-hydroxy-cyclopentylacetic acid), was isolated from pumpkin (*Cucurbita pepo* L.) seeds (Koshimizu et al., 1974). Cucurbic acid, which is an analogue of JA, inhibited the growth of rice seedling in a manner similar to JA.

Traumatic Acid

In 1935, Bonner and co-workers isolated a compound from plant extracts that stimulated callus growth in the parenchymous seed coat of immature been pods (Bonner et al., 1938). The active component was identified as traumatic acid (TA), trans-2-dodecenedioic acid (Fig. 3; English et al., 1939). Since the early work of Bonner, there have been several reports implicating TA as a wound compound, but also there have been several reports of inconsistent results. These inconsistencies led Lipetz (1970) and others to suggest that TA was not involved in the wound response. Part of the discrepancy in results might be explained by experimental protocol. Scott et al. (1961) showed that TA acid caused intense mitosis when injected into the hollow pith of castor bean internodes, and Strong and Kruitwangen (1967) showed that TA accelerated abscission in cotton plants in a manner that could not be reproduced by other growth regulators. In both cases, however, small amounts of TA were used but at high concentrations. The requirement for high concentrations of TA in several bioassays has been a consistent problem, as in the inhibition of rice seedling growth (Sekiya et al., 1986). Tanimoto and Harada (1984), however, demonstrated that submicromolar concentrations of TA stimulated adventitious bud formation in the epidermis of *Torenia* stem segments. A possible explanation for all of the inconsistencies in results may be that TA is only an analogue of the actual in vivo regulator.

Zimmerman and Coudron (1979), using the original bioassay of Bonner and English (1938), found that the actual active substance was probably 12-oxo-*trans*-10-dodecenoic acid (10-ODA; Fig. 3) and that the di-acid, TA, arose through nonenzymatic oxidation during extraction procedures. In fact, if strict isolation conditions were maintained, TA could not be isolated as a natural constituent of bean plants. Although the physiological activity of 10-ODA was not extensively investigated, both 10-ODA and TA were active in the runner bean bioassay. The possible involvement of 10-ODA in wound signaling needs further investigation, but such investigation may be difficult because of the unstable nature of the compound and its commercial unavailability (Zimmerman and Coudron, 1979). Leukocytes also produce an analogue of 10-ODA, 12-oxododeca-5,8,10-(Z,Z,E)-trienoic acid, from arachidonic acid (Glasgow et al., 1986). This short-chain aldehyde was shown to have significant biological activity and may be involved in intense inflammatory reactions.

LIPID OXIDATION AND STRESS SIGNALING
Lipid Oxidation During Stress

Major environmental stresses, including wounding, mechanical stimulation, and disease, can change membrane lipids and cause the release of free fatty acids (Galliard, 1978; Imaseki, 1985; Kuiper, 1985; Theologis and

Laties, 1980; 1981). This release of free fatty acids has been shown to be due to hydrolysis of cellular membrane esters by various phospholipases (Galliard, 1978). Hydrolysis of membrane lipids has also been found to occur with disease, drug, and anoxia stress in mammalian issues (Hostetler, 1985). In animal tissues, the A-type phospholipases have been implicated in release of free fatty acids and formation of lysophospholipids. Lipoxygenase and enzymes involved in metabolizing hydroperoxy fatty acids, apparently are constitutive in most animal and plant tissues, so that synthesis of eicosanoids in animals and the C_{18}-eicosanoid equivalents in plants requires only the availability of the free fatty acid substrates (Needleman et al., 1986; Vick and Zimmerman, 1987). Therefore, metabolites of linoleic and linolenic acid, generated by lipoxygenase action, are logical possibilities as second messengers of the imposed stress.

A major stress, such as wounding, in which the cells are actually disrupted, leads to a rapid loss of most of the esterified lipid in the plant cell due to autolysis (Galliard, 1978; Moreau and Isett, 1985). The released fatty acids are oxidized by lipoxygenase in the lysates, and this oxidation leads to the synthesis of fatty acid metabolites including several C_6 and C_9 aldehydes and alcohols (Hamilton-Kemp et al., 1987). Also, 10-ODA can be a product of tissue wounding (Hatanaka et al., 1977). Synthesis of 10-ODA or TA in cell lysates has been suggested as a mechanism for generating a wound signal that can be transported to undamaged tissues a few cell layers away from the actual trauma (Bonner and English, 1938; Imaseki, 1985; Lipetz, 1970; Zimmerman and Coudron, 1979). The connection between TA or 10-ODA to wounding is, therefore, quite plausible. Treshow (1955) studied the mechanism of mechanically stimulated tumor formation in green tomatoes and showed that TA specifically caused tumor formation when injected beneath the skin of the tomatoes.

Ryan and colleagues, however, have demonstrated another type of wound signal (Green and Ryan, 1972, 1973). Damaged tomato leaves accumulate proteinase inhibitors that apparently function in the defense against attack by insects. The wound stress causes autolysis of the cell wall in damaged cells, and this autolysis generates specific carbohydrate polymers that act as the wound signal (Ryan, 1974). Carbohydrate wound signals can also be generated by the invasion of a tissue by disease-causing organisms, and these elicitors can cause the synthesis of phytoalexins (Darvill and Albersheim, 1984; Erseek and Király, 1986). Phytoalexins are chemical agents used in the defense against disease. Elicitors clearly act as the chemical messengers of the disease or wounding stress. However, the wound signal still must be transduced in the target tissues in order to initiate the synthesis of phytoalexins or protease inhibitors (see Boller, Chapter 10). Involvement of peroxidation of fatty acids is again likely during the transduction process.

The invasion of fungi in plant tissues has been associated with the production of several C_6 aldehydes and alcohols that are obviously the result of lipoxygenase activity (Urbash, 1984), and in potato tubers, the actual elicitor of phytoalexin accumulation is not just a carbohydrate but also requires arachidonic acid (Zook et al., 1987).

The above discussion indicates that stress signaling during wounding and disease is a complex and, as yet, poorly understood phenomena in plant tissues. However, the situation in milder stresses, such as water and temperature stress, is even less clear. Again, turnover of membrane lipids may be involved in signal transduction. In animal membranes, simple depolarization by electrical stimulus or depolarizing agents is known to induce inositol phospholipid turnover (Nishizuka, 1984).

Production of the stress signal may not involve the actual release of free fatty acid from membranes prior to oxidation. In green tissues, a stress signal may be generated directly by changes in the chloroplast. Chloroplasts possess the potential for generating several forms of damaging oxygen species, some of which can initiate peroxidation of unsaturated lipids in membranes (Halliwell, 1984). Membrane lipids can also be directly peroxidized by certain bacterial toxins (Daub, 1982). These peroxidized lipids provide a potential source for hydroperoxy fatty acids, and the peroxidized fatty acid could be specifically excised from phospholipid by phospholipase A_2. Evidence for a similar situation has been found in mammalian tissues, where phospholipase A_2 has been shown to preferentially excise hydroperoxide-containing fatty acids from membrane lipids (Van Kuijk et al., 1987). The probable location in the chloroplast of hydroperoxide dehydrase and the other enzymes that participate in the synthesis of JA may reflect such a process (Vick and Zimmerman, 1987).

Nature of the Messengers

The lipoxygenase-formed metabolites of fatty acid released during stress are logical second messengers of the imposed stress. The similarity to mammalian systems is inescapable. In mammals, eicosanoid synthesis is triggered by release of arachidonic acid from membranes into the cytoplasm, and metabolism of arachidonic acid in mammals is often tied to stress-related phenomena such as hypersensitivity, allergy, inflammation, and even injury (Hostetler, 1985; Needleman et al., 1986; Ohuchi et al., 1985; Samuelsson, 1983; Vane, 1985; Van Kuijk et al., 1987). The nature of stress-related second messengers in plants is unknown, but a metabolite of linoleic or linolenic acid, which is similar to the eicosanoids in animals, is a likely candidate. Both JA and 10-ODA are distinct possibilities, but these compounds appear to function as extracellular messengers, rather than as second messengers. It is more likely that 10-ODA or TA acts as a local, diffusible signal of stress and that JA acts as a long-distance signal of the imposed stress by being transported in the phloem.

The nature of a possible fatty-acid-derived second messenger may be difficult to determine because compounds derived from hydroperoxy fatty acids can be highly unstable, with a short lifetime, even in vivo (Samuelsson, 1985). It should also be noted that the hydroperoxy derivatives of the fatty acids might act directly as second messengers without further processing, since such compounds have been shown to have pharmacological activity in animals (Needleman et al., 1986).

Senescence

The changes that occur in cellular membranes during senescence are in many ways similar to stress-induced changes, but these changes appear to occur much more gradually (Leshem et al., 1984; Sridhara and Leshem, 1986). In fact, Leshem (1987) has recently proposed that the syndrome of senescence in plants is intimately tied to a phospholipase-A_2-mediated cascade that resembles the phosphoinositide cascade of animals (Fig. 1). He proposes that senescence is triggered by entry of Ca^{2+} into the cytoplasm and that this Ca^{2+} activates phospholipase A_2. Phospholipase activation leads to release of linoleic and linolenic acids, which are then oxidized by lipoxygenase to produce oxy-free radicals, ethylene and JA. The process would be self-propagating and would eventually lead to senescent membranes. He also proposes that to counteract this *senescence cascade* there is a phytohormone-driven membrane regeneration. This proposition has merit in that it explains several diverse phenomena that occur during senescence, but the proposition appears to be an oversimplification of the physiological function of lipid peroxidation.

The proposal made by Leshem (1987) appears to require that lipoxygenase activity must be coupled to senescence, but Vick and Zimmerman (1984) have shown that the enzymes involved in processing hydroperoxy fatty acids into JA and 12-ODA, including lipoxygenase, are found in highest concentrations in young tissues, not senescent tissue. Also, evidence connecting JA to senescence promotion is, at best, weak. Finally, Peterman and Siedow (1985) found that if the epicotyl was removed from aging soybean cotyledons, senescence was halted and regreening occurred, and the level of at least one type of lipoxygenase increased during regreening, not during senescence.

Another problem with the senescence cascade, as proposed by Leshem (1987), is the fact that wounding of tissues should cause turnover of phospholipid and Ca^{2+} release in the cells, and therefore, initiate the senescence cascade. However, wounding often inhibits senescence in cells adjacent to the wound instead of stimulating it (Giridhar and Thimann, 1985). Therefore, it seems unlikely that the major function of phospholipase-dependent release of free fatty acids is promotion of senescence, rather than signal transduction. It should be noted that all stimulus–response mechanisms must have some mechanism for relaxation of the transduction process when the stimulus is removed. A likely situation in senescence is that some part of the relaxation process is impaired, and this impairment leads to continued degradation of cellular membranes. A hypothetical situation might be the decline in the synthesis of specific proteins that inhibit the activity of phosopholipase, such as the lipocortins which are inhibitors of phospholipase in animal tissues (Fauvel et al., 1987).

CONCLUSIONS

The question of whether plants utilize a signal transduction system similar to the phosphoinositide cascade (Fig. 1) found in animals remains unanswered. However, the circumstantial evidence for such a pathway in plants is steadily increasing (see Boss, Chapter 2). This chapter has attempted to address the evidence for the participation of PKC and lipoxygenase in a phosphoinositide pathway in plants. Although the evidence is preliminary, plants apparently do have a PKC with similar properties to the animal enzyme. Also, plants appear to have a variety of phospholipases that could be utilized to release free fatty acids into the cytoplasm, and some of these phospholipases have properties similar to those of the animal phospholipases. Finally, free fatty acids can be metabolized by a lipoxygenase-dependent pathway to compounds that are similar to the eicosanoids found in animals. Therefore, plants probably do utilize some variant of this branch of the phosphoinositide pathway. How similar the plant pathway is to the one found in animals remains to be seen.

A major difference between the animal and plant pathways is the apparent lack of arachidonic acid in plant membranes. Therefore, plants must utilize linoleic and linolenic acids for the synthesis of lipoxygenase metabolites. Even with this difference, several of the plant metabolites have structures reminiscent of the animal eicosanoids (Figs. 2–4). Two such compounds, JA and 10-ODA, apparently do play a role as plant growth regulators, but it is still too early to tell if a second messenger will come from this class of compounds. The unstable nature of many of the eicosanoids, however, should mandate a careful search for new lipoxygenase-derived metabolites from plant tissues that might serve as second messengers.

Another major difference between plants and animals that must be considered in any future research is the role of the chloroplast in signal transduction. There is strong evidence for the localization of at least one lipoxygenase and at least some of the enzymes that metabolize lipoxygenase products in the chloroplast (Vick and Zimmerman, 1987). In some ways, the investigations into signal transduction in plants has proceeded backward, with investigations into the nature of the transduction process coming before an understanding of the signals that are transduced. Although environmental signals, such as stress, are logical signals for transduction involving the release of free fatty acids, any signal, including the various plant growth regulators, may, in part, be transduced using this pathway. Therefore, an understanding of the metabolism of fatty acids may hold one of the elusive keys to understanding how plants transduce information into cellular biochemistry.

ACKNOWLEDGMENTS

Cooperative investigations of the United States Department of Agriculture, Agricultural Research Service, and the North Carolina Agricultural Research Service, Raleigh, NC 27695-7601. Paper No. 11384 of the Journal Series of the North Carolina Agricultural Research Service, Raleigh, NC 27695-7601.

Mention of a trademark or proprietary product does not constitute a guarantee or warranty of the product by the United States Department of Agriculture or the North Carolina Agricultural Research Service and does not imply its approval to the exlcusion of other products that may also be suitable.

REFERENCES

Albro PW, L Fishbein (1971): Isolation and characterization of 5,8,12-trihydroxy-trans-9-octadecenoic acid from wheat bran. Phytochemistry 10:631–636.
Aldridge DG, S Galt, D Giles, WB Turner (1971): Metabolites of Lasiodiplodia theobromae. J Chem Soc 1971:1623–1627.
Andersen HH, CJ Hartzell, B De (1986): Chemistry and structure of cyclooxygenase-derived eicosanoids: A historical perspective. In Pike JE, DR Morton, Jr (eds): "Advances on Prostaglandin, Thromboxane, and Leukotriene Research," Vol. 14. New York: Raven pp 1–43.
Anderson JM (1985a): Simultaneous determination of abscisic acid and jasmonic acid in plant extracts using high-performance liquid chromatography. J Chromatogr 330:347–355.
Anderson JM (1985b): Evidence for phloem transport of jasmonic acid. Plant Physiol 77S:75.
Anderson JM (1986): Fluorescent hydrazides for the high performance liquid chromatographic determination of biological carbonyls. Anal Biochem 152:146–153.

Anderson JM (1987): Jasmonic acid-dependent increase in specific polypeptides and polyphenoloxidase in three species of composites. Plant Physiol 83S:135

Anderson JM (1988): Jasmonic acid-dependent increase in the level of specific polypeptides in soybean suspension cultures and seedlings. J Plant Growth Reg (in press).

Anderson JM, H Charbonneau, HP Jones, RD McCann, MJ Cormier (1980): Characterization of the plant nicotinamide adenine dinucleotide kinase activator protein and its determination as calmodulin. Biochemistry 19:3113–3120.

Anderson JM, MJ Cormier (1978): Calcium-dependent regulator of NAD kinase in higher plants. Biochem Biophys Res Commun 84:595–602.

Ashendel CL (1985): The phorbol ester receptor: A phospholipid-regulated protein kinase. Biochim Biophys Acta 822:219–242.

Attrep KA, JM Mariani Jr, M Attrep, Jr (1973): Search for prostaglandin A_1 in onion. Lipids 8:484–486.

Baertschi SW, CD Ingram, TM Harris, AR Brash (1987): Absolute configuration of cis-12-oxo-phytodienoic acid of flaxseed: Implications for the mechanism of biosynthesis form the 13(S)-hydroperoxide of linolenic acid. Biochemistry 27:18–24.

Berger RG, A Kler, F Drawert (1987): C_6-Aldehyde formation from linolenic acid in fruit cells cultured in vitro. Plant Cell Tiss Organ Cult 8:147–151.

Berridge MJ (1984): Inositol trisphosphate and diacylglycerol as second messengers. Biochem J 220:345–360.

Bild GS, SG Bhat, CS Ramadoss, B Axelrod (1978): Biosynthesis of a prostaglandin by a plant enzyme. J Biol Chem 253:21–23.

Bild GS, CS Ramadoss, S Lim, B Axelrod (1977): Double dioxygenation of arachidonic acid by soybean lipoxygenase-1. Biochem Biophys Res Commun 74:949–954.

Bittner S, S Gazit, A Blumenfeld (1971): Isolation and identification of a plant growth inhibitor from avocado. Phytochemistry 10:1417–1421.

Bloch CB, PJGM De Wit, J Kuć (1984): Elicitation of phytoalexins by arachidonic and eicosapentaenoic acid: A host survey. Physiol Plant Pathol 25:199–208.

Blumenfeld A, S Gazit (1969): An endogenous inhibitor of auxins and kinetin. Isr J Bot 18:217–219.

Bonner J, J English Jr (1938): A chemical and physiological study of traumatin, a plant wound hormone. Plant Physiol 13:331–348.

Borgeat P, B Samuelsson (1979): Arachidonic acid metabolism in polymorphonuclear leukocytes: Effects of ionosphore A23187. Proc Natl Acad Sci USA 76:2148–2152.

Bostock RM, RA Laine, JA Kuć (1982): Factors affecting the elicitation of sequiterpenoid phytoalexin accumulation by eicosapentaenoic and arachidonic acids in potato. Plant Physiol 70:1417–1424.

Christov AM, SG Vaklinova (1987): Effects of prostaglandins E_2 and F_2 on the electrofusion of pea mesophyll protoplasts. Plant Physiol 83:500–504.

Claeys M, L Ustunes, G Laekeman, AG Herman, AJ Vlietinck, A Ozer (1986): Characterization of prostaglandin E-like activity isolated from plant source (Allium cepa). Prog Lipid Res 25:53–58.

Connolly TM, WJ Lawing Jr, PW Majerus (1986): Protein kinase C phosphorylates human platelet inositol triphosphate 5'-phosphomonoesterase, increasing the phosphatase activity. Cell 46:951–958.

Crabalona L (1967): Presence of levorotatory methyl jasmonate, methyl cis-2-2-(2-penten-1-yl)-3-oxocyclopentenyl acetate, in the essential oil of Tunisian rosemary. C R Acad Sci C264:2074–2076.

Curry SC, AG Galsky (1975): The action of prostaglandins on GA_3 controlled responses. I. Induction of barley endosperm acid phosphatase activity by prostaglandins E_1 and E_2. Plant Cell Physiol 16:799–804.

Cyong J-C, M Takahashi, K Hanabusa, Y Otsuka (1982): Guanosine $3',5'$-monophosphate in fruits of *Evodia rutaecarpa* and *E. officinalis*. Phytochem 21:777–778.

Darvill AG, P Albersheim (1984): Phytoalexins and their elicitors. A defense against microbial infection in plants. Ann Rev Plant Physiol 35:243–275.

Dathe W, H Ronsch, A Preiss, W Schade, G Sembdner, K Schreiber (1981): Endogenous plant hormones of the broad bean, *Vicia faba* L. (–)-jasmonic acid, a plant growth inhibitor in pericarp. Planta 153:530–535.

Daub ME (1982): Peroxidation of tobacco membrane lipids by the photosensitizing toxin, cercosporin. Plant Physiol 69:1361–1364.

Demole E, E Lederer, D Mercier (1962): Isolement et determination de la structure du jasmonate de methyle, constituant odoerant caracteristique de l'essence de jassmin. Helv Chim Acta 45:675–685.

Donnelly TE Jr, R Jensen (1983): Effect of fluphenazine on the stimulation of calcium-sensitive phospholipid-dependent protein kinase by 12-*O*-tetradecanoyl phorbol-13-acetate. Life Sci 33:2247–2253.

Dyer MI (1980): Mammalian epidermal growth factor promotes plant growth. Proc Natl Acad Sci USA 77:4836–4837.

Elliott DC, JD Skinner (1986): Calcium-dependent, phospholipid-activated protein kinase in plants. Phytochemistry 25:39–44.

Engels F, H Willems, FP Nijkamp (1986): Cyclooxygenase-catalyzed formation of 9-hydroxylinoleic acid by guinea pig alveolar macrophages under nonstimulated conditions. Fed Eur Biochem Soc Lett 209:249–252.

English J Jr, J Bonner, AJ Haagen-Smit (1939): Structure and synthesis of a plant wound hormone. Science 90:329

Érsek T, Z Király (1986): Phytoalexins: Warding-off compounds in plants? Physiol Plant 68:343–346.

Fauvel J, J-P Salles, V Roques, H Chap, H Rochat, L Douste-Blazy (1987): Lipocortin-like anti-phospholipase A₂ activity of endonexin. Fed Eur Biochem Soc Lett 216:45–50.

Favre B, G Turian (1987): Identification of a calcium- and phospholipid-dependent protein kinase (protein kinase C) in *Neurospora crassa*. Plant Sci 49:15–21.

Favus S, O Gonzalez, P Bowman, A Galasky (1977): Inhibition of crown-gall tumor formation on potato discs by cyclic-AMP and prostaglandins E₁ and E₂. Plant Cell Physiol 18:469–472.

Fletcher RA, T Venkatarayappa, V Kallidumbil (1983): Comparative effects of abscisic acid and methyl jasmonate in greening cucumber cotyledons and its application to a bioassay for abscisic acid. Plant Cell Physiol 24:1057–1064.

Ford-Hutchinson AW (1985): Leukotrienes as mediators of human disease. In Hayaishi O, S Yamamoto (eds): "Advances on Prostaglandin, Thromboxane, and Leukotriene Research," Vol 15. New York: Raven Press, pp 353–355.

Galliard T (1978): Lipolytik and lipoxygenase enzymes in plants and their action in wounded tissues. In Kahl G (ed): "Biochemistry of Wounded Plant Tissues." Berlin: Walter de Gruyter, pp 155–201.

Galliard T (1979): The enzymatic degradation of membrane lipids in higher plants. In Appelqvist L-A, C Lijenberg (eds): "Advances in the Biochemistry and Physiology of Plant Lipids." Amsterdam: Elsevier/North-Holland, pp 121–132.

Gellerman JL, WH Anderson, DG Richardson, H Schlenk (1975): Distribution of arachidonic acid and eicosapentaenoic acids in the lipids of mosses. Biochim Biophys Acta 388:277–290.

Giridhar G, KV Thimann (1985): Interaction between senescence and wounding in oat leaves. Plant Physiol 78:29–33.

Glasgow WC, TM Harris, AR Brash (1986): A short-chain aldehyde is a major lipoxygenase product in arachidonic acid-stimulated porcine leukocytes. J Biol Chem 261:200–204.

Green TR, CA Ryan (1972): Wound-Induced protease inhibitor in plant leaves: A possible defense mechanism against insects. Science 175:776–77.

Green TR, CA Ryan (1973): Wound-Induced protease inhibitor in tomato leaves: Some effects of light and temperature on the wound response. Plant Physiol 51:19–21.

Groenewald EG, JH Visser (1974): The effect of certain inhibitors of prostaglandin biosynthesis on flowering of *Pharbitis nil*. Z Pflanzenphysiol 71:67–70.

Groenewald EG, JH Visser (1978): The effect of arachidonic acid, prostaglandins and inhibitors of prostaglandin synthetase, in the flowering of excised *Pharbitis nil* shoot apices under different photoperiods. Z Pflanzenphysiol 88:423–429.

Haigh WG, R Safford, AT James (1969): Fatty acid composition and biosynthesis in ferns. Biochim Biophys Acta 176:647–650.

Halliwell B (1984): "Chloroplast Metabolism." Oxford: Clarendon, pp 180–206.

Hamberg M (1987): Mechanism of corn hydroperoxide isomerase: Detection of 12,13(S)-oxido-9-(Z), 11-octadecadienoic acid. Biochim Biophys Acta 920:76–84

Hamilton-Kemp TR, RA Andersen (1986): Volatiles from winter wheat: Identification of additional compounds and effects of tissue source. Phytochemistry 25:241–243.

Hamilton-Kemp TR, RA Andersen (1984): Volatiles from *Triticum aestivum*. Phytochemistry 23:1176–1177.

Hamilton-Kemp TR, RA Andersen, DF Hildebrand, JH Loughrin, PD Fleming (1987): Effects of lipoxygenase inhibitors on the formation of volatile compounds in wheat. Phytochemistry 26:1273–1277.

Hatanaka A, T Kajiwara, J Sekiya, Y Kido (1977): Formation of 12-oxo-trans-10-dodecenoic acid in chloroplasts from *Thea sinensis* leaves. Phytochemistry 16:1828–1829.

Hedqvist P, S-E Dahlen, U Palmerz (1985): Leukotrienes as mediators of airway anaphylaxis. In Hayaishi O, S Yamamoto (eds): "Advances on Prostaglandin, Thromboxane, and Leukotriene Research," Vol 15. New York: Raven Press, pp 345–348.

Hendry, G (1986): Why do plants have cytochrome P 450γ detoxification versus defense. New Phytol 102:239–247.

Hitchcock C, BW Nichols (1971): "Plant Lipid Biochemistry." New York: Academic Press, p 74.

Holub BJ, A Kuksis, W Thompson (1970): Molecular species of mono-, di-, and triphosphoinositides of bovine brain. J Lipid Res 11:558–564.

Hoshino T (1983): Identification of acetylcholine as a natural constituent of *Vigna* seedlings. Plant Cell Physiol 24:829–834

Hostetler KY (1985): The role of phospholipases in human diseases. In Kuo JF (ed): "Phospholipids and Cellular Regulation," Vol 1. Boca Raton, FL: CRC Press, pp 181–206.

Imaseki H (1985): Hormonal control of wound-induced responses. In Pharis RP, DM Reid (eds): "Hormonal Regulation of Development. III. Role of Environmental Factors, Encyclopedia of Plant Physiology, New Series," Vol 11. Berlin: Springer-Verlag, pp 485–512.

Irvine RF, RM Moor (1986): Micro-injection of inositol 1,3,4,5-tetrakisphosphate activates sea urchin eggs by a mechanism dependent on external Ca^{2+}. Biochem J 240:917–920

Jamieson GR, EH Reid (1972): The component fatty acids of some marine algal lipids. Phytochemistry 11:1423–1432.

Jamieson GR, EH Reid (1975): The fatty acid composition of fern lipids. Phytochemistry 14:2229–2232.

Janistyn B (1982a): Gas chromatographic-mass spectroscopic identification of prostaglandin $F_2\alpha$ in flowering *Kalanchoe blossfeldiana* v. Poelln. Planta 154:485–487.

Janistyn B (1982b): Gas chromatographic-mass spectroscopic identification and qualification of arachidonic acid in wheat-germ oil. Planta 155:342–344.

Kashman Y, I Neeman, A Lifshitz (1969): New compounds from avocado pear. Tetrahedron 25:4617–4631.

Kishimoto A, N Kajikawa, M Shiota, Y Nishizuka (1983): Proteolytic activation of calcium-activated, phospholipid-dependent protein kinase by calcium-dependent neutral protease. J Biol Chem 258:1156–1164.

Kolattukudy PE (1980): Cutin, suberin, and waxes. In Stumpf PK (ed): "The Biochemistry of Plants," Vol 4. New York: Academic Press, pp 571–645.

Koshimizu K, H Fukui, S Usuda, T Mitsui (1974): Plant growth inhibitors in seeds of pumpkin. In "Plant Growth Substances, 1973." Tokyo: Hirokawa, pp 86–92.

Kuhn H, Schewe T, SM Rapoport (1986): The stereochemistry of the reactions of lipoxygenases and their metabolites: Proposed nomenclature of lipoxygenases and related enzymes. In Meister A (eds): "Advances in Enzymology," Vol 58. New York: John Wiley & Sons, pp 273–311.

Kuiper PJC (1985): Environmental changes and lipid metabolism of higher plants. Physiol Plant 64:118–122.

Kumar N, MG Motto (1986): Volatile constituents of peony flowers. Phytochemistry 25:250–253.

Larqué-Saavedra A (1979): Studies on the effect of prostaglandins on four plant bioassay systems. Z Pflanzenphysiol 92:263–270.

Leshem YY (1987): Membrane phospholipid catabolism and Ca^{2+} activity in control of senescence. Physiol Plant 69:551–559.

Leshem YY, S Sridhara, JE Thompson (1984): Involvement of calcium and calmodulin in membrane deterioration during senescence of pea foliage. Plant Physiol 75:329–335.

Lipetz J (1970): Wound healing in plants. In Bourne GH, JF Danielli (eds): "International Review of Cytology," Vol. 27. New York: Academic Press, pp 1–28.

Lytle TF, JS Lytle, A Caruso (1976): Hydrocarbons and fatty acids of ferns. Phytochemistry 15:965–970.

Majerus PW, TM Connolly, H Deckmyn, TS Ross, TE Bross, H Ishii, VS Bansal, DB Wilson (1986): The metabolism of phosphoinositide-derived messenger molecules. Science 234:1519–1526.

Majerus PW, EJ Neufeld, M Laposata (1985): Mechanisms for eicosanoid precursor uptake and release by a tissue culture cell line. In: Bleasdale JE, J Eichberg, G Hauser (eds): "Inositol and Phosphoinositides: Metabolism and Regulation." Clifton, NJ: Humana, pp 443–457

Malvoisin E, F Wild, G Zwingelstein (1987): 12-O-Tetradecanoyl phorbol 13-acetate stimulates the myristylation of an 82 kDa protein in HL-60 cells, Fed Eur Biochem Soc Lett 215:175–178.

Martin TW, RB Wysolmerski, D Lagunoff (1987): Phosphatidylcholine metabolism in endothelial cells: Evidence for phospholipase A and a novel Ca^{2+}-independent phospholipase C. Biochim Biophys Acta 917:296–307.

Meyer A, O Miersch, C Buttner, W Dathe, G Sembdner (1984): Occurrence of the plant growth regulator jasmonic acid in plants. J Plant Growth Regul 3:1–8.

Michell R (1986): A second messenger function for inositol tetrakisphosphate. Nature 324:613.

Miersch O, A Meyer, S Vorkefeld, G Sembdner (1986): Occurrence of (+)-iso-jasmonic acid in Vicia faba L. and its biological activity. J Plant Growth Regul 5:91–100.

Moreau RA, TF Isett (1985): Autolysis of membrane lipids in potato leaf homogenates: Effects of calmodulin antagonists. Plant Sci 40:95–98.

Morré DJ, JT Morré, RT Varnold (1984): Phosphorylation of membrane-located proteins of soybean in vitro and response to auxin. Plant Physiol 75:265–268.

Musch MW, MI Siegel (1985): Antigenic stimulated release of arachidonic acid, lipoxygenase activity and histamine release in a cloned murine mast cell MC9. Biochem Biophys Res Commun 126:517–525.

Muto S, S Miyachi (1977): Properties of a protein activator of NAD kinase from plants. Plant Physiol 59:55–60.

Needleman P, J Turk, BA Jakschik, AR Morrison, JB Lefkowith (1986): Arachidonic acid metabolism. Annu Rev Biochem 55:69–102.

Nichols BW, RS Appleby (1969): The distribution and biosynthesis of arachidonic acid in algae. Phytochemistry 8:1907–1915.

Niedel JE, PJ Blackshear (1986): Protein kinase C. In JW Putney, Jr (ed): "Phosphoinositides and Receptor Mechanisms." New York: Alan R. Liss, pp 47–88.

Nishida R, TE Acree (1984): Isolation and characterization of methyl epijasmonate from lemon (Citrus limon). J Agric Food Chem 32:1001-1003.

Nishihira J, T Ishibashi (1986): A phospholipase C with a high specificity for platelet-activating factor in rabbit liver light mitochondria. Lipids 21:780–785.

Nishizuka Y (1984): Turnover of inositol phospholipids and signal transduction. Science 225:1365–1370.

O'Brian CA, WL Arthur, IB Weinstein (1987): The activation of protein kinase C by the polyphosphoinositides phosphatidylinositol 4,5-diphosphate and phosphatidylinositol 4-monophosphate. Fed Eur Biochem Soc Lett 214:339–342.

O'Brian CA, DS Lawrence, ET Kaiser, IB Weinstein (1984): Protein kinase C phosphorylates the synthetic peptide ARG-LYS-ALA-SER-GLY-PRO-VAL in the presence of phospholipid plus either Ca^{2+} or a phorbol ester tumor promoter. Biochem Biophys Res Commun 124:296–302.

Ohuchi K, M Watanabe, N Hirasawa, S Tsurufuji (1985): Role of arachidonic metabolites in allergic inflammation in rats. In Hayaishi O, S Yamamoto (eds): "Advances on Prostaglandin, Thromboxane, and Leukotriene Research," Vol 15. New York: Raven Press, pp 357–360.

Oláh Z, Z Kiss (1986): Occurrence of lipid and phorbol ester activated protein kinase in wheat cells. Fed Eur Biochem Soc Lett 195:33–37.

Panossian AG, GM Avetissian, VA Mnatsakanian, SG Batrakov, SA Vartanian, ES Gabrielian, EA Amroyan (1983): Unsaturated polyhydroxy acids having prostaglandin-like activity from Bryonia alba. II. Major components. Planta Med 47:17–25.

Parker CW (1985): 5-lipoxygenases. In Hayaishi O, S Yamamoto (eds): "Advances on Prostaglandin, Thromboxane, and Leukotriene Research," Vol 15. New York: Raven Press, pp 167–168.

Peterman TK, JN Siedow (1985): Behavior of lipoxygenase during establishment, senescence, and rejuvenation of soybean cotyledons. Plant Physiol 78: 690–695.

Redemann CT, L Rappaport, RH Thompson (1968): Phaseolic acid: A new plant growth regulator from bean seeds. In Wightman F, G Setterfield (eds): "Biochemistry and Physiology of Plant Growth Regulator Substances." Ottawa: Runge Press, pp 109–124.

Roblin G, J Bonmort (1984): Effects of prostaglandin E, precursors and some inhibitors of prostaglandin biosynthesis on dark- and light-induced leaflet movements in Cassia fasciculata Michx. Planta 160:109–112.

Ryan CA (1974): Assay and biochemical properties of the proteinase inhibitor-inducing factor, a wound hormone. Plant Physiol 54:328–332.

Samuelsson B (1983): Leukotrienes: Mediators of immediate hypersensitivity reactions and inflammation. Science 220:568–575.

Samuelsson B (1985): Leukotrienes and related compounds. In Hayaishi O, S Yamamoto (eds): "Advances on Prostaglandin, Thromboxane, and Leukotriene Research," Vol 15. New York: Raven Press, pp 1–9.

Samuelsson B, S Hammarström, M Hamberg, CN Serhan (1985): Structural determination of leukotrienes and lipoxins. In Pike JE, DR Morton (eds): "Advances on Prostaglandin, Thromboxane, and Leukotriene Research," Vol 14. New York: Raven Press, pp 45–71.

Saniewski M (1979): Questions about occurrence and possible roles of prostaglandins in the plant kingdom. Acta Hort 91:73–81.

Saniewski M, J Czapski (1983): The effect of methyl jasmonate on lycopene and β-carotene accumulation in ripening red tomatoes. Experimenta 39:1373–1374.

Saniewski M, H Urbanek, J Czapski (1987): Effects of methyl jasmonate on ethylene production, chlorophyll degradation, and polygalacturonase activity in tomatoes. J Plant Physiol 127:177–181.

Savart M, A Belamri, V Pallet, A Ducastaing (1987): Association of calpains 1 and 2 with protein kinase C activities. Fed Eur Biochem Soc Lett 216:22–26

Schafer A, F Bygrave, S Matzenauer, D Marmé (1985): Identification of a calcium- and phospholipid-dependent protein kinase in plant tissue. Fed Eur Biochem Soc Lett 187:25–28.

Schewe T, SM Rapoport, H Kuhn (1986): Enzymology and physiology of reticulocyte lipoxygenase: Comparison with other lipoxygenases. In Meister A (ed): "Advances in Enzymology," Vol 58. New York: John Wiley & Sons, pp 191–272.

Scott FM, BG Bystrom, V Sjaholm (1961): Anatomy of traumatic acid-treated internodes of *Ricinus communis*. Bot Gaz 122:311–314.

Sekiya J, T Kajiwara, A Hatanaka, S Ishida (1986): Inhibition of rice seed germination by long chain fatty acids. Phytochemistry 25:2733–2734.

Shimizu T, O Radmark, B Samuelsson (1984): Enzyme with dual lipoxygenase activities catalyzes leukotriene A₄ synthesis from arachidonic acid. Proc Natl Acad Sci USA 81:689–693.

Sircar JC, CF Schwender, EA Johnson (1983): Soybean lipoxygenase inhibition by nonsteroidal antiinflammatory drugs. Prostaglandins 25:393–396.

Smallman BN, A Maneckjee (1981): The synthesis of acetylcholine by plants. Biochem J 194:361–364.

Sridhara S, YY Leshem (1986): Phospholipid catabolism and senescence of pea foliage membranes: Parameters of Ca^{2+}:calmodulin:phospholipase A_z-induced changes. New Phytol 103:5–16.

Strong FE, E Kruitwangen (1967): Traumatic acid: An accelerator of abscission in cotton explants. Nature 215:1380–1381.

Tagagi T (1964): 5,11,14,-Eicosatrienoic acid in *Podocarpus nagi* seed oil. J Am Oil Chem Soc 41:516–519.

Takeda A, E Hashimoto, H Yamamura, T Shimazu (1987): Phosphorylation of liver gap junction protein by protein kinase C. Fed Eur Biochem Soc Lett 210:169–172.

Tanimoto S, H Harada (1984): Stimulation of adventitious bud initiation by cyclic AMP and traumatic acid in *Torenia* stem segments. Biol Plant 26:337–341.

Theologis A, GG Laties (1980): Membrane lipid breakdown in relation to wound-induced and cyanide-resistant respiration in tissue slices. Plant Physiol 66:890–896.

Theologis A, GG Laties (1981): Wound-induced membrane lipid breakdown in potato tuber. Plant Physiol 68:53–58.

Tomioka H, C Ra, I Iwamoto, T Sato, K Kodama, S Yoshida, M Tanaka, S Yoshida (1985): Arachidonic acid metabolism of human granulocytes from asthmatic patients. In Hayaishi O, S Yamamoto (eds): "Advances in Prostaglandin, Thromboxane, and Leuko-triene Research," Vol 15. New York: Raven Press, pp 339–343.

Treshow M (1955): Physiology and anatomical development of tomato fruit tumor. Am J Bot 42:198–202.

Tsurumi S, Y Asahi (1985): Identification of jasmonic acid in *Mimosa pudica* and its inhibitory effect in auxin- and light-induced opening of the pulvinules. Physiol Plant 64:207–211.

Ueda J, J Kato (1980): Isolation and identification of a senescence-promoting substance from wormwood (*Aremisia absinthium* L.) Plant Physiol 66:246–249.

Ueda J, J Kato (1981): Promotive effect of methyl jasmonate on leaf senescence in the light. Z Pflanzenphysiol 103:357–359.

Ueda J, J Kato (1982): Inhibition of cytokinin-iunduced plant growth by jasmonic acid and its methyl ester. Physiol Plant 54:249–252.

Ueda J, J Kato, H Yamane, N Takahashi (1981): Inhibitory effect of methyl jasmonate and its related compounds on kinetin-induced retardation of oat leaf senescence. Physiol Plant 52:305–309.

Urbash I (1984): Produktion pflanzlicher C_6-wundgase und ihre wirkung auf einige phy-topathologene pilze. Z Naturforsch 39c:1003-1007.

Van den Bosch H (1982): Phospholipases. In Hawthorne JN, GB Ansell (eds): "Phospho-lipids." Amsterdam: Elsevier, pp 313–358.

Vane JR (1985): The road to prostacyclin. In Hayaishi O, S Yamamoto (eds): "Advances on Prostaglandin, Thromboxane, and Leukotriene Research," Vol 15. New York: Raven Press, pp 11–19.

Van Kuijk FJGM, A Sevanian, GJ Handelman, EA Dratz (1987): A new role for phospholipase A_2: Protection of membranes from lipid damage. Trends Biochem Sci 12:31–34.

Vick BA, DC Zimmerman (1979): Distribution of a fatty acid cyclase system in plants. Plant Physiol 64:203–205.

Vick BA, DC Zimmerman (1983): The biosynthesis of jasmonic acid: A physiological role for plant lipoxygenase. Biochem Biophys Res Commun 111:470–477.

Vick BA, DC Zimmerman (1984): Biosynthesis of jasmonic acid by several plant species. Plant Physiol 75:458–461.

Vick BA, DC Zimmerman (1986): Characterization of 12-oxo-phytodienoic acid reductase in corn: The jasmonic acid pathway. Plant Physiol 80:202–295.

Vick BA, DC Zimmerman (1987): Oxidative systems for modification of fatty acids: The lipoxygenase pathway. In Stumpf PK, EE Conn, (eds): "The Biochemistry of Plants: Lipids," Vol 9. New York: Academic Press, pp 53–90.

Weidhase RA, H-M Kramell, J Lehmann, H-W Liebisch, W Lerbs, B Parthier (1987a): Methyl jasmonate-induced changes in the polypeptide pattern of senescing barley leaf segments. Plant Sci 51:177–186.

Weidhase RA, J Lehmann, H Kramell, G Sembdner, B. Parthier (1987b): Degradation of ribulose-1,5-bisphosphate carboxylase and chlorophyll in senescing barley leaf segments triggered by jasmonic acid methylester, and counteraction by cytokinin. Physiol Plant 69:161–166.

Wheeler JJ, WF Boss (1987): Polyphosphoinositides are present in plasma membranes isolated from fusogenic carrot cells. Plant Physiol 63:536–541.

Wolfe LS (1982): Eicosanoids: Prostaglandins, thromboxanes, leukotreienes, and other deriva-tives of carbon-20 unsaturated fatty acids. J Neurochem 38:1–14.

Yamane H, H Abe, N Takahashi (1982): Jasmonic acid and methyl jasmonate in pollens and anthers of three *Camellia* species. Plant Cell Physiol 23:1125–1127.

Yamane H, H Takagi, H Abe, T Yokota, N Takahashi (1981): Identification of jasmonic acid in three species of higher plants and its biological activities. Plant Cell Physiol 22:689–697.

Zimmerman DC, CA Coudron (1979): Identification of traumatin, a wound hormone, as 12-oxo-trans-10-dodecenoic acid. Plant Physiol 63:536–541.

Zimmerman DC, P Feng (1978): Characterization of a prostaglandin-like metabolite of linolenic acid produced by a flaxseed extract. Lipids 13:313–316.

Zimmerman DC, BA Vick (1983): Stimulation of adventitious rooting in mung bean seedlings by jasmonic acid and 12-oxo-phytodienoic acid. Plant Physiol 72S:108.

Zook MN, JS Rush, JA Kuć (1987): A role for Ca^{2+} in the elicitation of rishitin and lubimin accumulation in potato tuber tissue. Plant Physiol 84:520–525

Second Messengers in Plant Growth
and Development, pages 213–225
© 1989 Alan R. Liss, Inc.

9. The Vacuole: A Potential Store for Second Messengers in Plants

Alain M. Boudet and Raoul Ranjeva

Centre de Physiologie Végétale, Université Paul Sabatier, Unité Associée au C.N.R.S. No. 241, F-31062 Toulouse, France

INTRODUCTION

The plant vacuole is a unique organelle that occupies most of the cell volume in mature cells. One of the most likely explanations of the progressive development and/or conservation of the vacuolar compartment during the course of evolution is related to the nutritive strategies of higher plants, which need a large tissue surface area to trap very dilute aerial and terrestrial nutrients. Vacuoles filled with water are one of the least expensive ways of achieving such an extensive contact with the external medium, and they require limited energy and metabolite consumption to maintain. In addition to this putative primary function, many other roles for vacuoles, which have been discussed in recent reviews (Wagner, 1985, Matile, 1987), have gradually emerged particularly after the successful isolation of pure vacuoles on a large scale.

Among these possible functions, it is tempting to suggest that the large vacuole, in close contact with the cytoplasm, may release and respond to second messengers or effectors involved in the transduction of signals. Higher plants have to adapt themselves in a permanent way to their changing environment. Therefore, they need to detect and transduce different signals. The involvement of the vacuole in such processes could link this specific plant organelle with one of the basic functional strategies of plant cells. Such a role implies a transient release (or uptake) of chemical messages through the tonoplast following the perception at this membrane of primary or

Address reprint requests to Alain M. Boudet, Centre de Physiologie Végétale, Université Paul Sabatier, Unité Associée au C.N.R.S. No. 241, 118, route de Narbonne, F-31062 Toulouse Cédex, France.

secondary stimuli. The principle of this mechanism would be similar to those extensively studied for the release and uptake of Ca^{2+} from endoplasmic reticulum in animal cells.

The basic idea of the involvement of vacuoles in transduction processes was first developed on theoretical grounds and through indirect arguments (Boudet et al., 1984, 1987). However, recent experimental evidence substantiates this concept more directly (Schumaker and Sze, 1987; Ranjeva et al, 1988) even though the complete sequence of events from a primary stimulus to a biological response, including vacuolar release of messages, has yet to be described.

VACUOLES IN RELATION TO CYTOPLASMIC CONCENTRATION CHANGES OF POTENTIAL SECOND MESSENGERS
General Considerations

Due to the unbalanced sizes of the vacuole and the cytoplasm, it may be assumed that the efflux of compounds from the vacuole could increase the cytoplasmic concentration of solutes by almost one order of magnitude. From this very simple observation, it turns out that the traffic of molecules and ions through the tonoplast is more critical when the solutes leave the vacuole than when they enter it. If one assumes that the translocated compounds trigger (or inhibit) important biological responses or act as part of the transduction machinery of stimuli, then the movements of compounds through the tonoplast are crucial. It also appears that only compounds that reversibly cross the tonoplast are of interest. Consequently, one should only consider the exchangeable pools but not the overall population of a given solute.

As the ratios of cytoplasmic to vacuolar concentrations of a regulatory substance may change, it may be anticipated that, depending upon the actual parameters, more than one basic transport mechanism is involved. At a given time the system may work either against a gradient (as a pump), requiring energy, or down a gradient (diffusion or exchange). In the context of transduction processes, ion channels, which have been recently identified on the tonoplast (Hedrich et al., 1986), are frequently involved. On perception of a stimulus, the activity of the terminal membranes is modified. The changes lead, for example, to the opening of channels. As a result, the cytoplasmic concentrations of second messengers increase and stimulate the activity of the processes dependent on them. Once the stimulus has disappeared, the concentration of second messenger must be lowered to a resting level, necessitating extrusion from the cytoplasm: one possible mechanism involves plasmalemma-bound expelling systems; another possibility in plants might involve the reversible trapping of compounds in the vacuole. In this way vacuoles could be involved in the overall transduction process at two levels: 1) by releasing second messengers to the cytoplasm; 2) by accepting the excess of cytoplasmic second messengers after a signal-induced transient increase.

The concept of second messengers, extensively discussed by Blowers and Trewavas in Chapter 1, is just emerging in higher plants. It is particularly substantiated in the case of calcium (Hepler and Wayne, 1985), which is present in the vacuoles and could be released from them. However, other vacuolar solutes are likely to be available for modifying cellular metabolism in a catalytic way when released from the vacuoles. These include H^+ and inorganic phosphate (Pi), hormones, hormone precursors, and hormone-like substances (polyamines). Due to their putative importance and in relation to the data available, we will discuss only the role of the vacuole in the control of the cytoplasmic concentrations of Pi, H^+, and Ca^{2+}.

Phosphate

The results obtained by ^{31}P nuclear magnetic resonance (NMR) spectroscopy strongly suggest that a controlled flux of Pi to and from the vacuole (which contains most of the cellular Pi) maintains the cytoplasmic concentration of Pi at a constant level. However, the Pi transport systems, which have in part been elucidated for other organelles, are not known for the tonoplast. Roby et al. (1987) have emphasized the fact that phosphate remobilization from the vacuole is a rather slow process when cytoplasmic Pi decreases. Such an observation does not fit with the possible role of Pi release in rapid responses to external stimuli and raises the problem both of the time scale and extent of cytoplasmic concentration changes in transduction processes.

Different researchers, however, have demonstrated changes in cytoplasm/vacuole phosphate concentration in response to environmental stimuli. Black et al. (1982) have directly measured the Pi pool in isolated *Sedum* vacuoles over 24 hr. The Pi level is consistently lower in vacuoles isolated late at night (about two-thirds of the day concentration), but the protoplast Pi level does not change. The authors propose that a substantial amount of Pi moves out from the vacuole at night and might be involved in processes such as transport of triose phosphates out of the chloroplast during starch degradation. The Pi concentration in the vacuole is ~ 0.2 mM late in the day. Calculations based on respective vacuole and cytoplasm volumes and on the drop in vacuolar phosphate at night demonstrate that the cytoplasmic level of Pi can change by 1–2 mM between day and night. Independent of the mechanisms responsible for the vacuolar Pi efflux and for the potential effects of the released Pi, these results show that environmental changes can induce Pi redistribution between the vacuole and cytoplasm with a day/night rhythm.

Another interesting example of phosphate redistribution is provided by the results of Strasser et al. (1983). These authors have shown through ^{31}P NMR spectroscopy some modifications of cytoplasmic phosphate concentrations in response to an external stimulus. Addition of an elicitor preparation to parsley cell suspension cultures led to the temporary increase of vacuolar phosphate at the expense of cytoplasmic phosphate. The elicitor-induced synthesis of coumarinic phytoalexins in the parsley cells was ascribed, at least in part, to the temporarily decreased cytoplasmic phosphate concentration. In this case, the role of the vacuole in the transduction of messages would be to store mobile phosphate through activation of tonoplast phosphate transport systems. These results provide an interesting example of the potential role of cytoplasm/vacuole solute exchanges in stimulus response coupling.

In a general way, as limited changes in cytosolic Pi can rapidly and deeply affect different metabolic processes, including some of the earliest events in photosynthesis (Walker and Sivak, 1986), alterations of the Pi movements between the vacuole and the cytosol might modify cell metabolism.

Protons

Experimental evidence demonstrates the involvement of the vacuole in the cytoplasmic pH homeostasis (Torimitsu et al., 1984). Several mechanisms, including H^+ or other ion transport through the tonoplast, cooperate to maintain the cytoplasmic pH around 7–7.5.

Although the cytoplasmic pH is maintained at a relatively constant value over a long-time scale, evidence exists to demonstrate that transient cytoplasmic pH changes may occur over short periods of time. Using pH-sensitive microelectrodes, Steigner et al. (1986) have shown that small and reversible changes in cytoplasmic pH (0.2 pH units) may occur in the green alga *Eremosphaera viridis* in response to light/dark transitions: light-off induces a rapid acidification and light-on a rapid alkalinization.

In higher plants, little evidence supports the involvement of proton exchanges through the tonoplast in the transduction of external stimuli. One of the more interesting results was provided by Chrestin and co-workers (Chrestin et al., 1985; Chrestin et al., 1987) on a rather specific biological system: *Hevea* latex. Treatments of *Hevea* bark with ethrel, which promotes latex production induce a marked alkalinization of the latex cytosol concomitant with vacuolar acidification (Fig. 1). This inverse correlation between the pH values of the two compartments suggests the existence of some vectorial H^+ fluxes at the tonoplast of the vacuole-like structures of the latex: the lutoids. In the lutoids, two systems located at the tonoplast can transfer protons in an opposite direction:

1. a Mg-dependent ATPase working as an electrogenic proton pump and generating a high electrochemical proton gradient across the lutoidic tonoplast.
2. NADH cytochrome c (b) oxidoreductase, which induces H^+ efflux from intact fresh lutoids.

Following ethrel treatment, different processes—including an increase in the synthesis of the lutoidic ATPase, an increase in the concentration of the substrate ATP in the cytosol, and the appearance of some low-molecular-weight activators in the cytosol— cooperate to induce a marked stimulation

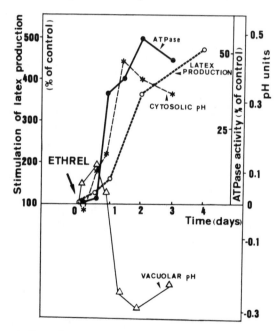

Fig. 1. Kinetics of effects of treatment of *Hevea* bark with ethrel on the production of natural rubber, on the pH changes of lutoids (vacuoles) and latex cytosol and on the tonoplast ATPase activity. (Redrawn from Chrestin et al., 1987, with permission of the publisher).

of the functioning of the lutoidic H^+ pumping ATPase and thus a significant cytosolic alkalinization (0.5 pH units) (Fig. 1). This results in the activation of some key enzymes in rubber synthesis, leading to an enhancement of rubber production. Such a sequence of events is in good agreement with a possible proton-development transduction mechanism of a hormonal stimulus involving the vacuole. However, the time scale for which a significant response to ethrel is observed is rather long (days) and could be related to the ethrel treatment procedure and/or the specificity of latex: a fluid cytoplasm that is expelled from wounded laticiferous "vessels" (articulated, anastomosed cells). For other systems, fine regulatory mechanisms of cytoplasmic pH have been shown to occur over short-term periods (Mathieu et al., 1986). Although these results and others reported above emphasize the relative flexibility of cytoplasmic pH in plants, these cytoplasmic pH changes probably occur more generally in a transient way over short periods of time.

It is important to stress that cytoplasmic pH changes can exert pleiotropic effects in animal cells and are particularly involved in the onset of cell proliferation and development (Swann and Whitaker, 1985; Moolenaar, 1986). For example, in human fibroblasts the epidermal growth factor induces a small (0.2 units) and rapid increase in the cytoplasmic pH by stimulation of the Na^+/H^+ exchanges at the plasmalemma. The small but relatively persistent alkaline pH shift may accelerate diverse pH-sensitive processes such as glycolysis, protein synthesis, and cytoskeletal organization. It is therefore proposed that the Na^+/H^+ exchanger may function as a transmembrane signal transducer involved in different biological processes. Such signals have been implicated in sea urchin egg activation, increased metabolism, and cell proliferation in response to growth factors. In a similar way, in higher plants proton transport systems at the tonoplast could be potential targets of primary stimuli or derived messengers in transduction processes. Thus, increased attention should be paid both to the regulation of the tonoplast-located proton transport systems, including the well-known ATPase, and to the pyrophosphatase, and the NADH redox system, which catalyzes H^+ efflux from the vacuole.

Calcium

In animal cells, the mobilization of calcium represents a primary mode of action for many external signals, including neurotransmitters, hormones, and growth factors. *Mobilization* is the term used for the action of regulatory or modulatory stimuli on both the entry of external calcium through the plasmalemma and/or the release of the cation from internal stores. Experiments based on permeabilized cells and microsomal fractions clearly support the last possibility showing that inositol-1,4,5-trisphosphate (IP_3) acts to mobilize calcium from internal compartments such as the endoplasmic reticulum (Berridge, 1986).

As for animal cells, it is now accepted that the cytoplasmic calcium in resting plant cells is about $10^{-6}-10^{-7}$ M (Moore and Ackerman, 1984; Gilroy et al., 1987). Such a low concentration is presumably controlled by active extrusion processes occurring both at the plasmalemma and the membrane of calcium-trapping organelles. Thus it has been established that a calcium–ATPase-like activity is associated with the tonoplast (Fukumoto and Venis, 1986). In addition, there is a "secondary pump," which allows a calcium/proton antiport fueled by the proton gradient generated by H^+-ATPase (Schumaker and Sze, 1985; Schumaker and Sze, 1986). These mechanisms may be involved in the protection of the cytoplasm against high levels of calcium.

The reverse situation should also be considered, that is, the vacuole as a reservoir of calcium. If the plant plasmalemma is considered to be a reference system, it has been shown that calcium channels exist on this terminal membrane. They have been characterized both by antagonist-binding experiments and calcium influx measurements (Andrejauskas et al., 1985; Graziana et al., 1988). These calcium channels seem to be regulated, like their counterpart in animal cells, by the depolarization of the membrane (interior less negative). However, voltage-dependent calcium channels similar to those present on the plasmalemma could hardly occur on the tonoplast, since the transmembrane potential difference is inverted (interior positive) (Barbier-Brygoo et al, 1985).

The possibility remains that vacuoles may release calcium in a voltage-independent manner, and IP_3 appears to be a good candidate as a chemical message. Different experimental data have shown that plants are able to form various IPs (Boss and Massel, 1985; Dillenschneider et al., 1986). Moreover, the IP_3-dependent release of calcium from sealed vesicles has been demonstrated (Drøbak and Ferguson, 1985; Reddy and Poovaiah, 1987). The membrane fraction used was not clearly defined and was referred to as microsomes. Nevertheless, Schumaker and Sze (1987) have extended these preliminary observations to tonoplast-derived vesicles. Here, addition of micromolar concentrations of IP_3 resulted in a transient loss of vesicular calcium within 15 sec (Fig. 2). The loss of calcium was immediately followed by an increase in vesicle-trapped calcium that can be inhibited by ethylene glycol bis (2-aminoethyl ether)-N,N,N',N' tetracetic acid (EGTA). The action of IP_3 is very specific and cannot be mimicked by other

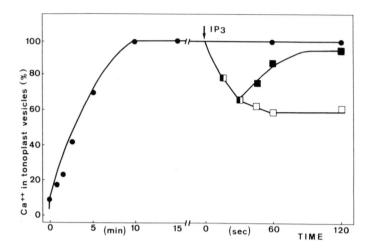

Fig. 2. Effects of inositol-1,4,5-trisphosphate (IP₃) release from oat root tonoplast vesicles preloaded with $^{45}Ca^{2+}$. ●——●, without IP₃; ■——■, 1.5 μM IP₃; □——□, 1.5 μM IP₃ + 0.7 mM EGTA.(Redrawn from Schumaker and Sze, 1987, with permission of the publisher.)

inositol derivatives. Moreover, IP₃ was inactive on the ATPase activity, confirming the idea that IP₃ does not affect the driving force for Ca^{2+} uptake. In parallel with the use of $^{45}Ca^{2+}$ loaded tonoplast vesicles, we used intact *Acer pseudoplatanus* vacuoles isolated according to Alibert et al. (1982). The efflux of endogenous vacuolar Ca^{2+} was monitored using a fluorescent probe, Quin 2, in the incubation medium. Under these conditions fluorescence is increased by factors stimulating the release of vacuolar calcium, since the probe does not cross the membranes and can only react with the external calcium. Using this experimental approach, it was shown that IP₃ controlled the Ca^{2+} efflux in a dose-dependent manner (Ranjeva et al., 1988) (Fig. 3). Taken together these results strongly suggest that a mechanism similar to those occurring in animal cells is also effective in plant cells for the controlled release of calcium from vacuoles.

This role of IP₃ is to be considered in context with a previous hypothesis concerning the role of calcium channel regulation in cellular responses to plant hormones (Pickard, 1984) and with the effects of 2,4-dichlorophenoxy-acetic acid (2,4-D) on phosphatidylinositol turnover in isolated soybean membranes (Morré et al., 1984). In a more general way, it can be postulated that plant hormones could exert some of their effects through a rise in cytosolic Ca^{2+} brought about by increased phosphoinositide turnover and subsequent release of IP₃.

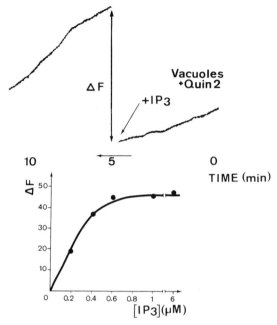

Fig. 3. Effects of inositol-1,4,5-trisphosphate (IP₃) on calcium efflux from intact vacuoles. Upper panel: Ca^{2+} release was estimated by spectrofluorometry. The addition of IP₃ to a vacuole suspension in presence of a fluorescent probe Quin 2 increased the fluorescence in a transient way. Addition of a Ca^{2+} ionophore or of a detergent induced a higher increase of fluorescence, confirming the occurrence of exchangeable Ca^{2+} within the vacuole. Lower panel: Effects of IP₃ concentration on Ca^{2+} efflux from isolated vacuoles.

CONCLUSIONS AND PROSPECTS

The analysis of solute fluxes through the tonoplast over long time periods clearly supports, in different situations, the involvement of vacuoles in cytoplasmic homeostasis (Bligny et al., 1984; Torimitsu et al., 1984). Apart from these movements, it can be assumed that more subtle and rapid migrations between the vacuolar and cytoplasmic compartments are likely to occur. They could have a transient effect on the cytoplasmic levels of potential effectors and thus act in transduction processes.

As far as calcium is concerned, it appears that the plant vacuole may serve as a reservoir of free calcium, as the endoplasmic reticulum does in animal cells, and that the same chemical message, IP₃ is able to release Ca^{2+} from these internal stores both in plants and animals. Such an analogy strongly supports the potential role of the vacuole in transduction processes (Fig. 4).

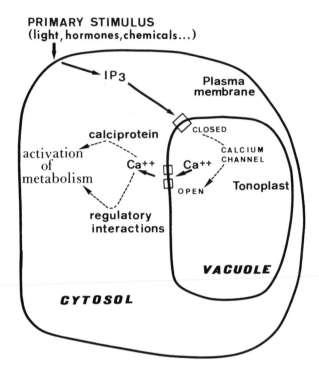

Fig. 4. Model of stimulus–response coupling involving the action of inositol-1,4,5-trisphosphate (IP₃) on vacuolar calcium release.

This basic concept can be extended to the controlled release of different vacuole-located metabolic effectors. In this way, characterization of the transport systems at the tonoplast is an important step toward further progress.

From the experimental viewpoint, tonoplast vesicles have been more widely used for transport studies than isolated vacuoles, since they are easier to handle and can be loaded with different labeled solutes. As vesicles of known orientations can be now obtained by free-flow electrophoresis (Canut et al., 1987), such fractions should be useful for future studies on the topography of transport systems. However, despite their fragility, isolated vacuoles can exhibit different advantages in some circumstances. The cytoplasmic side of the membrane (the right side) is unambiguously in contact with the tested effector, and the organelle keeps its natural pool of exchangeable solutes intact. Consequently, this experimental model is closer to the in vivo situation.

Future directions of research should include the characterization of the different transport systems and receptors (e.g., IP_3 receptors) at the tonoplast using biochemical techniques and recombinant DNA technology. They should also consider the topography of these proteins and their regulation. It is clear that this membrane houses Ca^{2+}/calmodulin-dependent enzymes and phosphorylated proteins (Teulières et al., 1985). The degree of phosphorylation of these transport systems could be an efficient and versatile means for their control (Ranjeva and Boudet, 1987).

In addition to the characterization of solute fluxes through the tonoplast and of the corresponding transport systems, further progress will require setting up methods that allow the in situ measurement of solute traffic. Such a technique basically exists for Pi and H^+ (NMR), even though it is not fully adapted to short-term measurements. Other techniques have to be designed in a similar way for other solutes. There is certainly a long way to go before directly demonstrating the involvement of the vacuole in the release of active effectors in response to environmental or cellular changes. Such a demonstration is difficult, as it implies switching from experiments on isolated vacuoles or tonoplast vesicles to studies on vacuoles in their cellular context. Technical improvements are occurring so rapidly, however, that this perspective may be considered realistic.

REFERENCES

Alibert G, A Carrasco, AM Boudet (1982): Changes in biochemical composition of vacuoles isolated from *Acer pseudoplatanus* L. during cell culture. Biochim Biophys Acta 721:22–29.
Andrejauskas E, R Hertel, D Marmé (1985): Specific binding of the calcium antagonist ([³H])verapamil to membrane fractions from plants. J Biol Chem 260:5411–5414.
Barbier-Brygoo H, R Gibrat, JP Renaudin, S Brown, JM Pradier, C Grignon, J Guern (1985): Membrane potential difference of isolated plant vacuoles: Positive or negative? II. Comparison of measurements with microelectrodes and cationic probes. Biochim Biophys Acta 819:215–224.
Berridge MJ (1986): Inositol triphosphate and calcium mobilization. In "Calcium and the Cell, Ciba Foundation Symposium 122.", John Wiley & Sons, Chichester: pp 39–57.
Black CC, NW Carnal, WH Kenyon (1982): Compartmentation and the regulation of CAM. In Ting IP, M Gibbs (eds): "Crassulacean Acid Metabolism." Baltimore: Waverly Press, pp 51–68.
Bligny R, F Rebeille, J Guern, JB Martin, R Douce (1984): Phosphate metabolism in intact plant cells. In Boudet AM, G Alibert, G Marigo, PJ Lea (eds): "Annual Proceedings of the Phytochemical Society of Europe," Vol 24, Oxford: Oxford University Press. pp 49–67.
Boss WF, M Massel (1985):Polyphosphoinositides are present in plant tissue culture cells. Biochem Biophys Res Commun 132:1018–1023.
Boudet AM, G Alibert, G Marigo (1984): Vacuoles and tonoplast in the regulation of cellular metabolism. In Boudet AM, G Alibert, G Marigo, PJ Lea (eds): "Annual Proceedings

224 / Boudet and Ranjeva

of the Phytochemical Society of Europe." Vol 24. Oxford: Oxford University Press, pp 29–47.

Boudet AM, G Alibert, G Marigo, R Ranjeva (1987): The vacuole: Possible role in signal transduction versus cytoplasmic homeostasis. In Marin B (ed): Plant Vacuoles: Their Importance in Plant Cell Compartmentation and Their Applications in Biotechnology. New York: Plenum, pp 455–465.

Canut H, AO Brightman, AM Boudet, DJ Morré (1987): Determination of sidedness of plasma membrane and tonoplast vesicles isolated from plant stems. In C Leaver, H S₃e (eds): Plant Membranes: Structure, Function, Biogenesis. New York: Alan R. Liss, Inc., pp 141–149.

Chrestin H, X Gidrol, H d'Auzac, JL Jacob, B Marin (1985): Cooperation of a "Davies Types" biochemical pH-stat and the tonoplastic bioosmotic pH-stat in the regulation of the cytosolic pH of Hevea latex. In Marin B (ed): "Biochemistry and Function of Vacuolar Adenosine-Triphosphatase in Fungi and Plants." New York: Springer-Verlag, pp 245–259.

Chrestrin H, X Gidrol, B Marin (1987): Transtonoplast pH changes as a signal for activation of the cytosolic metabolism in Hevea latex cells in response to treatments with ethrel (an ethylene releaser): Implications for the production of a secondary metabolite: Natural rubber. In Marin B (ed): "Plant Vacuoles: Their Importance in Plant Cell Compartmentation and Their Applications in Biotechnology." New York: Plenum, pp 467–476.

Dillenschneider M, AM Hetherington, A Graziana, G Alibert, P Berta, J Haiech, R Ranjeva (1986): The formation of inositol phosphate derivatives by isolated membranes from Acer pseudoplatanus is stimulated by guanine nucleotides. FEBS Lett 208:413–417.

Drobak BK, IB Ferguson (1985): Release of Ca²⁺ from plant hypocotyl microsomes by inositol-1,4,5-triphosphate. Biochim Biophys Res Commun 130:1241–1246.

Fukumoto M, MA Venis (1986): ATP-dependent Ca²⁺ transport in tonoplast vesicles from apple fruit. Plant Cell Physiol 27:491–497.

Gilroy S, WA Hughes, AJ Trewavas (1987): Calmodulin antagonists increase free cytosolic calcium levels in plant protoplasts in vivo. FEBS Lett 212:133–137.

Graziana A, M Fosset, R Ranjeva, AM Hetherington, M Lazdunski (1988): Ca²⁺ channel inhibitors that bind to plant cell membranes block Ca²⁺ entry into protoplasts. Biochemistry 27:764–768.

Hedrich R, UI Flügge, JM Fernandez (1986): Patch-clamp studies of ion transport in isolated plant vacuoles. FEBS Lett 204:228–232.

Hepler PK, RO Wayne (1985): Calcium and plant development. Annu Rev Plant Physiol 36:397–439.

Mathieu Y, J Guern, M Pean, C Pasquier, JC Beloeil, JY Lallemand (1986): Cytoplasmic pH regulation in Acer pseudoplatanus cells. II. Possible mechanisms involved in pH regulation during acid-load. Plant Physiol 82:846–852.

Matile P (1987): The sap of plant cells. New Phytol 105:1–26.

Moolenaar WH (1986): Regulation of cytoplasmic pH by Na⁺/H⁺ exchange. Trends Biochem Sci 11:141–143.

Moore AL, KE Ackerman (1984): Calcium and plant organelles. Plant Cell Environ 7:423–429.

Morré DJ, B Gripshover, A Monroe, JT Morré (1984): Phosphatidylinositol turnover in isolated soybean membranes stimulated by the synthetic growth hormone 2,4-dichlorophenoxyacetic acid. J Biol Chem 259:15364–15368.

Pickard BG (1984): Voltage transients elicited by sudden step-up of auxin. Plant Cell Environ 7:171–178.

Ranjeva R, AM Boudet (1987): Phosphorylation of proteins in plants: Regulatory effects and potential involvement in stimulus/response coupling. Ann Rev Plant Physiol 38:73–93.

Ranjeva R, A Carresco, AM Boudet (1988): Inositol trisphosphate stimulates the release of calcium from intact vacuoles isolated from *Acer* cells. FEBS Lett 230:137–141.

Reddy ASN, BW Poovaiah (1987): Inositol 1,4,5-triphosphate induced calcium release from corn coleoptile microsomes. J Biochem 101:569-573.

Roby C, JP Martin, R Bligny, R Douce (1987): Biochemical changes during sucrose deprivation in higher plant cells. J Biol Chem 262:5000–5007.

Schumaker KS, H Sze (1985): A Ca^{2+}/H^+ antiport system driven by the proton electrochemical gradient of a tonoplast H^+-ATPase from oat roots. Plant Physiol 79:1111–1117.

Schumaker KS, H Sze (1986): Calcium transport into the vacuole of oat roots: Characterization of H^+/Ca^{2+} exchange activity. J Biol Chem 261:12172–12178.

Schumaker KS, H Sze (1987): Inositol 1,4,5-triphosphate releases Ca^{2+} from vacuolar membrane vesicles of oat roots. J Biol Chem 262:3944–3946.

Steigner W, K Kohler, W Urbach, S Wilhelm (1986): Transient cytoplasmic pH changes after light-off and light-on in *Eremosphoaera viridis*. Plant Physiol Suppl 80:144.

Strasser H, KG Tietjen, K Himmelspach, U Mattern (1983): Rapid effect of an elicitor on uptake and intracellular distribution of phosphate in cultured parsley cells. Plant Cell Rep 2:140–143.

Swann K, M Whitaker (1985): Stimulation of the Na^+/H^+ exchanger of sea urchin eggs by phorbol ester. Nature 314:274–277.

Teulières C, G Alibert, R Ranjeva (1985): Reversible phosphorylation of tonoplast proteins involves tonoplast-bound calcium-calmodulin-dependent protein kinase(s) and protein phosphatase(s). Plant Cell Rep 4:199–201.

Torimitsu K, Y Yazaki, K Nagsuka, E Otha, M Sakata (1984): Effect of external pH on the cytoplasmic and vacuolar phs in mung bean root tip cells: a ^{31}P nuclear magnetic resonance study. Plant Cell Physiol 25:1403–1409.

Wagner GJ (1985): Vacuoles. In Liskins HF, JF Jackson (eds): "Modern Methods of Plant Analysis," Vol 1. Berlin: Springer-Verlag, pp 105–133.

Walker DA, MN Sivak (1986): Photosynthetic and phosphate: A cellular affair? Trends Biochem Sci 11:176–179.

Second Messengers in Plant Growth
and Development, pages 227–255
© 1989 Alan R. Liss, Inc.

10. Primary Signals and Second Messengers in the Reaction of Plants to Pathogens

Thomas Boller

Abteilung Pflanzenphysiologie, Botanisches Institut der Universität Basel,
CH-4056 Basel, Switzerland

INTRODUCTION

Plants are exposed to a vast number of potentially pathogenic microorganisms but are resistant to most of them. An important part of resistance is the induced biochemical defenses (Bell, 1981), for example, the formation of phytoalexins (Darvill and Albersheim, 1984; Kuć and Rush, 1985; Ebel, 1986; Dixon, 1986), the induction of antifungal hydrolases (Boller, 1985, 1987), and the so-called hypersensitive response, that is, the necrosis of a few plant cells around an invading microorganism (Dixon, 1986). The characterization of these defense reactions at the molecular level is presently a focus of interest.

Many observations indicate that specific molecules of the pathogen, the so-called elicitors, are the primary signals responsible for the induction of the defense reactions. The term *elicitor* was coined originally for inducers of phytoalexin biosynthesis (Keen, 1975), but it is now used more generally for inducers of various defense reactions (Callow, 1984; Dixon, 1986; Ebel, 1986). Elicitors have been studied extensively (reviews: West, 1981; Yoshikawa, 1983; Callow, 1984; Darvill and Albersheim, 1984; Dixon, 1986; Halverson and Stacey, 1986; Ralton et al., 1986, 1987). Molecules that suppress the defense response, sometimes referred to as suppressors, also have been studied.

Address reprint requests to Dr. Thomas Boller, Abteilung Pflanzenphysiologie, Botanisches Institut der Universität Basel, Hebelstrasse 1, CH-4056 Basel, Switzerland.

A number of reviews have emphasized the similarity of elicitors and hormones (Yoshikawa, 1983; Darvill and Albersheim, 1984; Dixon, 1986). In this context, the question of signal transduction between the pathogen-derived elicitors (or suppressors) and the defense responses is a matter of particular interest. Very little is known about the perception of elicitors by postulated receptors and the subsequent signal transduction. Do second messengers play a role in this process? The present article focuses on three aspects.

1. There is evidence that some primary pathogen-derived signals are enzymes. They probably act by way of the release of plant cell-wall-derived secondary signals, the so-called endogenous elicitors. Other primary pathogen-derived signals are insoluble structural elements of pathogens. These are modified and processed by plant enzymes, resulting in the release of soluble secondary signals. In both cases, signal transduction and release of the secondary signals occur extracellularly and are different from the established model of signal transduction by intracellular second messengers in animals.

2. There are good indications but no clear-cut evidence that the primary and secondary signals, produced extracellularly, elicit defense responses by way of intracellular second messengers.

3. Hitherto, attempts to demonstrate the participation of second messengers of the type known from animal studies have failed or have yielded equivocal results.

This survey leads to the conclusion that the search for intracellular second messengers in plant–pathogen interactions has not yet met with success but is promising, particularly if possibilities beyond those known from animal studies are included.

PATHOGEN-PRODUCED ELICITORS: PRIMARY SIGNALS INVOLVED IN PATHOGEN RECOGNITION

Resistance against microorganisms generally involves the induction of defense reactions. A given plant (or a plant cultivar) can induce defense reactions against microorganisms that are never pathogenic to the plant species (nonhost resistance, nonspecific resistance) as well as against microorganisms that are pathogenic to other plants (cultivars) of the same species (host resistance, cultivar-specific resistance).

In both cases, as a general concept, a plant cell recognizes an approaching microorganism by a primary signal, that is, by a specific physical or chemical change in its environment due to the presence of the microorganism. Physical changes are rarely discussed as potential primary signals although local pressure changes, for example, at the level of the plasma membrane, might be a specific alteration caused by many fungi. Chemical constituents of the pathogens are likely to represent the predominant primary signals recognized by plant cells.

Such chemical signals can be classified into extracellular compounds released by a pathogen and into structural elements of the pathogen's surface. The first can be recognized at a distance from the pathogen, the second only by probing the pathogen's surface. Both secreted and surface-bound primary signals may be involved in nonhost resistance and host resistance. It is reasonable to expect widespread fungal and bacterial components as signals for recognition in nonhost resistance. Highly specific compounds may function as signals in cultivar-specific host resistance. The cloning of avirulence genes has opened a new approach to the latter signals.

Extracellular Components Released by Pathogens

For pathogens that can be grown in liquid culture, extracellular components can readily be obtained in large quantities from culture filtrates, and most of the extracellular elicitors described in this section have been obtained in this way. However, a pathogen may secrete very different compounds in liquid culture and in its interaction with plants. The work of De Wit's group highlights this problem. Filtrates from pure cultures of various races of *Cladosporium fulvum,* a tomato pathogen displaying a high degree of race-cultivar specificity, contained potent unspecific but no specific elicitors (De Wit and Kodde, 1981). In contrast, the extracellular fluid produced by *Cladosporium fulvum* in its natural interaction with tomato contained highly active race-specific elicitors but no unspecific elicitors (De Wit and Spikman, 1982). Clearly, extraction of the intercellular fluid of diseased tissue yielded the biologically more relevant elicitors. This technique may prove useful in studies of other plant–pathogen interactions. A similar approach has been used by Cruickshank and Smith (1986), who looked for nonspecific elicitors formed in the interaction between pea endocarp and *Monilinia fructicola,* a nonpathogen of peas.

In work with pure cultures of pathogens, the type of elicitors formed may strongly depend on the composition of the growth medium and the age of the culture (De Wit and Kodde, 1981; Woodward and Pegg, 1986). It may be useful to employ growth conditions that approach the natural situation as closely as possible. One possibility is to germinate the spores of a pathogen for a short time on nutrient-poor media. Cell-free germination fluids obtained in this way from conidia of *Botrytis cinerea* (Thorpe and Hall, 1984; Bowen and Heale, 1987) or from zoospores of *Phytophthora infestans* (Chai and Doke, 1987) were found to contain potent elicitors.

Extracellular polysaccharides and glycoconjugates. Culture filtrates from *Phytophthora megasperma* f.sp. *glycinea*, a soybean pathogen, contained polysaccharides that nonspecifically elicited phytoalexin accumulation in soybean cotyledons (Ayers et al., 1976a). In culture filtrates from different races of the bean pathogen, *Colletotrichum lindemnthianum*, elicitor activity was found in the glycoprotein fraction, and the least virulent race yielded the most potent elicitors (Anderson, 1980a). The nonspecific elicitor activity obtained, as mentioned above, from culture fluids of *Cladosporium fulvum* resided in glycopeptides and glycoproteins (De Wit and Kodde, 1981). Culture filtrates of another tomato pathogen, *Verticillium albo-atrum*, contained polysaccharide elicitors of high or low molecular weight, depending on the culture conditions; only the high molecular weight type showed a weak indication of cultivar specificity (Woodward and Pegg, 1986).

Extracellular enzymes. To penetrate into plant tissues, pathogens require extracellular enzymes to degrade components of the plant cell walls. The pectic enzymes have been particularly well studied (review: Collmer and Keen, 1986). Some of them, for example, the polygalacturonase of *Rhizopus stolonifer* (Lee and West, 1981a) and of other fungi (Amin et al., 1986) and the endopolygalacturonic acid lyase from *Erwinia carotovora* (Davis et al., 1984), have been found to act as elicitors. Fungal enzyme mixtures used to prepare protoplasts contain pectic enzymes. This may be one reason for the fact that some plant protoplasts form phytoalexins constitutively (Mieth et al., 1986).

Lee and West (1981b) have examined the elicitor activity of the endo-polygalacturonase purified from *Rhizopus stolonifer* more closely. They used chemical treatments to partially inactivate its enzymatic activity and found that elicitor activity decreased in parallel. This strongly indicates that the elicitor activity of the enzyme is based on its catalytic activity. Subsequently, it was shown that the heat-labile pectic enzymes formed heat-stable elicitors when incubated with plant cell wall preparations (Bruce and West, 1982; Davis et al., 1984). These results are compatible with the following model: the pathogen-produced extracellular pectic enzymes are the primary signals for recognition. The pectin in the plant cell wall is the *receptor* for these primary signals. The signal–receptor interaction occurs extracellularly and produces plant-derived pectic fragments as secondary signals for the induction of the defense reactions. Such pectic fragments, called *endogenous elicitors*, are discussed in more detail in the section on extracellular second messengers.

Other enzymes might similarly function as primary signals. Several elicitors in the spore germination fluids of *Phytophthora infestans* were found to be heat labile and pronase sensitive, indicating that they were extracellular enzymes; none of them had pectolytic activity (Osman and Moreau, 1985). Extracellular enzymes of pathogens frequently are glycoproteins. They may act as primary signals independently of their catalytic activities. For example, extracellular invertases purified from different strains of *Phytophthora megasperma* f.sp. *glycinea* had a capacity to suppress phytoalexin accumulation in soybean in a race-specific manner (Ziegler and Pontzen, 1982). The inhibitory activity persisted in heat-inactivated enzymes but was lost upon periodate oxidation, indicating that the carbohydrate portion and not the catalytic site was important for the effect. Similarly, elicitor-active glycoproteins isolated from the culture fluid of *Phytophthora infestans* (Keenan et al., 1985) and from cell walls of *Puccinia graminis* (Moerschbacher et al., 1986) were heat stable.

Pathogen-produced toxins. Victorin, a toxin produced in liquid culture by the oat pathogen, *Helminthosporium victoriae,* is an elicitor of phytoalexins in oat (Mayama et al., 1986). Victorin, which consists of a group of small, related peptides, is a cultivar-specific toxin and elicitor: only cultivars carrying the Pc-2 gene are susceptible to *Helminthosporium victoriae* and to victorin. In susceptible cultivars, the compound acts as an elicitor of phytoalexins at 10 pg ml^{-1} and causes extensive necrosis at 100 pg ml^{-1}. The dominant Pc-2 gene, which confers susceptibility to victorin, also confers hypersensitive resistance to *Puccinia coronata* f.sp. *avenae.* It is tempting to hypothesize that victorin produces its toxic effect by mimicking a race-specific elicitor released by *Puccinia coronata,* which induces the hypersensitive reaction on oats with the Pc-2 gene (Mayama et al., 1986).

Race-specific elicitors. As mentioned above, race-specific elicitors were found in the intercellular fluid of susceptible tomato plants infected with *Cladosporium fulvum* (De Wit and Spikman, 1982). One of these elicitors was purified and shown to be a peptide with $M_r = 5,500$ (De Wit et al., 1985, 1986). Its presence in the intercellular fluid did not depend on the genotype of the tomato plant but strictly correlated with the presence of the avirulence gene "A9" in the fungus, indicating that it was a direct or indirect product of this gene. This peptide acted as a specific elicitor of necrosis in a tomato cultivar that carried a gene for hypersensitive resistance ("R9") against the races of the pathogen carrying the avirulence gene "A9."

Structural Elements of Pathogens

A particularly well-studied source of primary signals in plant–pathogen interactions is the pathogen cell wall. In addition, the plasma membrane of pathogens also can yield elicitors.

Pathogen cell walls as elicitors. Isolated cell walls of microorganisms have nonspecific elicitor activity in various systems (Hadwiger and Beckman, 1980; Bryan et al., 1985; Kurosaki et al., 1986). How can they be recognized by the plant? Plant cells are enclosed in walls with rather narrow pores (Ralton et al., 1986, 1987). The putative receptors on the plant plasma membrane probably cannot come into contact with insoluble pathogen cell walls but only with soluble fragments. Therefore, the first step in recognition of pathogen cell walls most likely consists in the release of soluble elicitor-active fragments by extracellular plant enzymes. Enzymes that release soluble elicitors from pathogen cell walls have been described in the pea (Hadwiger and Beckman. 1980) and in the soybean (Yoshikawa et al., 1981; Keen and Yoshikawa, 1983).

In terms of the primary signal-recognition model, this situation can be described as follows: The cell wall of the pathogen acts as a primary signal. The receptor of this signal is an extracellular plant enzyme directed against fungal structures. Signal–receptor interaction results in the release of soluble pathogen wall fragments that act as secondary extracellular signals, similar to the endogenous elicitors. Plant enzymes releasing elicitors from pathogen walls and their products are considered further in the section on extracellular second messengers.

Experimentally, preparations of soluble fragments from isolated microbial cell walls have been obtained most frequently by the use of a heat treatment or mild acid hydrolysis. The first example of such an elicitor was prepared from walls of *Phytophthora megasperma* var. *sojae* (Ayers et al., 1976b). This preparation was a complex mixture of glucan oligo- and polysaccharides and elicited phytoalexin formation nonspecifically in soybean as well as in nonhost plants (Cline et al., 1978). Similar glucan mixtures were isolated from cell walls of *Colletotrichum lindemuthianum* and *Fusarium oxysporum;* they were elicitors in bean but not in tomato (Anderson, 1980b). Elicitor preparations can be further fractionated, to purify individual elicitor-active components. Work in this direction has culminated in the isolation of a chemically pure heptasaccharide from cell walls of *Phytophthora megasperma,* which acts as a potent elicitor in soybean (Sharp et al., 1984). Interestingly, seven other structurally similar heptasaccharides had no elicitor activity and did not reduce the potency of the active compound, indicating a high specificity of the putative receptors for the elicitor.

Components of pathogen membranes as elicitors. Bostock et al. (1981, 1982) analyzed mycelial extracts from the potato pathogen, *Phytophthora infestans,* which elicited phytoalexin formation in potato. They found that most of the elicitor activity resided in lipids containing arachidonic acid and eicosapentaenoic acid. Most of the polyunsaturated fatty acids occurred in esterified form in membrane phospholipids of *Phytophthora infestans* (Bostock et al., 1982, Creamer and Bostock, 1986). Interestingly, the major storage protein of potatoes, patatin, is a lipid acyl hydrolase with the potential to release free fatty acids from some but not all phospholipids of *Phytophthora infestans* (Creamer and Bostock, 1986). It is tempting to wonder if the fungal membrane is the primary signal for recognition, if the plant lipid acyl hydrolase or other plant enzymes act as receptors for this signal, and if the signal–receptor interaction causes the production of free polyunsaturated fatty acids that act as secondary signals.

Products of Avirulence Genes as Potential Primary Signals

In several interactions between plants and pathogenic bacteria, resistance results from the interplay of specific genes in the two organisms according to the gene-for-gene hypothesis (Flor, 1956). Hypersensitive resistance occurs only in interactions of plants possessing a single dominant gene, the *resistance gene,* with pathogens possessing a corresponding dominant gene, the *avirulence gene.* In such an interaction, the product of the plant's resistance gene directly or indirectly "recognizes" the product of the avirulence gene; this recognition leads to the hypersensitive response. The product of the avirulence gene may represent a race-specific elicitor or an enzyme involved in its formation. As discussed above, one such race-specific elicitor has been isolated (De Wit et al., 1985, 1986). In other cases, as the interaction between lettuce and the biotrophic pathogen, *Bremia lactucae,* a search for race-specific elicitors was unsuccessful (Crucefix et al., 1987). Now, techniques of molecular biology have opened a new approach: several bacterial avirulence genes have been cloned (Staskawicz et al., 1984; Niepold et al., 1985; Gabriel et al., 1986). It will be exciting to examine the products of such avirulence genes and to find out whether they indeed act as race-specific elicitors or as enzymes catalyzing their production or release.

EXTRACELLULAR SECOND MESSENGERS INVOLVED IN THE REACTION OF PLANTS TO PATHOGENS

The interaction of extracellular enzymes of the pathogen with the plant cell wall or, conversely, the interaction of the cell wall of the pathogen with extracellular plant enzymes may frequently be a primary event in recognition. The interactions may result in the formation and release of soluble fragments, of plant or fungal origin, which act as elicitors and can be considered extracellular second messengers. The signal for systemic induced resistance is a different example of an extracellular second messenger. Its nature is still unknown.

Endogenous Elicitors: Soluble Fragments Released From Plant Cell Walls

Extracellular enzymes of pathogens can act as primary signals in the induction of defense responses. In particular, as discussed above, pectic enzymes have been found to act as heat-labile elicitors and to yield heat-stable elicitors when incubated with plant cell walls (Collmer and Keen, 1986). The heat-stable elicitor preparation released by the *Rhizopus stolonifer* endopolygalacturonase (Bruce and West, 1982) was fractionated; the most active fraction was a linear oligogalacturonide composed of 13 galacturonic acid residues (Jin and West, 1984). In a similar study, the heat-stable elicitor fraction formed by a bacterial endopolygalacturonic acid lyase from isolated soybean cell walls (Davis et al., 1984) was found to contain linear oligosaccharides composed of ≈10 galacturonic acid units as the most active elicitors (Davis et al., 1986a).

Heat-stable endogenous elicitors also can be formed in the absence of pathogen-derived enzymes. The first preparations of *endogenous elicitors* were obtained from homogenized bean tissue (Hargreaves and Bailey, 1978; Hargreaves and Selby, 1978). It has been postulated that the so-called abiotic elicitors—for example, ultraviolet light, heavy metal ions, or mechanical wounding—act by way of liberation of endogenous elicitors, which, as extracellular second messengers, induce defense reactions (Darvill and Albersheim, 1984). An endogenous elicitor was extracted from soybean cell walls by hot water or partial acid hydrolysis (Hahn et al., 1981). The most active fragment obtained from this elicitor preparation or from an acid hydrolysate of citrus pectin was an oligogalacturonate of 12 units (Nothnagel et al., 1983). A heat-labile endogenous elicitor, presumably an enzyme, was extracted from soybean stems (Lyon and Albersheim, 1982). It released heat-stable elicitor activity from soybean cell walls. These findings support the endogenous elicitor hypothesis and indicate that plant enzymes can liberate elicitor-active fragments from the plant's own cell walls.

Kurosaki et al. (1985) found that a crude pectin preparation extracted from carrot cells was not active as an elicitor of phytoalexins in carrot cells. However, digestion of the crude pectin by pectinase or trypsin yielded heat-stable elicitors. Further analysis indicated that both pectic fragments and glycopeptides acted as elicitors.

It should be noted that the data on endogenous elicitors are derived from model experiments in which the microbial pectic enzymes are obtained from pure liquid cultures and the endogenous elicitor is prepared in vitro. It remains to be shown that endogenous elicitors play a role in plant–pathogen interactions or in the abiotic elicitation of defense reactions in vivo. Fry

(1986) has isolated a possible biologically active cell wall fragment, a xyloglucan nonasaccharide, from spinach cell suspension culture. A similar search for the endogenous elicitors will be useful.

An interesting example illustrating extracellular movement of an endogenous elicitor released by the plant itself has been described by Dixon et al. (1983). They found that an abiotic elicitor, denatured ribonuclease, resulted in the induction of phenylalanine ammonia-lyase (PAL) and of phytoalexin production in suspension-cultured cells and in hypocotyl sections. This induction could be transmitted through a dialysis membrane to cells that were not in direct contact with the abiotic elicitor, indicating that a low molecular weight, extracellular factor of host plant origin acted as endogenous elicitor.

A complication of the endogenous elicitor concept arises because of the presence of inhibitors of microbial pectic enzymes in plants (Collmer and Keen, 1986). These might reduce the direct damage done by pectinase but also might interfere with the generation of the "second messenger," that is, the endogenous elicitor. Microbial pectic enzymes have been found to kill plant cells rapidly (e.g., Keon, 1985); the endogenous elicitors derived from pectin do not kill cells even at high concentrations (Darvill and Albersheim, 1984). Other plant cell-wall-derived signals might be responsible for the killing activity. An oligosaccharide with killing activity was obtained from an acid hydrolysate from soybean cell walls (Yamazaki et al., 1983). It will be interesting to learn whether such fragments can be generated by pectic enzymes. A further complication is the finding that pectic enzymes do not only cause rapid killing, but they also prevent it: A highly purified pectinase suppressed the hypersensitive response elicited by *Pseudomonas syringae* in tobacco leaves (Baker et al., 1986).

Soluble Fragments Released From Pathogen Cell Walls and Membranes

Insoluble cell walls of pathogens probably act as elicitors because extracellular plant enzymes release elicitor-active fragments. In this case, the pathogen cell wall is the primary signal for recognition. The plant enzymes involved can be likened to receptors of this primary signal. The soluble fragments released from the pathogen wall are extracellular second messengers similar to the endogenous elicitors. To now, this extracellular signal transduction chain has received only little attention.

Extracts from pea pods released soluble elicitors from cell walls of *Fusarium solani* (Hadwiger and Beckman, 1980). Neither the elicitor-releasing enzymes nor the fragments generated were analyzed further. However, fungal chitosan and plant chitosan-degrading enzymes were suggested to be important components of this interaction (Nichols et al., 1980). The smallest chitosan oligomer fully active as an elicitor was the heptamer (Kendra and Hadwiger, 1984). The presence of a chitosanase in pea tissue was initially suspected (Nichols et al., 1980), but this was not substantiated in later work (Mauch et al., 1984). It is still possible that the chitinase present in pea tissue (Mauch et al., 1984) releases fragments resembling those analyzed by Kendra and Hadwiger (1984) from partly deacetylated chitin.

Insoluble chitin appeared to act as an elicitor of lignification in wheat leaves, and it was proposed that this effect was mediated by soluble chitin fragments produced by plant chitinase (Pearce and Ride, 1982). Others did not observe an induction of lignification by chitin (Moerschbacher et al., 1986). Chito-oligosaccharides obtained by acid hydrolysis appeared to be inducers of chitinase in melon tissue (Roby et al., 1987). It should be noted that chitinase in bean leaves is localized in the vacuoles of the cells; thus, it cannot come into contact with extracellular pathogen cell walls (Boller and Vögeli, 1984). However, the pathogen-induced chitinase in cucumber (Métraux and Boller, 1986) is located in the cell wall (Boller and Métraux, 1988).

Soybean extracts released soluble elicitors from cell walls of *Phytophthora megasperma* f.sp. *sojae* (Yoshikawa et al., 1981). Further analysis led to the identification of elicitor-releasing enzymes: two isoenzymes of β-1,3-glucanase were purified and found to release elicitor-active molecules from the pathogen cell walls (Keen and Yoshikawa, 1983). The localization of these enzymes was not examined. Surprisingly, the elicitor-active soluble products generated by the β-1,3-glucanases appeared to be high molecular weight glucomannans (Keen et al., 1983). The glucomannans released by the β-1,3-glucanase from the cell walls of an incompatible race of the fungus were somewhat more potent elicitors than those from a compatible race, indicating some host specificity (Keen et al., 1983).

The cell walls of suspension-cultured soybean cells were found to contain a different type of β-glucan hydrolase (Cline and Albersheim, 1981a). This enzyme inactivated a soluble glucan elicitor of *Phytophthora megasperma* f.sp. *glycinea,* indicating that extracellular plant enzymes might be important not only in the release of soluble elicitors but also in their destruction (Cline and Albersheim, 1981b).

Plant chitinase and β-1,3-glucanase inhibit fungal growth directly (Schlumbaum et al., 1986; Mauch et al., 1988). The results discussed above indicate that they may also function in signal perception by releasing soluble fragments from the pathogen cell walls. In pea pods, chitinase and β-1,3-glucanase are formed constitutively at a low level but are induced in response to infection or treatment with pathogen cell walls (Mauch et al., 1984). It is tempting to speculate that this represents a positive feedback loop, increasing the cells' capacity to recognize the pathogen. It remains to be seen, however, whether or not the two hydrolases participate in the release of elicitor-active fragments in this system.

Chitinase and β-1,3-glucanase are directed against cell wall components present in many pathogenic fungi and might have a role in nonpathogen recognition and nonhost resistance (Boller, 1985, 1987). However, there might also be highly specific enzymes that release race-specific soluble fragments from microbial cell walls. Indications for such enzymes have been obtained in a study of the interaction of legumes with *Rhizobium* (Solheim and Fjellheim, 1984). Crude enzyme preparations from pea and clover roots differentially degraded the polysaccharides from *Rhizobium leguminosarum* and *trifolii,* the pea and clover symbionts, respectively.

The Signal for Systemic Induced Resistance

In some plants, a local necrotizing infection induces resistance against a second infection systemically (reviews: Kuć, 1982; Sequeira, 1983). Obviously, a signal moves from the site of local infection to other sites in the plant. Fungi, bacteria, and even viruses can elicit this response, provided they cause necrotizing lesions. It is most likely that the signal for systemic resistance is produced as a secondary signal in the infected leaf in response to the different primary signal(s) produced by the different pathogens. The signal is probably transported in the phloem and is not remobilized from or produced in protected leaves, as shown by grafting experiments (Dean and Kuć, 1986).

A systemic effect that has received much attention is the systemic induction of proteinase inhibitors in locally wounded tomato and potato plants (review: Ryan, 1984). It has been shown that accumulation of proteinase inhibitors can be induced in excised tomato shoots when chitosan fragments (Walker-Simmons and Ryan, 1984) or pectic fragments (Walker-Simmons et al., 1984) are supplied via the transpiration stream, indicating that second messengers similar to the fungal cell wall fragments and to the endogenous elicitor may play a role in this system. The smallest pectic fragment that was active was the dimer (Bishop et al., 1984). It should be noted, however, that a signal generated in a wounded leaf is highly unlikely to move into other leaves via the transpiration stream. The transpiration stream is expected to flow into the wounded leaf, not out of it. As in the case of systemic induced resistance, it is more probable that the signal moves in the phloem. It remains to be shown that elicitor-like oligosaccharides can be transported in the phloem.

The potato proteinase inhibitor gene was engineered into tobacco plants, which have no wound-inducible proteinase inhibitors (Sanchez-Serrano et al., 1987). Interestingly the potato gene was turned on systemically in the transgenic tobacco plants upon local wounding, indicating that potato and tobacco possess a similar second messenger for the systemic signaling of wounding.

INTRACELLULAR SECOND MESSENGERS IN PLANT–PATHOGEN INTERACTIONS?

An important part of the defense reaction is a rapid change in gene expression (Chappell and Hahlbrock, 1984; Cramer et al., 1985; Somssich et al., 1986; Lawton and Lamb, 1987). Two simple models, derived from well-established theories about the mode of action of animal hormones, could account for the elicitation of such defense responses by extracellular primary or secondary signal molecules. 1) The extracellular signal molecule itself is transported into the cell and interacts with a nuclear receptor to turn genes on. 2) The extracellular signal molecule does not enter the cell but interacts with a receptor at the plasma membrane. Elicitor–receptor interaction causes the formation or release of an intracellular second messenger, which turns the genes on. At present, there is no direct evidence for either model. The following discussion is divided into three parts.

In the first part, experiments are presented that give indications for or against a possible involvement of intracellular second messengers in the elicitation of defense responses. The weight of this admittedly indirect evidence is in favor of the second messenger model. The second part deals with compounds known to play a role as intracellular or extracellular second messengers in animals. Attempts to demonstrate their occurrence in plants and their involvement in the elicitation of defense responses are discussed. In the third part, the possibility is considered that plant hormones act as second messengers.

Indications for and Against Involvement of Intracellular Second Messengers in Elicitation of Defense Responses

Membrane receptors for elicitors. One approach to find receptors for elicitors is to look for elicitor-binding membrane proteins. In soybean, a membrane preparation was found to exhibit saturable binding ($K_d = 12 \ \mu M$) of ^{14}C-labeled mycolaminaran (Yoshikawa et al. 1983). However, mycolaminaran is a relatively weak elicitor in soybean, compared with the glucan elicitors derived from *Phytophthora megasperma* f.sp. *glycinea*. Binding studies with a partially purified, highly potent ^3H-labeled glucan elicitor from *Phytophthora megasperma* f.sp. *glycinea* disclosed a saturable, pronase-sensitive high-affinity binding site (K_d = 0.2 μM) on membrane fractions from soybean cells (Schmidt and Ebel, 1987). Binding studies were performed in the presence of inhibitors of β-glucosidases, to prevent degradation of elicitors. When the membrane fractions were separated by sucrose density centrifugation, the binding activity co-migrated with a plasma membrane marker. Bound elicitor was displaced by various polysaccharides, including mycolaminaran; the concentrations of the polysaccharides required for 50% inhibition of binding corresponded closely with the concentrations required to elicit phytoalexin biosynthesis at one-half of the maximal rate. These results provide a strong indication for the presence of a specific elicitor-recognition protein on the plasma membrane. It is tempting to suggest that this elicitor-recognition protein is a receptor involved in signal transduction in the same way as animal hormone receptors. However, it could also be a transport protein responsible for uptake of the elicitor, an enzyme responsible for elicitor processing, or a lectin-like membrane protein (Nothnagel and Lyon, 1986) without signal-transduction function.

Rapid membrane changes caused by elicitors. Several workers have shown a rapid change in plasma membrane properties in response to elicitors. An immediate depolarization of the plasma membrane potential in response to elicitors has been recorded (Katou et al., 1982; Pelissier et al., 1986). A rapid stimulation of K^+ efflux and concomitant H^+ influx has been found in tobacco cells (Atkinson et al., 1985) and cotton cells (Pavlovkin et al., 1986) treated with the pathogen, *Pseudomonas syringae,* or in tobacco cells treated with a pectate lyase from *Erwinia chrysanthemi* (Atkinson et al., 1986). Low and Heinstein (1986) used lipophilic fluorescent probes to record membrane changes in suspension-cultured cotton and soybean cells treated with elicitors. An abrupt change in fluorescence of the membrane-associated probes occurred after a lag time of 0–10 min. Strasser and Matern (1986) found elicitors to cause a decrease in phosphate uptake within 2–3 min, an effect that was reversible in the first 20 min of treatment.

These data indicate that elicitors can change membrane properties within minutes. It is tempting to speculate that such membrane changes are symptoms of the release of an intracellular second messenger. For example, if intracellular Ca^{2+} were a second messenger, the elicitor-receptor interaction might increase intracellular Ca^{2+} by a sudden increase of the Ca^{2+} conductance of the plasma membrane, and depolarization of the membrane might be a side effect of this process.

Multiple signals, multiplicity of responses. The hypersensitive response with its multiple facets is displayed against a wide variety of pathogens. Even viruses can elicit this response (Zaitlin and Hull, 1987). In the latter case, it is very likely that the primary signal for recognition is present intracellularly where viruses multiply. Furthermore, it is unlikely that signals derived from viruses directly activate the multiple defense responses. It is much more probable that initial recognition is transduced by way of intracellular plant-derived signals that activate the defense responses. Given the likelihood that viruses induce the defense reactions by intracellular second messengers, the extracellular signals generated by bacterial and fungal pathogens might well activate the same intracellular messenger system.

It has frequently been observed that different elicitors act synergistically and that various compounds modulate the sensitivity of plant cells to elicitors. This might be explained by any of several theories, especially that of the second messenger model, where various signals and effectors interact to modulate the release of the second messenger and thereby alter the response. For example, arachidonic acid and a glucan elicitor stimulate phytoalexin synthesis in potato at much smaller concentrations when used in combination (Kurantz and Osman, 1983; Maniara et al., 1984; Preisig and Kuć, 1985). Similarly, a fungal glucan elicitor and a plant-derived pectic oligosaccharide act synergistically in the induction of phytoalexin production in soybean cotyledons (Davis et al., 1986b) and in cultured parsley cells (Davis and Hahlbrock, 1987). In tobacco cells, in contrast, pectate lyase and its products suppress the hypersensitive response induced by *Pseudomonas syringae* (Baker et al., 1986). In potato protoplasts, water-soluble glucans extracted from different races of *Phytophthora infestans* have been found to suppress the hypersensitive reaction in a host-specific manner (Doke and Tomiyama, 1980; Doke, 1983b). The biological significance of this observation is questionable because of the large amount of glucan used in the experiment (Darvill and Albersheim, 1984; Dixon, 1986).

Brefeldin A, a toxin produced by the safflower pathogen, *Alternaria carthami,* partially inhibits elicitor-induced accumulation of phytoalexin in safflower cell cultures at submicromolar concentrations (Tietjen and Matern, 1984). In soybean and in cotton cells, citrate specifically reduces phytoalexin production and membrane alterations, measured by fluorescence transitions, in a parallel way (Apostol et al., 1987), indicating that citrate modulates a signal transduction process in the membrane.

In some instances, a number of different elicitor preparations have been found to elicit essentially the same multiple defense reactions (e.g., Kombrink and Halhbrock, 1986). This is compatible with the idea that different extracellular signals are recognized by different receptors, all of which stimulate production of the same intracellular messenger. However, as stressed by Dixon (1986), it is equally possible that individual elicitor preparations contain a number of different signals that differentially induce parts of the defense response. Thus, the similarity in the response elicited by different crude elicitor preparations might arise because each contains a number of different signals. For example, different elicitor preparations from *Colletotrichum lindemuthianum* differentially induced PAL, phytoalexin, formation and lignification (browning) in bean cells (Dixon et al., 1981; Hamdan and Dixon, 1986). It should also be noted that frequently only one or a few typical reactions have been measured in analyses of defense responses, for example, the induction of PAL and chalcone synthase (reviews: Dixon, 1986; Ebel, 1986). However, the elicitor preparations that induce these two enzymes to the same degree may induce very different levels of phytoalexins (Ebel et al., 1984; Kombrink and Hahlbrock, 1986).

Time course of the induction of defense reactions. Several workers have studied the sequence of induction of defense reactions in response to elicitors (reviews: Dixon, 1986; Ebel, 1986). Determinations of the activities of enzymes frequently seem to indicate that there are "early" and "late" responses (e.g., Chappell et al., 1984; Kombrink and Hahlbrock, 1986). However, it is possible that these responses are regulated with the same kinetics at the level of gene activation. As aptly discussed by Kombrink and Hahlbrock (1986), activity determinations do not take into account turnover of the enzymes and mRNAs. Proteins that turn over most quickly appear to be induced most rapidly; those with little turnover appear to be induced most slowly. Studies at the level of translatable or hybridizable mRNA or at the level of nuclear transcription indicate that induction of a large number of defense proteins occurs very rapidly, within 5–20 min, in response to elicitors (Cramer et al., 1985; Somssich et al., 1986; Lawton and Lamb, 1987). This means that second messengers, if they exist, must be released and rendered active within minutes.

Uptake of microbial components into plant cells. If elicitors act directly, rather than by way of second messengers, one should expect to find elicitors or, more generally, pathogen-derived molecules within plant cells. Indeed, electron microscopic evidence indicates the presence of antigenically reacting fungal molecules within plant cells.

Chitosan has been localized immunocytochemically in the nuclei of pea cells (Hadwiger et al., 1981). It should be noted, however, that preservation of the natural localization of chitosan is difficult. As a polycation, chitosan has a strong tendency to stick to nucleic acids; it might attach to nuclear sites only artifactually during fixation.

Antigens corresponding to fungal surface structures, so-called fimbriae, were detected in the cytoplasm of infected tobacco (Day et al., 1986) or *Vicia faba* (Svircev et al., 1986) using immunocytochemical techniques, indicating that fungal surface material can enter plant cells. In contrast, O'Connell et al. (1986) found no antigens of *Colletotrichum lindemnthianum* in infected bean cells.

Second Messengers Known From Animals

Cyclic nucleotides. Treatments with exogenously supplied cAMP or dibutyryl cAMP have been reported to induce phytoalexin production in sweet potato (Oguni et al., 1976) and in cultured carrot cells (Kurosaki et al., 1987). However, the occurrence of endogenous cAMP in plants has been a point of debate. Although cAMP has now been unequivocally demonstrated in plants, its levels are very low and difficult to reconcile with second messenger functions (review: Newton et al., 1986). For example, the level of cAMP assayed by radioimmunoassay was reported to be about 0.5 pmol g^{-1} fresh weight in uninfected *Nicotiana glutinosa;* a two- to fourfold increase occurred within 24–36 hr after infection with a virus causing hypersensitive reaction compared with mock-infected controls (Rosenberg et al., 1982).

The effect of elicitor treatment and fungal infection on cAMP levels was tested in soybean tissues (Hahn and Grisebach, 1983). The cAMP level in untreated soybean hypocotyls was 0.6 pmol g^{-1} fresh weight; in untreated suspension-cultured cells, it was below 0.1 pmol g^{-1} fresh weight, the limit of detection. There were no changes in the cAMP level in response to infection of hypocotyls or elicitor treatments of cultured cells. In contrast, in similar experiments with carrot cell cultures, the cAMP level was found to increase several fold within 30 min after an elicitor treatment (Kurosaki et al., 1987).

Calcium ions. The role of cytoplasmic calcium levels in regulation of plant metabolism is beginning to be appreciated (review by Kauss, 1987). As in animal cells, cytoplasmic Ca^{2+} levels are very low but may increase rapidly in response to various stimuli. Kauss and his group have studied the rapid induction of callose formation by chitosan in cultured soybean cells. They observed that chitosan increased the permeability of the cells (Young et al., 1982). Polylysine, histone, and DEAE-dextran had the same effect, indicating that chitosan acted unspecifically as a polycation in this experimental system. Chitosan and other polycations rapidly stimulated callose deposition in the cells (Köhle et al., 1985). In vitro, Ca^{2+} ions strongly stimulated the callose-forming enzyme, β-1,3-glucan synthase (Köhle et al, 1985).

On this basis, it was suggested that the chitosan treatment caused a rapid influx of calcium from the extracellular space into the cytoplasm and that increased cytoplasmic calcium stimulated β-1,3-glucan synthase (Köhle et al., 1985; Kauss, 1987). Kauss' group also observed that chitosan induced phytoalexin biosynthesis (Köhle et al., 1984) and hypothesized that cytoplasmic Ca^{2+} levels are part of the signal chain leading to induction of the enzymes for phytoalexin biosynthesis (Köhle et al., 1985; Kauss, 1987). It has been found that a reduction of the extracellular Ca^{2+} level reduces phytoalexin accumulation in response to elicitors in soybean cells (Ebel, 1986) and in carrot cells (Kurosaki et al., 1987). In these cases, it remains to be shown that cytoplasmic calcium levels are modulated also.

Inositol phosphates. Polyphosphoinositides have received much attention as second messengers in animal tissues. They also occur in plant tissues (Boss, Chapter 2). The level of polyphosphoinositides (PIP and PIP_2) has been examined in soybean and parsley cells treated with an elicitor of phytoalexins (Strasser et al., 1986). In both cell types, there were no consistent fluctuations in the level of PIP and PIP_2 between 30 sec and 20 min after elicitor treatment, indicating that they have no second messenger function in this system.

Diacylglycerol. Release of polyphosphoinositides from phospholipids in animal systems is accompanied by the production of diacylglycerols, which are also active as second messengers. The possibility that diacylglycerols play a role in plant–pathogen interactions remains to be examined.

Lipid metabolites. In animals, oxidation of arachidonic acid by lipoxygenases or cyclooxygenases gives rise to a number of molecules (leukotrienes, lipoxins, prostanoids) that have intracellular and extracellular signal functions (review: Needleman et al., 1986). It is intriguing that arachidonic acid is an elicitor of phytoalexins in the potato (Bostock et al., 1981, 1982). Arachidonic acid, but not linolenic acid, induces lignification and phytoalexin formation, indicating a degree of specificity of arachidonic acid; additionally, both arachidonic acid and linolenic acid have unspecific effects on potato tissue and protoplasts (Bostock et al., 1986; Davis and Currier, 1986). Arachidonic-acid-induced synthesis of rishitin was inhibited by salicylhydroxamic acid (SHAM) and by tetraethylthiuram disulfide (disulfiram), two compounds that inhibited both cyanide-insensitive respiration and lipoxygenase in potato (Stelzig et al., 1983). However, it remains to be shown that arachidonic acid acts by way of oxidized derivatives. It should also be noted that arachidonic acid exhibits no elicitor activity in a number of other plant species (Bloch et al., 1984).

Lipoxygenase, hydroperoxide isomerase, and hydroperoxide cyclase are present in plants and might be involved in the formation of second messengers derived from the plant's own unsaturated fatty acids (Vick and Zimmermann, 1981; Anderson, Chapter 8). Increases in lipoxygenase were observed during the hypersensitive response (Ruzicska et al., 1983) and after elicitor treatment (Ocampo et al., 1986). It remains to be seen whether lipoxygenase or related enzymes form specific products from plant lipids in response to elicitors and whether these products formed are related to induction of defense responses.

Superoxide anions. In mammals, phagocytes and other lymphocytes generate superoxide anions (O_2^-) as a response to injury and infection (reviews: Badwey and Karnovsky, 1980; Beaman and Beaman, 1984). This compound may be involved in the oxidative killing of phagocytized microorganisms, either directly or indirectly through formation of hydroxyl radicals and singlet oxygen; in addition, it may induce further defense reactions. Superoxide anion is generated also in plant cells. In work with the potato, it has been found that tissues (Doke, 1983a) and protoplasts (Doke, 1983b) generate O_2^- in a NADPH-dependent reaction when stimulated with cell wall elicitors of *Phytophthora infestans*. An NADPH-dependent O_2^--producing system was found in membrane-enriched fractions from infected or elicitor-treated tissue but not from uninfected tissue (Doke, 1985).

Infiltration of the tissue with superoxide dismutase retarded the hypersensitive reaction and reduced the accumulation of phytoalexins in potato tissue infected with incompatible races of *Phytophthora infestans,* indicating that O_2^- may have a role as a "trigger" of the defense reactions (Doke, 1983a). The exact role of O_2^- in defense responses remains to be determined. In a study of legume cotyledons treated with the abiotic elicitor, silver nitrate, it was found that scavengers of the hydroxyl radical and of singlet oxygen, but not scavengers of the superoxide anion, inhibited phytoalexin accumulation, indicating that radicals other than O_2^- may be important in induction of defense responses (Epperlein et al., 1986).

Oligoadenylates. Oligoadenylates are thought to play an important role in the induction of interferon-linked antiviral defense in animals. Sela's group has reported that exogenously supplied oligoadenylate inhibited virus multiplication in plants (Devash et al., 1982). They also reported that human interferon would protect tobacco from virus infections (Orchansky et al., 1982) and that interferon, as well as a plant-produced antiviral factor, would act by way of inducing oligoadenylate synthesis in plants (Reichman et al., 1983). They suggested that interferon-related resistance in animal cells had a close parallel in plant cells. Others have failed to detect oligoadenylates or associated proteins in virus-infected plants (Cayley et al., 1982) and were unable to detect activity of human interferon against plant virus infection (Antoniw et al., 1984). The significance of the work of Sela's group has been doubted (Pierpoint, 1983); it remains to be confirmed.

Plant Hormones

Ethylene. One of the early reactions of plants to infections or elicitor treatments is the increased formation of ethylene (review: Boller, 1982). This appears to be due to the rapid induction of aminocyclopropane-1-

carboxylate (ACC) synthase, the key enzyme for ethylene biosynthesis (Chappell et al., 1984). Does ethylene have a role as a second messenger, controlling at least part of the disease response?

It has been shown by inhibitor studies that endogenously formed ethylene is not involved in induction of PAL and phytoalexin synthesis in the soybean (Paradies et al., 1980). This is not too surprising since exogenously applied ethylene does not induce these reactions. Defense reactions that can be induced both by infections and exogenous ethylene include the formation of chitinase and β-1,3-glucanase (Boller et al., 1983; Mauch et al., 1984) and the deposition of hydroxyproline-rich glycoproteins (Roby et al., 1985).

The question of whether ethylene acted as a second messenger in the induction of these reactions in response to elicitors or infection was investigated. Using a specific inhibitor of ethylene biosynthesis, aminoethoxyvinylglycine (AVG), ethylene biosynthesis in infected or elicitor-treated tissue was shown to be inhibited below the level of uninfected controls (Mauch et al., 1984; Roby et al., 1985, 1986). In the case of pea pods treated with pathogens or elicitors, it was found that chitinase and β-1,3-glucanase were strongly induced despite inhibition of ethylene biosynthesis, indicating that ethylene was a symptom of stress but not a necessary signal for the induction of the hydrolases (Mauch et al., 1984). Similar results were obtained for the induction of hydroxyproline-rich glycoproteins (Roby et al., 1985) and of chitinase (Roby et al., 1986) in infected or elicitor-treated melon plants, although AVG partially inhibited the responses.

The relationship of ethylene production and of PAL induction in parsley cells treated with an elicitor was also examined (Chappell et al., 1984). In this case, exogenous ethylene did not induce PAL. However, suppression of ethylene formation by AVG partly inhibited PAL induction in elicitor-treated cells, an effect that was fully reversed by addition of ACC but only partly by addition of ethylene.

Taken together, these data indicate that stress ethylene production may modulate the regulation of certain defense responses but does not act as a second messenger in their induction by elicitors.

Auxin and cytokinin. The case of crown gall disease caused by *Agrobacterium tumefaciens* may illustrate the diversity of signal transduction chains found in plant–pathogen interactions. It has long been known that *Agrobacterium* releases a signal that elicits increased formation of indoleacetic acid and cytokinin in plant tissues. These hormones can be considered second messengers, which cause autonomous tumor growth of the tissue, in crown gall disease. It is now known that the bacterial *signal* is a piece of DNA, the so-called T-DNA; that the *receptor* is the plant nuclear DNA and its transcription machinery; and that *signal transduction* involves expression of the transferred bacterial genes which code for key enzymes in the biosynthesis of auxin and cytokinin (review: Nester et al., 1984).

CONCLUSION

The example of auxin and cytokinin induction by *Agrobacterium*, summarized in the previous paragraph, is meant as a mild provocation. The general concept that primary signals are produced by pathogens, recognized by plant receptors, and transduced into specific responses by way of intracellular second messengers is readily applicable to this particular case; however, the individual components of the system are radically different from the model of signal transduction by intracellular second messengers in animal cells.

Attempts to transpose the model of second messengers derived from animal cells more or less exactly to plant–pathogen interactions have so far met with little success. However, a number of possibilities remain to be tested. One candidate for an intracellular second messenger of elicitor action is the calcium ion. In addition, it might be useful to look for new signals in plants. Particularly fruitful stimuli for new research might come from a comparison of the signal exchange of plants and pathogens with that between plants and symbionts (Lugtenberg, 1986; Halverson and Stacey, 1986). In the case of the symbiosis between legumes and *Rhizobium*, genetic evidence indicates that several signals derived from the bacterium are necessary for the correct establishment and functioning of the symbiosis (Halverson and Stacey, 1986).

The initial recognition of many pathogens occurs in the extracellular space. Here too, the model of an interaction between primary pathogen-derived signals and plant receptors and of the ensuing release of second messengers is readily applicable; again, it is very different in its details from the second messenger model derived from hormone action in animal cells.

Recognition of a pathogen may occur at the level of interactions between enzymes of one organism and structural compounds of the other. When soluble elicitors are released from the cell walls or membranes of pathogens, the plant enzymes liberating the elicitors function like receptors. When pathogen enzymes act as elicitors, the plant cell wall is the functional equivalent of a receptor. In either case, at least a part of the specificity of plant–pathogen interactions may reside in the specificity of the extracellular enzymes and of their substrates. It will be a challenge to find out which soluble elicitors and suppressors are produced in specific plant–pathogen interactions in vivo and how such naturally occurring signal molecules act in the regulation of the various defense responses.

REFERENCES

Amin M, F Kurosaki, A Nishi (1986): Extracellular pectinolytic enzymes of fungi elicit phytoalexin accumulation in carrot suspension culture. J Gen Microbiol 132:771–777.

Anderson AJ (1980a): Differences in the biochemical compositions and elicitor activity of extracellular components produced by three races of a fungal plant pathogen, *Colletotrichum lindemuthianum*. Can J Microbiol 26:1473–1479.

Anderson AJ (1980b): Studies on the structure and elicitor activity of fungal glucans. Can J Bot 58:2343–2348.

Antoniw JF, RF White, JP Carr (1984): An examination of the effect of human α-interferon on the infection and multiplication of tobacco mosaic virus in tobacco. Phytopathol Z 109:367–371.

Apostol I, PS Low, P Heinstein, RD Stipanovic, DW Altman (1987): Inhibition of elicitor-induced phytoalexin formation in cotton and soybean cells by citrate. Plant Physiol 84:1276–1280.

Atkinson MM, J-S Huang, JA Knopp (1985): The hypersensitive response of tobacco to *Pseudomonas syringae* pv. *pisi:* Activation of a plasmalemma K^+/H^+ exchange mechanism. Plant Physiol 79:843–847.

Atkinson MM, CJ Baker, A Collmer (1986): Transient activation of plasmalemma K^+ efflux and H^+ influx in tobacco by a pectate lyase isozyme from *Erwinia chrysanthemi*. Plant Physiol 82:142–146.

Ayers AR, J Ebel, F Finelli, N Berger, P Albersheim (1976a): Host–pathogen interactions. IX. Quantitative assays of elicitor activity and characterization of the elicitor present in the extracellular medium of cultures of *Phytophthora megasperma* var. *sojae*. Plant Physiol 57:751–759.

Ayers AR, J Ebel, BS Valent, P Albersheim (1976b): Host–pathogen interactions. X. Fractionation and biological activity of an elicitor isolated from the mycelial walls of *Phytophthora megasperma* var. *sojae*. Plant Physiol 57:760–765.

Badwey JA, ML Karnovsky (1980): Active oxygen species and the functions of phagocytic leukocytes. Annu Rev Biochem 49:695–726.

Baker CJ, MM Atkinson, MA Roy, A Collmer (1986): Inhibition of the hypersensitive response in tobacco by pectate lyase. Physiol Mol Plant Pathol 29:217–225.

Beaman L, BL Beaman (1984): The role of oxygen and its derivatives in microbial pathogenesis and host defense. Annu Rev Microbiol 38:27–48.

Bell AA (1981): Biochemical mechanisms of disease resistance. Annu Rev Plant Physiol 32:21–81.

Bishop PD, G Pearce, JE Bryant, CA Ryan (1984): Isolation and characterization of the proteinase inhibitor-inducing factor from tomato leaves: Identity and activity of poly- and oligogalacturonide fragments. J Biol Chem 259:13172–13177.

Bloch CB, PJGM De Wit, J Kuć (1984): Elicitation of phytoalexins by arachidonic and eicosapentaenoic acids: A host survey. Physiol Plant Pathol 25:199–208.

Boller T (1982): Ethylene-induced biochemical defenses against pathogens. In Wareing PF (ed): "Plant Growth Substances 1982." New York: Academic Press, pp 303–312.

Boller T (1985): Induction of hydrolases as a defense reaction against pathogens. In Key JL, T Kosuge (eds): "Cellular and Molecular Biology of Plant Stress." New York: Alan R. Liss, pp 247–262.

Boller T (1987): Hydrolytic enzymes in plant disease resistance. In Kosuge T, EW Nester (eds): "Plant–Microbe Interactions: Molecular and Genetic Perspectives," Vol 2. New York: Macmillan, pp 385–413.

Boller T, A Gehri, F Mauch, U Vögeli (1983): Chitinase in bean leaves: Induction by ethylene, purification, properties and possible function. Planta 157:22–31.

Boller T, JP Métraux (1988): Extracellular localization of chitinase in cucumber. Physiol Mol Plant Pathol 33, in press.

Boller T, U Vögeli (1984): Vacuolar localization of ethylene-induced chitinase in bean leaves. Plant Physiol 74:442–444.

Bostock RM, JA Kuć, RA Laine (1981): Eicosapentaenoic and arachidonic acids from *Phytophthora infestans* elicit fungitoxic sesquiterpenes in the potato. Science 212:67–69.

Bostock RM, RA Laine, JA Kuć (1982): Factors affecting the elicitation of sesquiterpenoid phytoalexin accumulation by eicosapentaenoic and arachidonic acids in potato. Plant Physiol 70:1417–1424.

Bostock RM, DA Schaeffer, R Hammerschmidt (1986): Comparison of elicitor activities of arachidonic acid, fatty acids and glucans from *Phytophthora infestans* in hypersensitivity expression in potato tuber. Physiol Mol Plant Pathol 29:349–360.

Bowen RM, JB Heale (1987): Four "necrosis-inducing" components, separated from germination fluid of *Botrytis cinerea* induce active resistance in carrot tissue. Physiol Mol Plant Pathol 30:55–66.

Bruce RJ, CA West (1982): Elicitation of casbene synthetase activity in castor bean: The role of pectic fragments of the plant cell wall in elicitation by a fungal endopolygalacturonase. Plant Physiol 69:1181–1188.

Bryan IB, WG Rathmell, J Friend (1985): The role of lipid and non-lipid components of *Phytophthora infestans* in the elicitation of the hypersensitive response in potato tuber tissue. Physiol Plant Pathol 26:331–341.

Callow JA (1984): Cellular and molecular recognition between plants and fungal pathogens. In Linskens HF, J Heslop-Harrison (eds): "Encylopedia of Plant Physiology, New Series, Vol 17, Cellular Interactions." Berlin: Springer, pp 212–237.

Cayley PJ, RF White, JF Antoniw, NJ Walesby, IM Kerr (1982): Distribution of the ppp $(A2'p)_nA$-binding protein and interferon-related enzymes in animals, plants and lower organisms. Biochem Biophys Res Commun 108:1243–1250.

Chai HB, N Doke (1987): Activation of the potential of potato leaf tissue to react hypersensitively to *Phytophthora infestans* by cytospore germination fluid and the enhancement of this potential by calcium ions. Physiol Mol Plant Pathol 30:27–37.

Chappell J, K Hahlbrock (1984): Transcription of plant defense genes in response to UV light or fungal elicitor. Nature 311:76–78.

Chappell J, K Hahlbrock, T Boller (1984): Rapid induction of ethylene biosynthesis in cultured parsley cells by fungal elicitor and its relationship to the induction of phenylalanine ammonia-lyase. Planta 161:475–480.

Cline K, P Albersheim (1981a): Host–pathogen interactions. XVI. Purification and characterization of a β-glucosyl hydrolase/transferase present in the cell walls of soybean cells. Plant Physiol 68:207–220.

Cline K, P Albersheim (1981b): Host–pathogen interactions. XVII. Hydrolysis of biologically active fungal glucans by enzymes isolated from soybean cells. Plant Physiol 68:221–228.

Cline K, M Wade, P Albersheim (1978): Host–pathogen interactions. XV. Fungal glucans which elicit phytoalexin accumulation in soybean also elicit the accumulation of phytoalexins in other plants. Plant Physiol 62:918–921.

Collmer A, NT Keen (1986): The role of pectic enzymes in plant pathogenesis. Annu Rev Phytopathol 24:383–409.

Cramer CL, TB Ryder, JN Bell, CJ Lamb (1985): Rapid switching of plant gene expression induced by fungal elicitor. Science 227:1240–1243.

Creamer JR, RM Bostock (1986): Characterization and biological activity of *Phytophthora infestans* phospholipids in the hypersensitive response of potato tuber. Physiol Mol Plant Pathol 28:215–225.

Crucefix DN, PM Rowell, PFS Street, JW Mansfield (1987): A search for elicitors of the hypersensitive reaction in lettuce downy mildew disease. Physiol Mol Plant Pathol 30:39–54.

Cruickshank IAM, MM Smith (1986): Elution profiles of pisatin elicitor activity from extracts of *Monilinia fructicola, Pisum sativum* and diffusates from their interaction. J Phytopathol 116:48–59.

Darvill AG, P Albersheim (1984): Phytoalexins and their elicitors: A defense against microbial infection in plants. Annu Rev Plant Physiol 35:243–275.

Davis DA, WW Currier (1986): The effect of phytoalexin elicitors, arachidonic and eicosapentaenoic acids, and other unsaturated fatty acids on potato tuber protoplasts. Physiol Mol Plant Pathol 28:431–441.

Davis KR, AG Darvill, P Albersheim, A Dell (1986a): Host–pathogen interactions. XXX. Characterization of elicitors of phytoalexin accumulation in soybean released from soybean cell walls by endopolygalacturonic acid lyase. Z Naturforsch 41c:39–48.

Davis KR, AG Darvill, P Albersheim (1986b): Host–pathogen interactions. XXXI. Several biotic and abiotic elicitors act synergistically in the induction of phytoalexin accumulation in soybean. Plant Mol Biol 6:23–32.

Davis KR, K Hahlbrock (1987): Induction of defense responses in cultured parsley cells by plant cell wall fragments. Plant Physiol 85:1286–1290.

Davis KR, GD Lyon, AG Darvill, P Albersheim (1984): Host–pathogen interactions. XXV. Endopolygalacturonic acid lyase from *Erwinia carotovora* elicits phytoalexin accumulation by releasing plant cell wall fragments. Plant Physiol 74:52–60.

Day AW, RB Gardner, R Smith, AM Svircev, WE McKeen (1986): Detection of fungal fimbriae by protein A-gold immunocytochemical labelling in host plants infected with *Ustilago heufleri* or *Peronospora hyoscyami* f.sp. *tabacina*. Can J Microbiol 32:577–584.

Dean RA, J Kuć (1986): Induced systemic protection in cucumbers: The source of the "signal." Physiol Mol Plant Pathol 28:227–233.

Devash Y, S Biggs, I Sela (1982): Multiplication of tobacco mosaic virus in tobacco leaf disks is inhibited by (2'5') oligoadenylate. Science 216:1415–1416.

De Wit PJGM, MB Buurlage, KE Hammond (1986): The occurrence of host-, pathogen- and interaction-specific proteins in the apoplast of *Cladosporium fulvum* (syn. *Fulvia fulva*) infected tomato leaves. Physiol Mol Plant Pathol 29:159–172.

De Wit PJGM, AE Hofmann, GCM Velthuis, JA Kuć (1985): Isolation and characterization of an elicitor of necrosis isolated from intercellular fluids of compatible interactions of *Cladosporium fulvum* (syn. *Fulvia fulva*) and tomato. Plant Physiol 77:642–647.

De Wit PJGM, E Kodde (1981): Further characterization and cultivar specificity of glycoprotein elicitors from culture filtrates and cell walls of *Cladosporium fulvum* (syn. *Fulvia fulva*). Physiol Plant Pathol 18:297–314.

De Wit PJGM, G Spikman (1982): Evidence for the occurrence of race and cultivar-specific elicitors of necrosis in intercellular fluids of compatible interactions of *Cladosporium fulvum* and tomato. Physiol Plant Pathol 21:1–11.

Dixon RA (1986): The phytoalexin response: Elicitation, signaling and control of host gene expression. Biol Rev 61:239–291.

Dixon RA, PM Dey, MA Lawton, CJ Lamb (1983): Phytoalexin induction in French bean: Intercellular transmission of elicitation in cell suspension cultures and hypocotyl sections of *Phaseolus vulgaris*. Plant Physiol 71:251–256.

Dixon RA, PM Dey, DL Murphy, IM Whitehead (1981): Dose responses for *Colletotrichum lindemuthianum* elicitor-mediated enzyme induction in French bean cell suspension cultures. Planta 151:272–280.

Doke N (1983a): Involvement of superoxide anion generation in the hypersensitive response of potato tuber tissue to infections with an incompatible race of *Phytophthora infestans* and to the hypal wall components. Physiol Plant Pathol 23:345–357.

Doke N (1983b): Generation of superoxide anion by potato tuber protoplasts during the hypersensitive response to hyphal wall components of *Phytophthora infestans* and specific inhibition of the reaction by suppressors of hypersensitivity. Physiol Plant Pathol 23:359–367.

Doke N (1985): NADPH-dependent O_2^- generation in membrane fractions isolated from wounded potato tubers inoculated with *Phytophthora infestans*. Physiol Plant Pathol 27:311–322.

Doke N, K Tomiyama (1980): Suppression of the hypersensitive response of potato tuber protoplasts to hyphal wall components by water soluble glucans isolated from *Phytophthora infestans*. Physiol Plant Pathol 16:177–186.

Ebel J (1986): Phytoalexin synthesis: The biochemical analysis of the induction process. Annu Rev Phytopathol 24:235–264.

Ebel J, WE Schmidt, R Loyal (1984): Phytoalexin synthesis in soybean cells: Elicitor induction of phenylalanine ammonia-lyase and chalcone synthase mRNAs and correlation with phytoalexin accumulation. Arch Biochem Biophys 232:240–248.

Epperlein MM, AA Noronha-Dutra, RN Strange (1986): Involvement of the hydroxyl radical in the abiotic elicitation of phytoalexins in legumes. Physiol Mol Plant Pathol 28:67–77.

Flor HH (1956): The complementary genetic systems in flax and flax rust. Adv Genet 8:29–54.

Fry SC (1986): In-vivo formation of xyloglucan nonasaccharide: A possible biologically active cell-wall fragment. Planta 169:443–453.

Gabriel DW, A Burges, GR Lazo (1986): Gene-for-gene interactions of five cloned avirulence genes from *Xanthomonas campestris* pv. *malvacearum* with specific resistance genes in cotton. Proc Natl Acad Sci USA 83:6415–6419.

Hadwiger LA, JM Beckman (1980): Chitosan as a component of pea–*Fusarium solani* interactions. Plant Physiol 66:205–211.

Hadwiger L, JM Beckman, MJ Adams (1981): Localization of fungal components in the pea–*Fusarium* interaction detected immunochemically with anti-chitosan and anti-fungal cell wall antisera. Plant Physiol 67:170–175.

Hahn MG, AG Darvill, P Albersheim (1981): Host–pathogen interactions. XIX. The endogenous elicitor, a fragment of a plant cell wall polysaccharide that elicits phytoalexin accumulation in soybeans. Plant Physiol 68:1161–1169.

Hahn MG, H Grisebach (1983): Cyclic AMP is not involved as a second messenger in the response of soybean to infection by *Phytophthora megasperma* f.sp. *glycinea*. Z Naturforsch 38c:578–582.

Halverson LJ, G Stacey (1986): Signal exchange in plant–microbe interactions. Microbiol Rev 50:193–225.

Hamdan MAMS, RA Dixon (1986): Differential biochemical effects of elicitor preparations from *Collectotrichum lindemuthianum*. Physiol Mol Plant Pathol 28:329–344.

Hargreaves JA, JA Bailey (1978): Phytoalexin production by hypocotyls of *Phaseolus vulgaris* in response to constitutive metabolites released by damaged plant cells. Physiol Plant Pathol 13:89–100.

Hargreaves JA, C Selby (1978): Phytoalexin formation in cell suspensions of *Phaseolus vulgaris* in response to an extract of bean hypocotyls. Phytochemistry 17:1099–1102.

Jin DF, CA West (1984): Characteristics of galacturonic acid oligomers as elicitors of casbene synthetase activity in castor bean seedlings. Plant Physiol 74:989–992.

Katou K, K Tomiyama, H Okamoto (1982): Effects of hyphal wall components of *Phytophthora infestans* on membrane potential of potato tuber cells. Physiol Plant Pathol 21:311–317.

Kauss H (1987): Some aspects of calcium-dependent regulation in plant metabolism. Annu Rev Plant Physiol 38:47–72.

Keen NT (1975): Specific elicitors of plant phytoalexin production: Determinants of race specificity in pathogens? Science 187:74–75.

Keen NT, M Yoshikawa (1983): β-1,3-Endoglucanase from soybean releases elicitor-active carbohydrates from fungus cell walls. Plant Physiol 71:460–465.

Keen NT, M Yoshikawa, MC Wang (1983): Phytoalexin elicitor activity of carbohydrates from *Phytophthora megasperma* f.sp. *glycinea* and other sources. Plant Physiol 71:466–471.

Keenan P, IB Bryan, J Friend (1985): The elicitation of the hypersensitive response of potato tuber tissue by a component of the culture filtrate of *Phytophthora infestans*. Physiol Plant Pathol 26:343–355.

Kendra DF, LA Hadwiger (1984): Characterization of the smallest chitosan oligomer that is maximally antifungal to *Fusarium solani* and elicits pisatin formation in *Pisum sativum*. Exp Mycol 8:276–281.

Keon JPR (1985): Cytological damage and cell wall modification in cultured apple cells following exposure to pectin lyase from *Monilinia fructigena*. Physiol Plant Pathol 26:11–29.

Köhle H, W Jeblick, F Poten, W Blaschek, H Kauss (1985): Chitosan-elicited callose synthesis in soybean cells as a Ca^{2+}-dependent process. Plant Physiol 77:544–551.

Köhle H, DH Young, H Kauss (1984): Physiological changes in suspension-cultured soybean cells elicited by treatment with chitosan. Plant Sci Lett 33:221–230.

Kombrink E, K Hahlbrock (1986): Responses of cultured parsley cells to elicitor from phytopathogenic fungi: Timing and dose dependency of elicitor-induced reactions. Plant Physiol 81:216–221.

Kuć J (1982): Induced immunity to plant disease. Biosci Rep 32:854–860.

Kuć J, JS Rush (1985): Phytoalexins. Arch Biochem Biophys 236:455–472.

Kurantz MJ, SF Osman (1983): Class distribution, fatty acid composition and elicitor activity of *Phytophthora infestans* mycelial lipids. Physiol Plant Pathol 22:363–370.

Kurosaki F, M Amin, A Nishi (1986): Induction of phytoalexin production and accumulation of phenolic compounds in cultured carrot cells. Physiol Mol Plant Pathol 28:359–370.

Kurosaki F, Y Tsurusawa, A Nishi (1985): Partial purification and characterization of elicitors for 6-methoxymellein production in cultured carrot cells. Physiol Plant Pathol 27:209–217.

Kurosaki F, Y Tsurusawa, A Nishi (1987): The elicitation of phytoalexins by Ca^{2+} and cyclic AMP in carrot cells. Phytochemistry 26:1919–1923.

Lawton MA, CJ Lamb (1987): Transcriptional activation of plant defense genes by fungal elicitor, wounding and infection. Mol Cell Biol 7:335–341.

Lee S-C, CA West (1981a): Polygalacturonase from *Rhizopus stolonifer,* an elicitor of casbene synthetase activity in castor bean (*Ricinus communis* L.) seedlings. Plant Physiol 67:633–639.

Lee S-C, CA West (1981b): Properties of *Rhizopus stolonifer* polygalacturonase, an elicitor of casbene synthetase activity in castor bean (*Ricinus communis* L.) seedlings, Plant Physiol 67:640–645.

Low PS, PF Heinstein (1986): Elicitor stimulation of the defense response in cultured plant cells monitored by fluorescent dyes. Arch Biochem Biophys 249:472–479.

Lugtenberg B (ed) (1986): "Recognition in Microbe–Plant Symbiotic and Pathogenic Interactions." Berlin: Springer.

Lyon GD, P Albersheim (1982): Host–pathogen interactions. XXI. Extraction of a heat-labile elicitor of phytoalexin accumulation from frozen soybean stems. Plant Physiol 70:406–409.

Maniara G, R Laine, J Kuć (1984): Oligosaccharides from *Phytophthora infestans* enhance the elicitation of sesquiterpenoid stress metabolites by arachidonic acid in potato. Physiol Plant Pathol 24:177–186.

Mauch F, LA Hadwiger, T Boller (1984): Ethylene: Symptom, not signal for the induction of chitinase and β-1,3-glucanase in pea pods by pathogens and elicitors. Plant Physiol 76:607–611.

Mauch F, B Mauch-Mani, T Boller (1988): Antifungal hydrolases in pea tissue. II. Inhibition of fungal growth by combinations of chitinase and β-1,3-glucanase. Plant Physiol (in press).

Mayama S, T Tani, T Ueno, SL Midland, JJ Sims, NT Keen (1986): The purification of victorin and its phytoalexin elicitor activity in oat leaves. Physiol Mol Plant Pathol 29:1–18.

Métraux JP, T Boller (1986): Local and systemic induction of chitinase in cucumber plants in response to viral, bacterial and fungal infections. Physiol Mol Plant Pathol 28:161–169.

Mieth H, V Speth, J Ebel (1986): Phytoalexin production by isolated soybean protoplasts. Z Naturforsch 41c:193–201.

Moerschbacher B, KH Kogel, U Noll, HJ Reisener (1986): An elicitor of the hypersensitive lignification response in wheat leaves isolated from the rust fungus *Puccinia graminis* f.sp. *tritici*. I. Partial purification and characterization. Z Naturforsch 41c:830–838.

Needleman P, J Turk, BA Jakschik, AR Morrison, JB Lefkowith (1986): Arachidonic acid metabolism. Annu Rev Biochem 55:69–102.

Nester EW, MP Gordon, RM Amasino, MF Yanofsky (1984): Crown gall: A molecular and physiological analysis. Annu Rev Plant Physiol 35:387–413.

Newton RP, EG Brown (1986): The biochemistry and physiology of cyclic AMP in higher plants. In Chadwick CM, DR Garrod (eds): "Hormones, Receptors and Cellular Interactions in Plants." Cambridge: Cambridge University Press, pp 115–153.

Nichols EJ, JM Beckman, LA Hadwiger (1980): Glycosidic enzyme activity in pea pod tissue and pea–*Fusarium solani* interactions. Plant Physiol 66:199–204

Niepold F, D Anderson, D Mills (1985): Cloning determinants of pathogenesis from *Pseudomonas syringae* pathovar *syringae*. Proc Natl Acad Sci USA 82:406–410.

Nothnagel EA, JL Lyon (1986): Structural requirements for the binding of phenylglycosides to the surface of protoplasts. Plant Physiol 80:91-98.

Nothnagel EA, M McNeil, P Albersheim, A Dell (1983): Host-pathogen interactions. XXII. A galacturonic acid oligosaccharide from plant cell walls elicits phytoalexins. Plant Physiol 71:916–926.

Ocampo CA, B Moerschbacher, HJ Grambow (1986): Increased lipoxygenase activity is involved in the hypersensitive response of wheat leaf cells infected with avirulent rust fungi or treated with fungal elicitor. Z Naturforsch 41c:559–563.

O'Connell RJ, JA Bailey, IR Vose, CJ Lamb (1986): Immunogold labelling of fungal antigens in cells of *Phaseolus vulgaris* infected by *Colletotrichum lindemuthianum*. Physiol Mol Plant Pathol 28:99–105.

Oguni I, K Suzuki, I Uritani (1976): Terpenoid induction in sweet potato roots by cyclic-3′,5′-adenosine monophosphate. Agric Biol Chem 40:1251–1252.

Orchansky P, M Rubinstein, I Sela (1982): Human interferons protect plants from virus infection. Proc Natl Acad Sci USA 79:2278–2280.

Osman S, R Moreau (1985): Potato phytoalexin elicitors in *Phytophthora infestans* spore germination fluids. Plant Sci 41:205–209.

Paradies I, JR Konze, EF Elstner, J Paxton (1980): Ethylene: Indicator but not inducer of phytoalexin synthesis in soybean. Plant Physiol 66:1106–1109.

Pavlovkin J, A Novacky, CI Ullrich-Eberius (1986): Membrane potential changes during bacteria-induced hypersensitive reaction. Physiol Mol Plant Pathol 28:125–135.

Pearce RB, JP Ride (1982): Chitin and related compounds as elicitors of the lignification response in wounded wheat leaves. Physiol Plant Pathol 20:119–123.

Pelissier B, JB Thibaud, C Grignon, MT Esquerré-Tugayé (1986): Cell surfaces in plant–microorganism interactions. VII. Elicitor preparations from two fungal pathogens depolarize plant membranes. Plant Sci 46:103–109.

Pierpoint WS (1983): Is there a phytointerferon? Trends Biochem Sci 8:5–7.

Preisig CL, JA Kuć (1985): Arachidonic-acid-related elicitors of the hypersensitive response in potato and enhancement of their activities by glucans from *Phytophthora infestans* (Mont.) de Bary. Arch Biochem Biophys 236:379–389.

Ralton JE, BJ Howlett, AE Clarke (1986): Receptors in host–pathogen interactions. In Chadwick CM, DR Garrod (eds): "Hormones, Receptors and Cellular Interactions in Plants." Cambridge: Cambridge University Press, pp 281–318.

Ralton JE, MG Smart, AE Clarke (1987): Recognition and infection processes in plant pathogen interactions. In Kosuge T, EW Nester (eds): "Plant–Microbe Interactions: Molecular and Genetic Perspectives," Vol 2. New York: Macmillan, pp 217–252.

Reichman M, Y Devash, RJ Suhadolnik, I Sela (1983): Human leukocyte interferon and the antiviral factors (AVG) from virus-infected plants stimulate plant tissues to produce nucleotides with antiviral activity. Virology 128:240–244.

Roby D, A Gadelle, A Toppan (1987): Chitin oligosaccharides as elicitors of chitinase activity in melon plants. Biochem Biophys Res Commun 143:885–892.

Roby D, A Toppan, MT Esquerré-Tugayé. (1985): Cell surfaces in plant microorganism interactions. V. Elicitors of fungal and of plant origin trigger the synthesis of ethylene and of cell wall hydroxyproline-rich glycoprotein in plants. Plant Physiol 77:700–704.

Roby D, A Toppan, MT Esquerré-Tugayé (1986): Cell surfaces in plant-microorganism interactions. VI. Elicitors of ethylene from *Colletotrichum lagenarium* trigger chitinase activity in melon plants. Plant Physiol 81:228–233.

Rosenberg N, M Pines, I Sela (1982): Adenosine 3'-5'-cyclic monophosphate: Its release in a higher plant by an exogenous stimulus as detected by radioimmunoassay. Fed Eur Biochem Soc Lett 137:105–107.

Ruzicska P, Z Gombos, GL Farkas (1983): Modification of the fatty acid composition of phospholipids during the hypersensitive reaction in tobacco. Virology 128:60–64.

Ryan CA (1984): Systemic responses to wounding. In Kosuge T, EW Nester (eds): "Plant–Microbe Interactions: Molecular and Genetic Perspectives," Vol. 1. New York: Macmillan, pp 307–320.

Sanchez-Serrano JJ, M Keil, A O'Connor, J Schell, L Willmitzer (1987): Wound-induced expression of a potato proteinase inhibitor II gene in transgenic tobacco plants. Eur Mol Biol Org J 6:303–306.

Schlumbaum A, F Mauch, U Vögeli, T Boller (1986): Plant chitinases are potent inhibitors of fungal growth. Nature 324:365–367.

Schmidt WE, J Ebel (1987): Specific binding of a fungal glucan phytoalexin elicitor to membrane fractions from soybean *Glycine max*. Proc Natl Acad Sci USA 84:4117–4121.

Sequeira L (1983): Mechanisms of induced resistance in plants. Annu Rev Microbiol 37:51–79.

Sharp JK, M McNeil, P Albersheim (1984): The primary structure of one elicitor-active and seven elicitor-inactive hexa (β-D-glucopyranosyl)-D-glucitols isolated from the mycelial walls of *Phytophthora megasperma* f.sp. *glycinea*. J Biol Chem 259:11321–11336.

Solheim B, KE Fjellheim (1984): Rhizobial polysaccharide-degrading enzymes from roots of legumes. Physiol Plant 62:11–17.

Somssich I, E Schmelzer, J Bollmann, K Hahlbrock (1986): Rapid activation by fungal elicitor of genes encoding "pathogenesis-related" proteins in cultured parsley cells. Proc Natl Acad Sci USA 83:2427–2430.

Staskawicz BJ, D Dahlbeck, NT Keen (1984): Cloned avirulence gene of *Pseudomonas syringae* pv. *glycinea* determines race-specific incompatibility on *Glycine max* (L.) Merr. Proc Natl Acad Sci USA 81:6024–6028.

Stelzig DA, RD Allen, SK Bhatia (1983): Inhibition of phytoalexin synthesis in arachidonic acid-stressed potato tissue by inhibitors of lipoxygenase and cyanide-resistant respiration. Plant Physiol 72:746–749.

Strasser H, C Hoffman, H Grisebach, U Matern (1986): Are polyphosphoinositides involved in signal transduction of elicitor-induced phytoalexin synthesis in cultured plant cells? Z Naturforsch 41c:717–724.

Strasser H, U Matern (1986): Minimal time requirement for lasting elicitor effects in cultured parsley cells. Z Naturforsch 41c:222–227.

Svircev AM, RB Gardiner, WE McKeen, AW Day, RJ Smith (1986): Detection by protein A-gold of antigens to *Botrytis cinerea* in cytoplasm of infected *Vicia faba*. Phytopathol 76:622–626.

Thorpe JR, JL Hall (1984): Chronology and elicitation of changes in peroxidase and phenylalanine ammonia-lyase activities in wounded wheat leaves in response to inoculation by *Botrytis cinerea*. Physiol Plant Pathol 25:363–379.

Tietjen KG, U Matern (1984): Induction and suppression of phytoalexin biosynthesis in cultured cells of safflower, *Carthamus tinctorius* L., by metabolites of *Alternaria carthami* Chowdhury. Arch Biochem Biophys 229:136–144.

Vick BA, DC Zimmermann (1981): Lipoxygenase, hydroperoxide isomerase, and hydroperoxide cyclase in young cotton seedlings. Plant Physiol 67:92–97.

Walker-Simmons M, D Jin, CA West, L Hadwiger, CA Ryan (1984): Comparison of proteinase inhibitior-inducing activities and phytoalexin elicitor activities of a pure fungal endopolygalacturonase, pectic fragments, and chitosans. Plant Physiol 76:833–836.

Walker-Simmons M, CA Ryan (1984): Proteinase inhibitor synthesis in tomato leaves: Induction by chitosan oligomers and chemically modified chitosan and chitin. Plant Physiol 76:787–790.

West CA (1981): Fungal elicitors of the phytoalexin response in higher plants. Naturwissenschaften 68:447–457.

Woodward S, GF Pegg (1986): Rishitin accumulation elicited in resistant and susceptible isolines of tomato by mycelial extracts and filtrates from cultures of *Verticillium alboatrum*. Physiol Mol Plant Pathol 29:337–347.

Yamazaki N, SC Fry, AG Darvill, P Albersheim (1983): Host–pathogen interactions. XXIV. Fragments isolated from suspension-cultured sycamore cell walls inhibit the ability of the cells to incorporate [^{14}C]leucine into proteins. Plant Physiol 72:864–869.

Yoshikawa M (1983): Macromolecules, recognition, and the triggering of resistance. In Callow JA (ed): "Biochemical Plant Pathology." Chichester: John Wiley & Sons, pp 267–298.

Yoshikawa M, NT Keen, M-C Wang (1983): A receptor on soybean membranes for a fungal elicitor of phytoalexin accumulation. Plant Physiol 73:497–506.

Yoshikawa M, M Matama, H Masago (1981): Release of a soluble phytoalexin elicitor from mycelial walls of *Phytophthora megasperma* var. *sojae* by soybean tissues. Plant Physiol 67:1032–1035.

Young DH, H Köhle, H Krauss (1982): Effect of chitosan on membrane permeability of suspension-cultured *Glycine max* and *Phaseolus vulgaris* cells. Plant Physiol 70:1449–1454.

Zaitlin M, R Hull (1987): Plant virus-host interactions. Annu Rev Plant Physiol 38:291–315.

Ziegler E, R Pontzen (1982): Specific inhibition of glucan-elicited glyceollin accumulation in soybean by an extracellular mannan-glycoprotein of *Phytophthora megasperma* f.sp. *glycinea.* Physiol Mol Plant Pathol 20:321–331.

Second Messengers in Plant Growth
and Development, pages 257–287
© 1989 Alan R. Liss, Inc.

11. Acetylcholine as a Signaling System in Plants

E. Hartmann and R. Gupta

Department of Botany, Johannes Gutenberg University Mainz, 6500 Mainz, Federal Republic of Germany (E.H.) and Department of Botany, University of Delhi, Delhi, India (R.G.)

INTRODUCTION

Acetylcholine (ACh) is an important neurochemical involved in the transmission of signals at junctions between nerves and between nerves and muscles (Nachmansohn, 1976). Hans Fischer (1971) stated in his book about the pharmacology of transmitter substances: "The use of ACh may be one of the oldest phylogenetic attempts to transduce a stimulus in the living material from one cell to another cell. This hypothesis is supported by finding sensitivity for ACh already in lower invertebrates and maybe still earlier in the plant kingdom." (translated from German).

A physiological role for ACh in plants was first envisaged in 1970, when Jaffe reported evidence for ACh as a primary biochemical messenger of photoactivation of *phytochrome*, a photoreceptor that controls nonphotosynthetic action of light in plant development.

The studies that followed during the 1970s raised the hope that plant biologists were on the threshold of discovering another aspect of the unity of plant and animal life at the molecular level. ACh was found in all the plants that were investigated. However, the enzymes of ACh metabolism are less widely distributed, and much of the initial evidence supporting the role of ACh as a mediator of phytochrome action has not been repeated in other laboratories. Thus, the initial enthusiasm did not last long, and the natural role of ACh in plants remains undiscovered.

Address reprint requests to Dr. E. Hartmann, Universität Mainz, Institut Allgemeine Botanik, Saarstrasse 21, 6500 Mainz, Federal Republic of Germany.

This paper summarizes briefly the concepts of the ACh system in animals and the reports on detection in plants of ACh and the enzymes of its metabolism. Reports on the exogenous application of ACh also are reviewed. The possible role of ACh in plants is highlighted from a heuristic point of view.

Bioelectrical Potentials

A charge difference exists across cell membranes. It is created by unequal diffusion of various ions across the membranes and by the activity of energy-driven ion pumps. Normally, the interior of all plant and animal cells is electrically negative with respect to the outside. The potential difference across the cell membrane at which the net flow of current is zero is defined as *resting potential*. In response to various stimuli, certain specific receptors on the cell membrane undergo conformational change, which, in turn, may open certain ion channels (and not others) in the membrane and, consequently, change the electrical polarity of the membrane. The localized change in the polarity of membrane thus induced in the vicinity of the receptor, leading to a quick transient potential change, is an *action potential*. However, the action potential is "fired" only if the magnitude of signal received by the cell has crossed a particular threshold. Therefore, action potential is an "all-or-none" type of response. Although the primary stimulus for firing the action potential could be physical, chemical, or electrical, during transmission the nature of the stimulus may change from one form to another, for example, physical to chemical, or chemical to electrical, and vice versa. The language of the message that is being transmitted depends on the frequency of the action potential (Stevens, 1979).

The action potential in nerve cells is a depolarization based on the prompt opening of Na^+ channels, leading to an influx of Na^+ ions driven by an electrochemical gradient. The process continues in a self-amplifying way until the membrane potential has shifted from its negative resting value (about -70 mV) to a positive Na^+ equilibrium potential (about $+50$ mV). The Na^+ channels close after about 1 msec (inactivation). In many types of neurons the recovery process is hastened by voltage-gated K^+ channels, which open in response to membrane depolarization.

Bioelectrical Potentials in Plants

Using techniques comparable to those used by neurophysiologists, external stimuli can be demonstrated to transiently depolarize or hyperpolarize plant membranes. Analogous to animal systems, spontaneous bioelectrical transients, which depend on exceeding a certain threshold, also are called *action potentials.* The work on bioelectrical potentials in plants has been the subject of several reviews and treatises (Sibaoka, 1966; Higinbotham, 1973; Pickard, 1973; Lüttge and Higinbotham, 1979; Simons, 1981).

The basic molecular processes during action potentials in plants, however, are still poorly understood. The complex morphological structure of plant cells consisting of large vacuoles, tonoplast, plasmalemma, and cell wall does not facilitate membrane potential measurements. Apart from experiments with large coenocytic algae, where the microelectrodes can be introduced selectively, potential differences in most plant cells represent the sum of electropotentials between the vacuole and the external medium. Nevertheless, it is well established that the ionic basis of action potentials in plant cells is given by K^+ and Cl^- and not by K^+ and Na^+. In many examples Ca^{2+} apparently is needed to establish membrane excitability.

A good communication and coordination system must exist not only in trees that may stand hundreds of feet above ground and send roots to great depth in soil but also in all plants. There are numerous reports that show propagation of stimulus-dependent membrane potentials (Brinckmann and Lüttge, 1974; Spanswick, 1972; Zawadzki and Trebacz, 1982, 1985). While a truly rapid communication system, like that found in nerves, is obviously not present in plants, cells of plants that exhibit rapid movements such as *Dionea muscipula* (Venus fly trap) and *Mimosa pudica* (touch-me-not) can propagate electrical stimuli as efficiently as nerves of lower animals. The physiological role of electrical signals in such communication processes, however, is still obscure.

Concept of Neurotransmitter Action

According to the current understanding, nerves interact with one, or sometimes with hundreds of other nerves, for example, at junctions called synapses, and exchange information. An impulse passed on to the next nerve at the synapse may be either excitatory or inhibitory of the firing of the receiving neuron. At many synapses, the exchange of information is mediated through certain chemicals that are termed neurotransmitters or neurohormones. These chemicals are kept in readiness at the nerve ending to be fired at the next nerve. As an action potential reaches the synapse at an axonal terminus, its effect causes an uptake of Ca^{2+}, which in turn facilitates the release of neurotransmitters from the presynaptic membrane, that is, membrane of the nerve ending (Llinas, 1982). Thus, the neurotransmitter molecules are released in the space between the membranes of the two interacting nerves and bind to the receptors on the membrane of the receiving nerve (i.e., the postsynaptic membrane), resulting in a conformational change.

In a resting state, that is, when no impulse is received, the inside of the postsynaptic membrane, like that of all plant and animal cells, is electrically negative to the outside and contains a higher concentration of K^+ and a lower concentration of Na^+ than that present outside. At the excitatory synapse, the conformational change in the neurotransmitter receptors opens ion channels, resulting in a large inflow of Na^+, which leads to depolarization or reversal of the polarity of the negatively charged inner membrane. The event triggers an action potential or a wave of depolarization all along the length of the nerve, again causing the release of the neurotransmitter at the next nerve ending.

Immediately after the passage of the action potential, the nerve membrane lies in a depolarized state and is unfit to receive the next impulse. The neurotransmitter receptor is restored to its original conformation by the removal of neurotransmitter molecules, either by certain enzymes that convert the neurotransmitter into an inactive form or by a system of active recycling. The negative polarity of the axonal membrane is also restored by selective closure of Na^+ channels and simultaneous opening of K^+ channels, to allow a large outflow of K^+ (Keynes, 1979).

In contrast to the depolarization at the excitatory synapse, an impulse at the inhibitory synapse causes *hyperpolarization*—increased negative voltage—of the postsynaptic membrane. This leads to difficulty in causing reversal of the excessively negative voltage when an excitatory impulse is received from another neuron. Thus the generation of bioelectrical potential at the postsynaptic membrane depends on the net voltage change caused by the excitatory and the inhibitory impulses received from different neurons (Eccles, 1965).

Acetylcholine and Other Neurotransmitters

Several neurotransmitters function at different synapses that take part in diverse types of communication and integration phenomena. ACh is the earliest and best known neurotransmitter. Synapses that release ACh are called cholinergic. The ACh system consists of 1) ACh molecules; 2) an enzyme to synthesize ACh, that is, choline acetyltransferase (ChAT); 3) an enzyme to hydrolyze ACh, that is, acetylcholinesterase (AChE); and 4) ACh receptor (AChR), the protein that mediates the molecular action of ACh. ACh molecules are synthesized at the nerve terminals from choline and acetyl-coenzyme A (CoA) and are stored in the synaptic vesicles. Each vesicle contains 10,000 ACh molecules to be fired at the next neuron. After release from the presynaptic membrane, ACh molecules move to the AChR at the outer surface of the postsynaptic membrane; once the action potential has been triggered, ACh is removed from the AChR by AChE, which hydrolyzes it to choline and acetic acid (Lester, 1977; Dunant and Israel, 1985). Other principal neurotransmitters, namely, norepinephrine, serotonin, and dopamine, act similarly through their respective receptors and enzymes (Axelrod, 1974).

History of Interest in the Role of Acetylcholine in Plants

The possibility of a "fundamental similarity of the mechanism of bioelectrogenesis in plant and animal cells" in terms of role of neurotransmitters was first suggested by Dettbarn (1962), while reporting the presence of an enzyme in *Nitella flexilis,* that could hydrolyze acetyl-β-methylcholine, a specific substrate of neuronal AChE. Dettbarn noted that although the speed of propagation of electrical stimulus in *Nitella* was about 1,000 times slower than in the nerves, the AChE activity was only one-tenth of that found in the giant axon of the squid.

A very exciting, early hypothesis about ACh functions in plants was published by Bennet-Clark in 1956. He tried to explain the auxin-mediated salt accumulation of plants by a biochemical pathway involving ACh, choline esterases, and the turnover of phosphatides. This hypothesis was not pursued although there were early indications that auxin effects might be regulated via a phosphatidylinositol (PI) turnover (Morré, et al., 1984).

Cumming and Wagner (1968) drew attention to the similarity of bioelectrical effects in plants and animals in response to light. They sought to invoke the theory of Nachmansohn (1966) of induction of conformational changes in excitable membranes in animals by chemicals like ACh to suggest the possibility that a similar mechanism controls phytochrome-regulated membrane permeability and ion flow. In the same year, Tanada reported a rapid electrotactic response of root tips of barley and mung bean mediated by phytochrome (Tanada, 1968a,b). In this phenomenon, the so-called "Tanada effect," the root tips of etiolated seedlings could be induced to stick to a negatively charged glass surface on exposure to red light, and made to unstick on a subsequent far-red light exposure. Tanada (1968a,b) postulated that development of charges on exposure of plant to light was among the primary events of phytochrome action. They were manifested in the same time period as that required to bring about conformational changes in the phytochrome molecule and provided evidence for Hendricks and Borthwick's (1967) theory of phytochrome action being a result of a change in membrane permeability.

Tanada's work on mung bean root tips was followed by that of Jaffe (1968), who actually measured the magnitude of the changes in membrane potential induced by phytochrome. Subsequent investigations of the mung bean plant for possible involvement of ACh in mediating the action of phytochrome revealed not only the presence of ACh (Jaffe, 1970) but also of two enzymes: cholinesterase, responsible for ACh hydrolysis (Riov and Jaffe, 1973a,b), and ChAT, responsible for synthesis of ACh (Jaffe, 1976). Jaffe's work provided a new orientation to phytochrome research. All old reports on the presence of ACh in plants that were scattered in pharmacological journals since 1914 were unearthed and viewed in a new perspective.

ACETYLCHOLINE AND ITS METABOLISM IN PLANTS

Distribution and Presence in Plants

Reports of presence. The literature on the presence of ACh in plants has been reviewed earlier by Guggenheim (1958), Whittaker (1963), Fluck and Jaffe (1974a), and Gupta (1983). The discovery of ACh in a biological material was reported by Ewins (1914), who isolated it from ergot, a fungus. In the 1920s and the 1930s, as details of ACh action in animals were being unraveled, several pharmacologists reported the presence of ACh in both fresh and fermenting plant juices and in bacteria (Guggenheim, 1958, Whittaker, 1963). However, Oury and Bacq (1937) could not detect ACh in 36 species of fungi, including ergot, and attributed the reported presence of ACh in plants to bacterial contamination. Due to the awareness that bacteria could produce ACh, a number of other reports on the presence of ACh in plants either could not be confirmed or were suspect (Whittaker, 1963). Most of the studies up until the 1960s reported very high concentrations of ACh, based on responses obtained in animal bioassays (Whittaker, 1963). In the 1970s ACh was detected in mung bean (Jaffe, 1970) and *Albizzia julibrissin* (Satter et al., 1972) by the clam heart bioassay and paper chromatography and in moss callus by employing both the frog heart bioassay and paper chromatography (Hartmann, 1971). Hartmann and Kilbinger (1974a) confirmed the presence of ACh in moss callus by using gas–liquid chromatography. ACh was detected in *Amaranthus caudatus, Cucurbita pepo, Helianthus annuus, Phaseolus vulgaris, Pisum sativum, Sinapis alba,* and *Spinacia oleracea* (Hartmann and Kilbinger, 1974b). Hoshino (1983b) has successfully extended the technique of field desorption mass spectroscopy to identify ACh in a sample of ACh-like substances that were isolated and purified from etiolated seedlings in *Vigna sesquipedalis.* Gupta (1983) has summarized all available reports on the presence or absence of ACh in plants, along with its concentration. The list of ACh-containing plants consists of 3 fungi, 1 bryophyte, 1 gymnosperm, and 36 angiosperms, in addition to 2 species of bacteria and fermenting juices of many plants. It is now widely believed that aside from being present in nerves of animals and in bacteria and plants that ACh is present in non-neuronal locations in all animal phyla (Sastry and Sadavongvivad, 1979).

Distribution of acetylcholine in different plant parts. Acetylcholine is generally present in all parts of the plants that have been investigated. Leaf hairs in *Urtica urens* (Emmelin and Feldberg, 1947) and seeds of *Artocarpus integra* (Lin, 1955, 1957) contain higher amounts of ACh than other parts of the plant. In *Phaseolus aureus,* growing regions of roots and shoot are particularly rich in ACh (Jaffe, 1970). Hartmann and Kilbinger (1974b) have reported that, invariably, all plants contained more ACh in the above-ground than in the below-ground parts.

Regulation of acetylcholine level by light. Acetylcholine levels have been reported to be controlled by phytochrome in a number of plants. Red light enhances while far-red light causes a decrease of ACh concentration in *Phaseolus aureus* (Jaffe, 1970, 1972), in a moss callus from a hybrid of *Funaria hygrometrica* × *Physcomitrium piriforme* (Hartmann, 1971; Hartmann and Kilbinger, 1974a) (Fig. 1), and in *Pinus sylvestris* (Kopcewicz et al., 1977). No ACh could be detected in etiolated seedlings of *Pisum sativum* and *Sinapis alba,* in contrast to its presence in light-grown plants (Hartmann

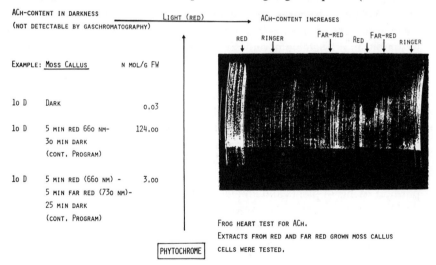

Fig. 1.   Light regulation of acetylcholine (ACh) content in moss callus. The moss callus was grown for 10 days under different light conditions and in darkness. The light programs (red light or red–far red light) were given continuously. The ACh content was measured by gas chromatography. The pharmacological frog heart test of ACh is also shown.

and Kilbinger, 1974b). In *Lemna gibba* G3 the contents of ACh are mediated by the photoperiodic regime. More ACh was found in plants grown under long-day conditions than in plants grown under continuous light or short-day conditions (Hoshino and Oota, 1978). ACh synthesis in membrane vesicles from soybean was stimulated by light (Jaffe, 1976). Incorporation of labeled precursors into ACh occurred only in light (Hartmann, 1979). Nevertheless, Satter et al. (1972) failed to observe any difference in the ACh level between light and dark in leaves of *Albizzia julibrissin*. On the other hand, a rise in ACh level in the dark, as well as a rise in the light in *Pinus sylvestris,* is evident from the data presented by Kopcewicz et al. (1977).

Cholinesterases

In nerves, ACh is hydrolyzed to choline and acetic acid by the enzyme AChE (E.C. 3.1.1.7). However, another enzyme pseudocholinesterase (E.C. 3.1.1.8) present at non-neuronal locations also hydrolyzes ACh, although at a slower rate. Both the enzymes are inhibited by carbamates, such as eserine and neostigmine, and by organophosphates, such as diisopropylfluorophosphate (DEP). According to a list compiled by Gupta (1983), the absence of ACh hydrolysis has been recorded in 65 species of plants, and the capacity of ACh hydrolysis has been reported in 103 species of green plants, 2 bacteria, and 1 slime mold. However, the conclusive proof for the presence of cholinesterase based on kinetic properties and response to inhibitors is available in only six species, of which four belonged to the Leguminosea—*Pseudomonas fluorescens* (Fitch, 1963a,b; Laing et al., 1967), *Phaseolus aureus* (Riov and Jaffe, 1973a,b), *Solanum melongena* (Fluck and Jaffe, 1975), *Pisum sativum* (Kasturi and Vasantharajan, 1976), *Phaseolus vulgaris* (Mansfield et al., 1978; Ernst and Hartmann, 1980), and *Cicer arietinum* (Gupta and Maheshwari, 1980). Lectins, plant proteins mainly isolated from legumes, agglutinate red blood cells by binding to glycoproteins on their surface (see Kauss, 1981). One of the glycoproteins on the surface of red blood cells is AChE (Juliano, 1973). Lectins can bind to purified AChE of plants and animals without affecting its catalytic activity. AChE can also bind to lectin-coupled-polyacrylamide beads (Bon and Rieger, 1975) and may also be bound to lectins *in vivo*. Therefore, it would be interesting to study the not yet completely understood physiological role of lectins in context with AChE in legumes and other plants.

Distribution and localization of cholinesterase in plants. AChE activity is present in all parts of the seedlings in mung bean (Fluck and Jaffe, 1974a) and Bengal gram (Gupta, 1983). The AChE activity in shoots has been found to decrease with the growth of seedlings. Almost 90% of the AChE activity in seedlings of *Cicer arietinum* is found in the roots (Gupta, 1983). Cell fractionation and electron microscopic studies in mung bean roots indicate that 95% of the AChE activity is associated with the cell wall (Fluck and Jaffe, 1974a). Maheshwari et al. (1982) observed AChE activity in the cell wall and nuclei of the root cells of *Pisum sativum*.

Lees and Thompson (1975) have reported increasing association of AChE with the plasmalemma and cell wall with growth of seedlings of *Phaseolus vulgaris*. A possible localization of AChE in the plasmalemma of bean tissues was extensively studied by Hartmann and co-workers. Hock (1983) found no AChE activity in protoplasts from beans. The activity was exclusively found in the cell wall residue. Bean hypocotyl tissue contained 94% of the enzyme activity in the cell wall fraction, 5% in the 100,000 g pellet, traces in other pellets, but there was no activity in the cytosolic fraction. The same pattern was found for bean callus. This enzymatic distribution pattern also was comparable to leaf tissue but with activity significantly increased in the chloroplast fraction (7%). Roshchina (1986) also reported AChE activity in pea chloroplasts. In contrast with the AChE activity pattern described for shoots and leaves, root tissue from beans had only 58% of the activity in the wall and 42% in other compartments (Table I).

The assumption that parts of the plasmalemma and not the cell wall were the main sites of AChE localization was not supported by studies with protoplasts. The preliminary immunological evidence that had suggested a localization of AChE on the surface of naked bean protoplasts (Hartmann et al., 1980, 1981) was found to be due to a nonspecific reaction of the polyclonal antibodies against AChE, which also contained antibodies against ChAT (Hartmann, unpublished).

Unfortunately, it was not possible to investigate the kinetics of restoration of AChE activity in cell walls of bean protoplasts because protoplasts from this species did not regenerate their cell walls (Hartmann and Hock, 1985). This question was, however, successfully researched with a soybean suspension, *Glycine max* callus (Kieffer, 1986). In this tissue AChE activity also was found exclusively in the cell wall fraction, but in regenerating protoplasts no correlation with the cell wall formation could be detected. The cell wall regeneration was rapid and could be monitored by calcofluor fluorescence after 16–24 hr. Yet no AChE activity was found before the fourth day of regeneration. It started at this time with high transient enzymatic activity, which decreased up to the twelfth day of growth. After 12 days of culture a second peak of activity could be observed. This peak was followed by a phase of declining activity.

The enzyme activity seemed to be correlated in some way with the cell division activity of the regenerating protoplasts and microcallus, respec-

TABLE I. Degradation and Synthesis of Acetylcholine by Enzymes Isolated From Bean *(Phaseolus vulgaris)* Seedlings

Degradation:

$$CH_3-\overset{\overset{\textstyle O}{\|}}{C}-O-CH_2-CH_2-N^+(CH_3)_3 \overset{AChE}{\rightleftharpoons} CH_3-COO^- + HOCH_2-CH_2-N^+(CH_3)_3$$

Acetylcholine   Acetate   Choline

Acetylcholinesterase
Molecular weight: 65,000 daltons
Apparent affinity constant (K_m): 0.46 mM
Highest enzyme activity: pH 8.0, 30–36°C
Substrate specificity: acetylcholine > propionylcholine > buturylcholine
Inhibitors: neostigmine > physostigmine > organophosphates > choline
Localization:
Enzyme activity determined from 7-day-old bean seedlings 1) hypocotyl hooks; 2) leaves; 3) roots. Protoplasts from this tissue showed no enzyme activity; x = no measurements

	Amount (%)			Specific activity $(10^{-10}$ kat.g^{-1} protein)		
Fraction	1	2	3	1	2	3
Cell wall	94	90	58	153.48	82.2	2,290.5
Cytosol	Nil	Nil	42	Nil	Nil	62.5
100,000g pellet	5	3	x	0.93	0.53	x
Chloroplasts	1	7	x	0.92	2.6	x

Synthesis:

$$HOCH_2-CH_2-N^+(CH_3)_3 + Acetyl-CoA \overset{ChAT}{\rightleftharpoons} ACh + CoA$$

Cholineacetyltransferase
Molecular weight: about 80,000
Apparent affinity constant (K_m): about 0.2 mM
Highest enzyme activity   (homogenate) pH 7.5, 35°C
                           (purified – less active) pH 5.5, 35°C
Localization:
Enzyme activity determined from 7-day-old bean seedlings 1) hypocotyl hooks; 2) protoplasts; x = no measurements

	Amount (%)		Specific activity $(10^{-9}$ kat.g^{-1} protein)	
Fraction	1	2	1	2
Cell wall	21	x	31.4	x
Cytosol	57	67	7.7	8.7
100,000g pellet	22	33	8.1	12.2

tively. The cytosolic activity did not decline and reached 42% of the total AChE activity after 23 days of growth. While the distribution of AChE was not measured at each stage of cell wall regeneration and callus growth, after several weeks of growth the callus had a AChE activity distribution pattern similar to the mother tissue, with only traces of cytosolic AChE activity.

These results indicated that callus regenerated from protoplasts slowly regained the normal AChE distribution. The physiological switch that controls the distribution pattern of AChE activity between cytosol and cell wall is unknown. The application of the AChE inhibitor, neostigmine (10^{-4}M), reduced cell wall regeneration. This effect was much more enhanced by the additional application of ACh (10^{-6} M), whereas ACh alone had no effect. Participation in the regulation and the maintenance of membrane flow or symplastic transport is still an attractive hypothesis for the physiological meaning of a plant ACh system, but there is so far only preliminary evidence supporting this speculation.

Cholineacetyltransferase

The first report of ChAT activity in plants came from Barlow and Dixon (1973). They found ChAT in the leaf hairs of stinging nettle *(Urtica dioica)*. The presence of ChAT also has been claimed in extracts of pea buds, cauliflower, and bean hypocotyl hooks (Biro, 1978). The enzyme was isolated and further characterized from bean seedlings *(Phaseolus vulgaris)* by Hoffmann (1982) and Hock (1983).

The reaction product ACh was identified using ^{14}C-choline as tracer. A purification of the enzyme was not possible because it became extremely unstable during purification. Experiments undertaken to stabilize the enzyme during preparation failed. Perhaps there were unidentified factors in homogenates that protect the enzyme, and the loss of these factors led to a fast inactivation of ChAT activities (Hoffmann, 1982). The enzyme was distributed in all parts of the bean cells. The highest ChAT activity was detected in the cytosolic fraction (57%), followed by the cell wall fraction (21%). The microsomal fraction (100,000g pellet) contained 22% of the activity. Based on specific activities, the ChAT activity in the cell wall was four times that of the cytosol. AChE and ChAT were thus found in the same cellular fractions (Hock, 1983). ChAT has also been isolated from *Urtica dioica, Helianthus annuus, Pisum sativum, Spinacia oleracea,* and *Oscillatoria agardhi* (Smallman and Maneckjee, 1981).

Jaffe (1976) reported the isolation of a membrane vesicle fraction rich in endoplasmic reticulum and containing high phytochrome content from soybean hypocotyls. On irradiation with red light, the vesicles synthesized 100 times more ACh-like substance than during incubation in dark or under far-red irradiation. A critical characterization of the ACh-like substances, thus produced, has not yet been reported. Its characterization would be desirable because, as discussed earlier, plants can synthesize chemicals other than ACh, which interfere with its extraction and assay (Hartmann, 1979; Smallman and Maneckjee, 1981).

Hartmann (1979) attempted to find out the pathway of ACh biosynthesis in plants by studying incorporation of labeled tracers in seedlings of *Phaseolus vulgaris*. Acetate was found to be a better precursor than choline for the synthesis of ACh. Most of the labeled choline could be traced in membrane fractions. Labeling of ACh was observed only in light but not in dark. No movement of radioactivity could be observed from glucose to ACh in a 24-hr period.

Acetylcholine Receptor

Although an acetylcholine receptor (AChR) has not been isolated from plants or any other non-nervous system, the presence of ACh binding activity in extracts of roots of *Phaseolus aureus* (Fluck and Jaffe, 1974b) and of a protein with a strong affinity for ACh in the cell wall residual fraction of *Phaseolus vulgaris* (Borbe, 1978; Hartmann et al., 1981; Weber, 1981) has been indicated (Table II). These reports give *Phaseolus aureus* and *Phaseolus vulgaris* the unique distinction of systems that potentially contain all components of a functional ACh system (Fig. 2).

ACETYLCHOLINE IN SIGNALING SYSTEMS

Acetylcholine as a Messenger of Phytochrome Action

Tanada effect: change in electrical polarity of root tips. Root tips of etiolated or light-grown barley and mung bean seedlings adhere, within a few

TABLE II. Characterization of Acetylcholine (ACh) Binding Sites in Cell Wall Extracts From Bean Hypocotyl Hooks Phaseolus Vulgaris*

Specific binding $Q = \dfrac{ACh\dagger}{ACh\dagger - ACh}$	$P < .05$
Cell wall protein fraction	6.04 ± 1.83 (n = 12)
Denaturated cell wall fraction	3.63 ± 0.89 (n = 5)
100,000g pellet	0.98 ± 0.40 (n = 5)

* The experiments were performed with ^{14}C labeled (ACh$^+$). The binding is calculated as quotient of overall binding (ACh$^+$) divided by the nonspecific binding (ACh$^+$ − ACh) based on competition experiments with nonradioactive (ACh 10^{-4} M).

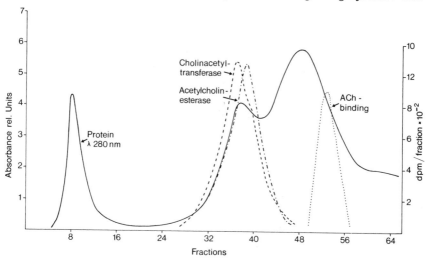

Fig. 2. Elution profile of the cell wall proteins from bean seedlings *(Phaseolus vulgaris)*. The proteins were eluted from the Sepharose 6B CL (Pharmacia) column by a 10-mM KH₂PO₄/NaOH buffer pH 7.0 with 18 ml/hr. The fractions were tested for enzyme and binding activity. The cell wall protein extract of beans seems to contain all components of an acetylcholine (ACh) system.

seconds of light exposure, to the bottom of a beaker that has been charged negatively by pretreatment with anions. This phenomenon exhibits rapid photoreversibility by red and far-red light (Tanada, 1968a,b).

A direct measurement of the electrical properties revealed that root tips became about 1 mV more electropositive on red-light exposure and reversed to a negative polarity on subsequent far-red-light exposure (Jaffe, 1968). The Tanada effect offered a special system for a study of rapid events in phytochrome action (Jaffe, 1968, 1970.)

The conditions for studying adhesion and release of root tips in the Tanada effect was characterized in detail by Yunghans and Jaffe (1970). The mechanism of development of a positive charge on root tips was presented by showing H^+ efflux from the roots following irradiation. The efflux could be reversed by subsequent far-red-light exposure. The action of red light could be mimicked by dipping the root tips in a solution of ACh (Jaffe 1970, 1972; Yunghans and Jaffe, 1972). Ca^{2+} ions and ATP were found to be obligatory for the release of root tips from the glass surface. Later, it was claimed that red light also caused the migration of Ca^{2+} from the nuclear envelope into the cytoplasm (Jaffe, 1972).

Rapid photoreversible adhesion of the root tips of mung bean to charged surfaces was confirmed by Racusen and Miller (1972), Bürcky and Kauss

(1974), and Racusen and Etherton (1975). However, doubts have been raised about the specific involvement of ACh in this effect. Tanada (1972) pointed out that abnormally high concentrations of ACh, 5×10^{-3} were required in causing adherence of root tips to charged glass surface. He presented data to show that even Na^+ at 3×10^{-4} M could induce the same degree of adherence as 3×10^{-4} M ACh and that K^+ (10^{-3} M) could specifically induce release of root tips, where adherence was initially caused by Na^+ or ACh. In bean hypocotyl hooks, high ACh concentrations inhibited the light-induced hyperpolarization. The same effect was given by increasing the K^+ concentration. The fact that applying ACh together with potassium reduced potassium uptake indicated that ACh and K^+ ions were competing for ion carrier systems (Hartmann, 1977). Since no ACh effect could be observed below 10^{-4} M, it was concluded that, contrary to Jaffe's proposal, ACh does not act as a hormone or transmitter but behaves as a cation interfering with the function of K^+ in the detachment process. Later, Tanada (1973) proposed that the hormones indoleacetic acid (IAA) and abscisic acid (ABA) caused the adherence and release, respectively, at concentrations of 10^{-9} M or less. Other plant hormones were ineffective.

Mediation of ATP Levels

Roshchina and Muhkin (1985) reported results about the influence of ACh on the regulation of ATP/NADPH ratio in chloroplasts and the ion permeability of the chloroplast thylakoid membranes. ACh was also shown to mimic red light in the phytochrome-mediated rapid uptake of O_2 and the decrease of ATP in bean root tips. Yunghans and Jaffe (1972) speculated that rapid changes in O_2 consumption and ATP utilization might play a role in the short- and the long-term action of light, primarily by using the bond energy of ATP for the active transport of ions, for example, the uptake of acetate, as shown later by Jaffe and Thoma (1973). However, rapid changes of ATP levels in bean roots could not be repeated by Bürcky and Kauss (1974).

In contrast, White and Pike (1974) and Kirshner et al. (1975) demonstrated that a brief red irradiation caused an increase in ATP levels, while the exogenous application of as low as 10 pM ACh in the dark causes a marked decrease. Fast changes in ATP levels mediated by phytochrome were shown by Hartmann et al. (1980). But the authors never found that the physiological effects of red light could be mimicked by ACh in either bean tissue or in moss callus. ACh became physiologically active after or during the irradiation. Light was necessary for ACh to have any effect. Thus, contrary to the observations of Yunghans and Jaffe (1972), ACh in these examples did not mimic red light.

Friederich and Mohr (1975) have shown that ATP levels and the energy charge of mustard seedlings *(Sinapis alba)* remain virtually unchanged even under light treatments that mediate drastic morphogenic and respiratory changes. They concluded that ATP levels cannot be a link in any causal chain of reactions originating from P_{fr} (far-red absorbing, physiologically active form of phytochrome) and leading to photomorphogenesis. A truly meaningful discussion of this important question as to whether or not phytochrome and/or ACh affect ATP levels is difficult, even if data clearly show a change in ATP level. As Marmé (1977) has pointed out, a wide range of cellular activities may be controlled by a single light stimulus. These activities also may contribute to the net ATP pool without any compelling reflection on the primary reaction controlled by phytochrome. Conclusive evidence, for or against, would therefore require studies on direct control of ATPase activity.

Acetylcholine in leaf movements. Leaves on many plants move from the normal to the drooping position and vice versa. Leaflets exhibit opening or closure movements in response to light, touch, electrical stimulus, chemicals, or temperature. Leaf movements also are regulated by endogenous rhythms. The role of electric potentials in leaf movements has been extensively reviewed by Umrath (1959), Satter and Galston (1981), and Simons (1981). The mechanism of movements has been studied in some detail in the leguminous plants, *Albizzia julibrissin, Mimosa pudica,* and *Samanea saman.* In the touch-sensitive or other rapid turgor-regulated movements, such as in *Mimosa pudica,* the primary stimulus changes the membrane potential in the sensing tissue. The signal is then propagated as an action potential to pulvinules and pulvini, which act as motor organs at the base of leaflets and leaves, respectively. Phytochrome-mediated nyctinastic or rhythmic movements are initiated also by a change in the membrane potential, although no action potential is yet known to be generated or propagated. The common point is that all stimuli finally cause large changes in ion fluxes, leading to turgor changes in the motor tissues.

Toriyama and Jaffe (1972) drew an analogy between the rapid ionic movements in nerves and muscles and those occurring in the motor cells of leaves, and they speculated about the involvement of ACh in leaf movements in *Mimosa pudica.* However, investigations by Satter et al. (1972) have negated the possibility of the involvement of ACh in leaf movements in *Mimosa pudica* as well as in *Albizzia julibrissin, Samanea saman,* and *Phaseolus multiflorus.* Although ACh could be detected in *Albizzia* (though not in the other plants), no differences in its titer were found between the lamina and the pulvinule. The ACh concentration in pulvinules did not change either during the phytochrome-mediated leaflet movements or during the rhythmic movements. Furthermore, exogenous applications of 10^{-5}–10^{-3}M ACh—up to 400 times the endogenous concentration in pulvinules—had no effect on leaflet movement either in red light, far-red light, or dark. Kumaravel et al. (1979), however, claimed to have established the presence of ACh and its involvement in leaf movements of *Samanea saman*. Although the contents of ACh of leaves in the closed stage were said to be higher than in the open condition during the day, no detailed report has yet been published regarding this aspect.

Contractive effects of ACh on plant cells also are known. Toriyama (1975) reported that the immature protoplasts of sieve tubes in root tips of mung bean were sensitive to the neurotransmitters, ACh and norepinephrine. Whereas ACh caused contraction of the protoplasts, atropine and norepinephrine inhibited this response. Later, Toriyama (1978) demonstrated that certain cells of the pericycle in the root tip yielded protoplasts, which exhibited a phytochrome-mediated reversible response. Based on the staining properties, the author conjectured that the ACh-sensitive cells contained some contractile proteins. Nevertheless, the significance of the above observations is difficult to assess at present, and there are no further reports about this effect.

Acetylcholine in flowering. ACh has been reported to cause changes in bioelectrical potentials in leaf surface accompanying photoperiodic induction in *Spinacia oleracea* and *Perilla nankinensis* (Greppin et al., 1973; Greppin and Horwitz, 1975). Phytochrome-mediated changes in membrane potentials also have been reported in *Lemna paucicostata* 6746 (Löpert et al., 1978) and *Lemna gibba* G1 (Kandeler et al., 1980). No evidence is available, however, for such potentials in signaling photoperiodic induction in *Lemna* or for an effect of ACh on biopotentials in these plants. The effect of ACh in flowering in *Lemna* has been studied in detail. Kandeler (1972) reported inhibition of flowering in a long-day plant *Lemna gibba* G1 by ACh and eserine, an inhibitor of ACh hydrolysis. Flowering in *Lemna paucicostata* was promoted by ACh. These results have been supported by Oota and co-workers in their studies on *Lemna gibba* (Oota and Hoshino, 1974; Oota, 1977; Hoshino, 1979). However, in a rather extensive study on several strains of *Lemna*, Khurana (1982) reported that ACh or inhibitors of AChE do not affect flowering.

Light-induced seed germination. Kostir et al. (1965) have reported an increase in the germinating power of seeds of some economically important plants by using ACh (Hadacova et al., 1981). Similarly, Holm and Miller (1972) have also found that ACh (10^{-4} M) mimics red light by promoting or inducing seed germination in *Agropyron repens*, *Echinchloa crusgalli*, *Che-*

nopodium album, Brassica kaber, and *Setaria viridis.* Eserine $(10^{-4}$ M) magnified the effect of ACh. Our own results (Gupta, unpublished) as well as those of Briggs (Marmé, 1977) do not support any effect of ACh on seed germination over a wide range of ACh concentrations and pH values.

Pollen germination and pollen tube growth. Pollen tube elongation in peanut, *Arachis hypogaea,* has been claimed to be influenced by phytochrome as well as by blue light (Chhabra and Malik, 1978). Red light promotes but blue light retards pollen tube elongation. ACh at a concentration of about 5.5×10^{-5} M promotes pollen tube elongation by mimicking red light (Chhabra and Malik, 1978). On the other hand, Gharyal (personal communication) finds ACh to be inhibitory to pollen tube elongation in *Lathyrus sativus.* Eserine, the cholinesterase inhibitor, also inhibits elongation of pollen tubes of *Lathyrus sativus,* but neostigmine has no effect. Inhibition by eserine of pollen tube elongation of *Crinum asiatum* (Martin, 1972) and lack of any effect of neostigmine in *Lathyrus latifolia* (Fluck and Jaffe, 1974a) have been recorded earlier. Acetylcholine or light do not influence pollen germination in *Pisum sativum, Cajanus cajan,* and *Lathyrus odoratus* (Gharyal, personal communication). It should be noted that pollen germination and tube elongation in many plants are very sensitive to the medium composition, particularly to the type and concentration of ions. No report on the effect of ACh has demonstrated specificity, since proper controls for effects of choline, acetic acid, or conjugate ions, for example, Na^+ and Cl^-, are lacking.

Phytochrome-mediated changes in enzyme activities. Saunders and McClure (1973) claimed that on dipping barley plumules into 10^{-4} M ACh for 5 min, the inductive effect of red light on activity of phenylalanine ammonia lyase (PAL) and flavonoid level was partly blocked, although ACh did not influence enzyme activity in dark alone. Such a claim, however, requires confirmation because, in contrast to the observed effect of ACh, similar treatment with either carbamylcholine (an AChR agonist), atropine (an AChR antagonist), or eserine (an AChE antagonist) was ineffective. Earlier, Kasemir and Mohr (1972) failed to find involvement of ACh in the phytochrome-mediated anthocyanin level—an indicator of PAL levels—in hypocotyls of *Sinapis alba.*

Under inductive photoperiodic conditions, peroxidase activity in leaves of long-day spinach shows rapid stimulation by far-red light (within 2 min), which was reversed by subsequent red light. Peroxidase activity in plants exposed to short days exhibited a diurnal rhythmicity in its sensitivity to far-red stimulation. The activity increased during the photoperiod and decreased during the dark period (Penel and Greppin, 1974). Interestingly,

ACh was reported to mimic red-light by immediately inhibiting the far-red stimulated peroxidase activity (Penel and Greppin, 1973). The effects of light were dependent on photoperiodic conditions. Because ACh caused rapid modification of light effects, the modulation of peroxidase activity was considered to be related to early events of floral induction (Penel and Greppin, 1973). There also appeared to be a striking analogy, with respect to the effect of ACh, between the inhibition of peroxidase and the stimulation of bioelectrical response of leaves (Greppin and Horwitz, 1975). Jones and Sheard (1975) and Ramaswamy (1979) have failed to observe any red-light-stimulating effect of ACh on nitrate reductase levels in either the pea or in wheat, although the activity of nitrate reductase in both the plants was influenced by phytochrome.

No difference in the activity of cholinesterase has been observed in Bengal gram seedlings grown in dark or light (Gupta, 1983), in mung bean roots irradiated with red or far-red light (Jaffe, 1976), in particulate matter from mung bean roots irradiated by red or far-red light (Jaffe, 1976), or in purified enzyme from pea roots irradiated by red or far-red light (Kasturi, 1979). However, Kasturi (1979) has reported phytochrome control of de novo synthesis of AChE in pea roots by demonstrating incorporation of labeled amino acids into enzyme protein. Although white light, red, and far-red light were all promotive for ACh synthesis, far-red light specifically caused a marked stimulation that was reversed by red light.

Acetylcholine in Blue Light Effects

Hartmann (1975) has reported a potential difference of about -30 mV between the surface of the bean hypocotyl hook and the cut base. Although the red–far-red reversible hook opening (a typical phytochrome-mediated photomorphogenic response) could be demonstrated, the bioelectrical changes were not related to phytochrome. Instead, blue light induced hyperpolarization (increased negative voltage) by about 7 mV. Later, Hartmann (1977) reported that ACh (10^{-9}–10^{-3} M) could partially suppress the blue light effect.

ACh in the presence of 5×10^{-6} M eserine also has been reported to mimic blue light in inducing conidiation in dark-grown mycelia of *Trichoderma viride* (Gressel et al., 1971). However, slightly lower or higher concentrations of eserine, for example, 3×10^{-6} M or 10^{-5} M were ineffective. ACh alone does not simulate blue light over a wide range of concentrations (10^{-5}–10^{-1} M). ACh alone, up to 10^{-2} M, or in combination with 10^{-3} M eserine, did not promote blue light induction of carotenoid biosynthesis in *Fusarium aqueductum* (Rau, 1980).

Acetylcholine and Plant Growth

ACh has been reported to influence the growth of several plants either alone or in combination with light and hormones, for example, it mimics red light to inhibit secondary root formation in etiolated seedlings of *Phaseolus aureus* (Jaffe, 1970, 1972) but inhibits secondary root formation only in light-grown seedlings of *Pisum sativum* (Kasturi, 1978). ACh inhibits the growth of tap root of *Lens culinaris* (Penel et al., 1976). It also inhibits elongation of hypocotyls in *Vigna sesquipedalis* (Hoshino, 1983a). ACh reverses the retardation caused by CCC of root growth in *Dolichos lablab* (Tung and Raghavan, 1968) and of leaf growth (Dekhuijzen, 1973). It also caused reversal of inhibition by ethrel of elongation of fern protonema (Bähre, 1975, 1977) and growth of soybean hypocotyls (Mukherjee, 1980). ACh is reported to promote growth of wheat seedlings (Dekhuijzen, 1973), of excised coleoptile segments of both oats (Evans, 1972, 1973) and wheat (Lawson et al., 1978), hypocotyl segments of the cucumber (Verbeke and Vendrik, 1977), and of intact hypocotyls of the soybean (Mukherjee, 1980).

ACh, however, does not influence phytochrome-mediated rapid growth inhibition in seedlings of *Tropaeolum majus* (Reed and Bonner, 1971), and it has no effect on growth of etiolated seedlings of *Sinapis alba* (Kasemir and Mohr, 1972), *Pisum sativum* (Kasturi, 1978), or *Cicer arietinum* (Gupta, 1983). In soybean leaves, high ACh concentration (11×10^{-3} M) had a strong inhibitory effect on ethylene production and mimicked the red-light effects. Atropine increased the ethylene production dramatically. This effect was reduced by ACh. Neostigmine had no effect (Jones and Stutte, 1986). It might be speculated that growth effects of ACh could be attributed to its effects on ethylene biosynthesis (Yang and Hoffmann, 1984).

However, unequivocal evidence for ACh having either a promotive or inhibitory effect on plant growth phenomena is not available. In some examples a concentration as high as 5×10^{-2} M ACh has been used in claiming a physiological response to ACh. In most, promotion or inhibition is not more than 20–30%, which is not very far from the standard deviations observed in untreated controls.

Problems of Exogenous Application of ACh

Some general problems in using ACh as an effector in plant physiology should be discussed in this context. There might be, in principle, a barrier to using exogenously applied ACh for elucidating its physiological significance. If ACh is an important agonist, it would be of great importance for the cells to regulate its internal ACh concentration very strictly.

The ACh-hydrolyzing enzyme AChE has been shown to be present on the surface and in the cell wall of plants (Fluck and Jaffe, 1974a; Jaffe, 1976;

Maheshwari et al., 1982), causing strong catabolism of ACh (Hartmann, 1978). Therefore, it may be reasoned that high concentrations of ACh (about 10^{-3} M) may be required to cause a response. However, responses induced with high ACh concentration are difficult to interpret, and they should be considered as being without any biological significance unless the effects of ACh can be distinguished from the reactions of its metabolic products, choline and acetic acid. Up to 90% of ACh may be hydrolyzed before its uptake by the plants (Hartmann, 1978). In most instances, however, degradation products of ACh were not used, inhibitors or stimulators of ACh action and metabolism were not tried, and dose-response curves were not reported.

In animals, too, AChE is juxtaposed to the AChR on the postsynaptic membrane and hydrolyzed about one-third of the ACh molecules even before they could bind to the receptor. Within a millisecond the concentration of ACh is reduced to 1% of that released by an impulse. The concentration of ACh in the synaptic cleft at the time of arrival of a nerve impulse has been estimated by various investigators as being from 5×10^{-4} M to more than 10^{-3} M (Lester, 1977). On the other hand, it is known that high transmitter concentrations and prolonged exposure to ACh cause the channels to enter a desensitized state, analogous to the naturally occurring inactivation of Na^{+} channels (Sakmann et al., 1980). The desensitization of ACh receptors occurs when ACh is not rapidly removed from its binding sites by AChE (Changeaux et al., 1984). Jones and Stutte (1984) observed a similar desensitization effect of ACh on ethylene production, when soybean tissue was pretreated with neostigmine.

Acetylcholine and Phospholipid Metabolism

This commentary so far has focused on a common idea—that ACh may act in plants in a manner analogous to its major effect in animal systems, that is, in transmission of signals. But the only possibility discussed is that ACh acts on membrane receptors, which directly results in the opening of some specific ion channels. This, in turn, would change the ionic composition of the intracellular fluid and lead to a series of metabolic changes. However, ACh also is believed to control membrane permeability in animal tissues by regulating the phospholipid metabolism of the membranes. Thus, ACh may enhance the enzymatic hydrolysis of PI and increase the content of phosphatidic acid in the membranes. Consequently, there would be a significant increase in phospholipid turn-over

This action of ACh, known as either the "phospholipid effect" or "phosphatidylinositol effect," is by far the best studied and most accepted nonsynaptic role of ACh in animals. It is mediated by the muscarinic cholinergic receptor on the cell membrane, and thus, it is inhibited by atropine (Michell, 1975; Putney, 1981). Hartmann et al. (1980) have reported that ACh inhibits ^{32}P incorporation into phospholipids of etiolated bean hypocotyls (Fig. 3). The phytochrome system also mediates phospholipid turnover (Fig. 4). In experiments with bean hook preparations labeled with ^{32}P, ACh (10^{-6} M) had a significant influence on the turnover of phosphatidylethanolamine (PE) and PI (Fig. 5). These authors thus claim similarities of the phospholipid effect in animals and plants (Hartmann and Pfaffmann, 1986). Hartmann and Schleicher (1977) have also isolated a choline kinase from bean seedlings. The in vitro activity of this enzyme could be stimulated by at-

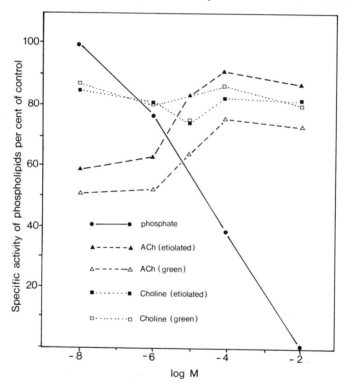

Fig. 3. Influence of acetylcholine (ACh) on the incorporation of phosphate (^{32}P) in the phospholipids of etiolated and green hypocotyl hooks of beans *(Phaseolus vulgaris)*. The hooks were incubated in the ^{32}P solution for 1 hr, and then the phospholipids were extracted by the standard method using methanol:chloroform (1:2). Choline was used as control. ACh did not inhibit phosphate incorporation at higher concentration. There was no difference between high ACh and high choline concentration.

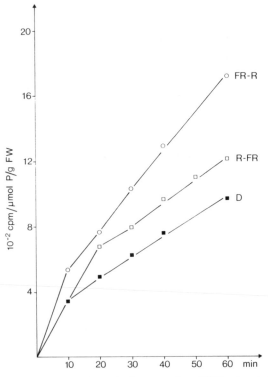

Fig. 4.   Effect of red and far red light on the incorporation of ^{32}P in phospholipids of etiolated bean *(Phaseolus vulgaris)* hypocotyl hooks. The radiolabeling and the extraction procedure was the same as in Figure 3. The phosphate (^{32}P) incorporation is expressed as specific labeling of phospholipid–phosphate on a fresh weight (FW) basis. The hooks were irradiated with 10 min red or 10 min red followed by 10 min far-red. The red-light-stimulated incorporation of phosphate in phospholipids was reversed by far-red light.

ropine and inhibited by eserine. An effect of ACh on choline kinase activity, however, was not found.

This is an area worthy of more intensive research. There are an increasing number of publications reporting the involvement of phospholipid metabolism in signal-transducing processes in plants (Anderson, Chapter 8; and Boss, Chapter 2). Nevertheless, involvement of ACh in the phospholipid effect, at least in a manner similar to that in animals, remains to be investigated. The interpretation of available data are complicated for several reasons: 1) atropine, an antagonist of ACh action, acts in the same way as ACh and, puzzlingly, much more effectively; 2) neostigmine and eserine, which should stimulate ACh action by increasing endogenous ACh levels,

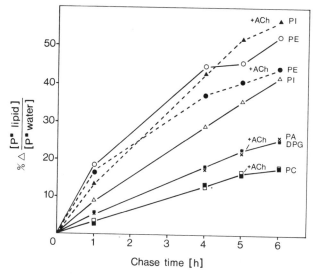

Fig. 5. Hypocotyl hooks of beans *(Phaseolus vulgaris)* were pulse-labeled for 2 hr with $^{32}P^*$ (P^*) in darkness. The turnover of single phospholipids was determined during chase time without (solid lines) and with (dashed lines) ACh. The turnover of phosphatidylethanolamine (PE) is inhibited in contrast to the stimulation of phosphatidylinositol (PI) turnover by ACh. Red light had comparable effects (data not shown). After the ^{32}P pulse (time 0) PE represented 43%; phosphatidylinositol (PI) 29%; phosphatidylcholine (PC) 12%; all other lipids 16% of ^{32}P-labeled lipid [P*lipid].

are reported to weaken the ACh effect; and 3) carbamylcholine, an ACh agonist, has no effect.

NONCHOLINERGIC AMINES AS REGULATORY MOLECULES IN PLANTS

Besides ACh, several other biogenic amines occur in plants, for example, norepinephrine, serotonin, dopamine, and glutamine. Of these, norepinephrine (noradrenaline) is an established neurotransmitter, whereas the other chemicals function in the nervous system but do not fulfill all the criteria for being neurotransmitters.

Dopamine and norepinephrine are catecholamines derived from tyrosine along the same metabolic pathway. There is no report on the presence in plants of epinephrine (adrenaline), which is a methylated derivative of norepinephrine. Epinephrine functions as a neurotransmitter in amphibians but as an endocrine hormone in mammals, illustrating that the same chemical may participate in cellular communication in different ways in different organisms (Silver, 1974).

Norepinephrine is claimed to influence flowering in *Lemna gibba* G3 (Oota and Nakashima, 1978) and in *Lemna paucicostata* (Khurana et al., 1987), where the presence of a functional norepinephrine–adenyl cyclase system is also indicated by use of agonists and antagonists of norepinephrine receptor (β-adrenergic receptor) and inhibitors of adenyl cyclase and phosphodiesterase. Dopamine and norepinephrine are reported to influence gibberellin-induced growth of lettuce hypocotyls (Kamisaka, 1979; Kamisaka and Shibata, 1982). Dopamine is present in certain varieties of the sugar beet, *Beta vulgaris,* and confers resistance to infection by the fungus *Cercospora beticola* (Hecker et al., 1970; cited in Smith, 1977).

There is striking structural similarity between the neurochemical, serotonin, and the plant growth hormone, indoleacetic acid (IAA), both of which are derivatives of tryptophan. It should be interesting, therefore, to compare the mechanism of action of the two regulators. While cAMP mediates the action of serotonin at certain synapses (Nathanson and Greengard, 1977), there is no evidence for its participation in IAA action. On the contrary, IAA and cAMP have been shown to exert opposing effects on differentiation of protonema cell types in the moss *Funaria hygrometrica* (Handa and Johri, 1976, 1979). In addition, IAA and cAMP are known to have opposite effects on the flowering of *Lemna gibba* G3 (Hoshino, 1979).

SUMMARY AND GENERAL CONCLUSIONS

During the last two decades the presence of ACh in various plants has been well established. In addition to bioassays, ACh has been analyzed by paper and gas–liquid chromatography, as well as by mass spectrometry. In contrast, the presence of choline acetyltransferase, the enzyme that catalyzes the synthesis of ACh, has been shown so far only in a few plants and cholinesterase appears to be found largely in members of the family Leguminosae. If ACh does act on the plant membranes, the way it does in animal cells, it would be important to draw attention to the fact that the acetylcholine receptor would become desensitized by prolonged exposure to high concentrations of the agonists (Lester, 1977).

The action of ACh in nerves and muscles involves participation of Ca^{2+} (see Llinas, 1982). The release of neurotransmitters from the presynaptic nerve endings also has been shown to be stimulated by calmodulin, a calcium binding protein (DeLorenzo et al., 1979). There is some evidence to support the idea that Ca^{2+} ions may modulate phytochrome action in plants (Roux et al. 1986). Thus, the original hypothesis that ACh may mediate phytochrome action still remains appealing to explain the "rapid" bioelectric

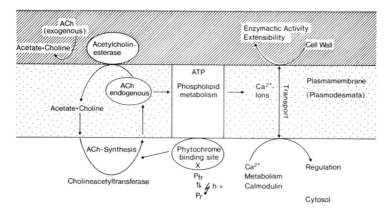

Fig. 6.   Summary of reported acetylcholine (ACh) effects and integration of these effects in a speculative model of the possible roles of ACh in plant plasma membranes.

response at the plasmalemma. Acetylcholine physiology may be one of the multiple responses influenced by phytochrome, although it remains to be proven whether or not the link is one of association rather than causal. Figure 6 summarizes potential mechanisms for involvement of ACh in the physiological responses discussed in this chapter.

REFERENCES

Axelrod J (1974): Neurotransmitters. Sci Am 230:59–71.
Barlow RB, ROD Dixon (1973): Cholineacetyltransferase in the nettle *Urica dioica* L. Biochem J 132:15–18.
Bähre R (1975): Zur Regulation des Protonemawachstums von *Athyrium filix-femina* (L.). II. Acetylcholine als Äthylenantagonist. Z Pflanzenphysiol 76:248–251.
Bähre R (1977): Zur Regulation des Protonemawachstums von *Athyrium filix-femina* (L.) Roth. IV. Wirkung cholinerger Substanzen in Gegenwart eines Antiauxins (PCIB). Z Pflanzenphysiol 81:278–282.
Bennet-Clark, TA (1956): Salt accumulation and mode of action of auxin: A preliminary hypothesis. In Wain, RL, F Wightman (eds): London: Butterworths pp 284–291.
Biro RL (1978): Choline acetylation and its phytochrome control in etiolated pea extracts. Master Thesis, Ohio University.
Bon S, F Rieger (1975): Interactions between lectins and electric eel acetylcholinesterase. Fed Eur Biol Soc Lett 53:282–286.
Borbe H (1978): Untersuchungen über das Vorkommen eines Acetylcholin-Rezeptors in Pflanzen. Master thesis, University of Mainz.
Bürcky K, H Kauss (1974): Veränderung im Gehalt an ATP und ADP in Wurzelspitzen der Mungobohne nach Hellrotbelichtung. Z Pflanzenphysiol 73:184–186.
Brinckmann E, U Lüttge (1974): Lichtabhängige Membranpotentialschwankungen und deren interzelluläre Weiterleitung bei panaschierten Photosynthese-Mutanten von *Oenothera*. Planta 119:47–57.

Changeaux JP, A Devillers-Thiery, P Chemouilli (1984): Acetylcholine receptor: An allosteric protein. Science 225:1335–1345.

Chhabra N, CP Malik (1978): Influence of spectral quality of light on pollen tube elongation in *Arachis hypogaea*. Ann Bot 42:1109–1117.

Cumming BG, E Wagner (1968): Rhythmic processes in plants. Annu Rev Plant Physiol 19:381–416.

Dekhuijzen HM (1973): The effect of acetylcholine on growth and on growth inhibition by CCC in wheat seedlings. Planta 111:149–156.

DeLorenzo RJ, SC Freedman, WB Yohe, SC Maurer (1979): Stimulation of Ca^{2+} -dependent neurotransmitter release and presynaptic nerve terminal protein phosphorylation by calmodulin and a calmodulin-like protein isolated from synaptic vesicles. Proc Natl Acad Sci USA 76:1838–1842.

Dettbarn WD (1962): Acetylcholinesterase activity in *Nitella*. Nateral 194:1175–1176.

Dunant Y, M Israel (1985): The release of acetylcholine. Sci Am 252:40–48.

Eccles J (1965): The synapse. Sci Am 212:56-66.

Emmelin N, W Feldberg (1947): The mechanism of the sting of the common nettle *(Urtica urens)*. J Physiol 106:440–455.

Ernst M, E Hartmann (1980): Biochemical characterization of an acetylcholine-hydrolyzing enzyme from bean seedlings. Plant Physiol 65:447–450.

Evans ML (1972): Promotion of cell elongation in *Avena* coleoptiles by acetylcholine. Plant Physiol 50:414–416.

Evans ML (1973): Rapid stimulation of plant cell elongation by hormonal and non-hormonal factors. Bioscience 23:711–718.

Ewins AJ (1914): Acetylcholine: A new active particle of ergot. Biochem J 8:44–49.

Fischer H (1971): "Vergleichende Pharmakologie von Überträgersubstanzen in Tier-systematischer Darstellung: Handbuch der Experimentelle Pharmakologie," Vol 25, New York: Springer Verlag.

Fitch WM (1963a): Studies on a cholinesterase of *Pseudomonas fluorescens*. I. Enzyme induction and the metabolism of acetylcholine. Biochemistry 2:1217–1221.

Fitch WM (1963b): Studies on a cholinesterase of *Pseudomonas fluorescens*. II. Purification and properties. Biochemistry 2:1221–1227.

Fluck RA, JM Jaffe (1974a): Cholinesterases from plant tissues. III. Distribution and subcellular localization in *Phaseolus areus* Roxb. Plant Physiol 53:752–758.

Fluck RA, JM Jaffe (1974b): The acetylcholine system in plants. Curr Adv Plant Sci 5:1–22.

Fluck RA, JM Jaffe (1975): Cholinesterases from plant tissues. VI. Preliminary characterization of enzymes from *Solanum melongena* L. and *Zea mays* L. Biochem Biophys Acta 410:130–134.

Friederich KE, H Mohr (1975): Adenosine 5′ -triphosphate content and energy charge during photomorphogenesis of the mustard seedling *Sinapis alba* L. Photochem Photobiol 22:49--53.

Greppin H, B Horwitz (1975): Floral induction and the effect of red and far-red preillumination on the light-stimulated bioelectric response of spinach leaves. Z Pflanzenphysiol 75:243--249.

Greppin H, BA Horwitz, PL Horwitz (1973): Light-stimulated bioelectric response of spinach leaves and photoperiodic induction. Z Pflanzenphysiol 68:336–345.

Gressel J, L Strausbauch, E Galun (1971): Photomimetic effect of acetylcholine on morphogenesis in *Trichoderma*. Nature 232:648–649.

Guggenheim M (1958): Die biogenen Amine in der Pflanzenwelt. In Ruhland W (ed): "Handbuch der Pflanzenphysiologie," Vol 8. Berlin: Springer Verlag, pp 889-988.

Gupta R (1983): Studies on cholinesterase in Bengal gram and probable role of acetylcholine in plants. Ph.D. Thesis, University of Delhi.

Gupta R, SC Maheshwari (1980): Preliminary characterization of a cholinesterase from roots of Bengal gram *Cicer arietinum* L. Plant Cell Physiol 21:1675–1680.

Hadacova V, J Hofman, RM De Almedia, K Vackova, M Kutacek, E Klozova (1981): Cholinesterase and choline acetyl-transferase in seeds of *Allium altaicum* (Pall.) Reyse. Biol Plant 23:220–227.

Handa AK, MM Johri (1976): Cell differentiation by 3', 5'-cyclic AMP in a lower plant. Nature 259:480–482.

Handa AK, MM Johri (1979): Involvement of cyclic adenosine 3',5'-monophosphate in chloronema differentiation in protonema cultures of *Funaria hygrometrica*. Planta 144:317–324.

Hartmann E (1971): Über den Nachweis eines Neurohormones beim Laubmooskallus und seine Beeinflussung durch das Phytochrom. Planta 101:159–165.

Hartmann E (1975): Influence of light on the bioelectric potential of the bean *(Phaseolus vulgaris)* hypocotyl hook. Physiol Plant 33:266–275.

Hartmann E (1977): Influence of acetylcholine and light in the bioelectric potential of bean *(Phaseolus vulgaris* L.) hypocotyl hook. Plant Cell Physiol 18:1203–1207.

Hartmann E (1978): Uptake of acetylcholine by bean hypocotyl hooks. Z Pflanzenphysiol 86:303–311.

Hartmann E (1979): Attempts to demonstrate incorporation of labeled precursors into acetylcholine by *Phaseolus vulgaris* seedlings. Phytochemistry 18:1643–1646.

Hartmann E, I Grasmück, K Hock (1981): Investigation about an acetylcholine-system in bean seedlings. In XIII International Botanical Congress, Sydney, p 250 (abstract).

Hartmann E, I Grasmück, N Lehrbach, R Müller (1980): The influence of acetylcholine and choline on the incorporation of phosphate into phospholipids of etiolated bean hypocotyl hooks. Z Pflanzenphysiol 97:377-389.

Hartmann E, K Hock (1984): Fatty acids in Protoplasts. In PE Pilet (ed): "The Physiological Properties of Plant Protoplasts." Berlin, Heidelberg, New York, Tokyo: Springer Verlag, pp 190–199.

Hartmann E, H Kilbinger (1974a): Gas–liquid-chromatographic determination of light-dependent acetylcholine concentrations in moss callus. Biochem J 137:249–252.

Hartmann E, H Kilbinger (1974b): Occurrence of light-dependent acetylcholine concentrations in higher plants. Experientia 30:1387–1388.

Hartmann E, H Pfaffmann (1986): Involvement of phospholipid-metabolism in signal transduction and development in plants. In Furuya M (ed): "Phytochrome and Plant Photomorphogenesis." Okazaki, Japan: Yamada Science Foundation, p 107.

Hartmann E, W Schleicher (1977): Isolierung und Charakterisierung einer Cholinkinase aus *Phaseolus vulgaris* L. Keimlingen. Z Pflanzenphysiol 83:69–80.

Hartmann E, K Schmid, M Nestler (1980): Effects of different light qualities on the light-induced biopotential changes and the ATP-content of etiolated hypocotyl hooks of beans (*Phaseolus vulgaris* L.) In De Greef J (ed): "Photoreceptors and Plant Development," Antwerpen: Antwerpen University Press, pp 335–350.

Hendricks SB, HA Borthwick (1967): The function of phytochrome in regulation of plant growth. Proc Natl Acad Sci USA 58:2125–2130.

Higinbotham N (1973): Electropotentials of plant cells. Annu Rev Plant Physiol 24:25–46.

Hock KF (1983): Untersuchungen zur Kompartimentierung der Enzyme des pflanzlichen Acetylcholin-Metabolismus mit Hilfe pflanzlicher Protoplasten. Ph.D. thesis, University of Mainz.

Hoffman N (1982): Lokalisierung, Reinigung und biochemische Charakterisierung der Cholinacetyltransferase aus Hypocotylhaken von *Phaseolus vulgaris* L. Master thesis, University of Mainz.

Holm RF, MR Miller (1972): Hormonal control of weed seed germination. Weed Sci 20:209–212.

Hoshino T (1979): Simulation of acetylcholine action by β-indoleacetic acid in inducing diurnal change of floral response to chilling under continuous light in *Lemna gibba* G3. Plant Cell Physiol 20:43–50.

Hoshino T (1983a): Effect of acetylcholine on the growth of the *Vigna* seedlings. Plant Cell Physiol 24:829–834.

Hoshino T (1983b): Identification of acetylcholine as a natural constituent of *Vigna* seedlings. Plant Cell Physiol 24:829–834.

Hoshino T, Y Oota (1978): The occurrence of acetylcholine in *Lemna gibba* G3. Plant Cell Physiol 19:769–776.

Jaffe MJ (1968): Phytochrome-mediated bioelectric potentials in mung seedlings. Science 162:1016–1017.

Jaffe MJ (1970): Evidence for the regulation of phytochrome-mediated processes in bean roots by the neurohumor, acetylcholine. Plant Physiol 46:768–777.

Jaffe MJ (1972): Acetylcholine as a native metabolic regulator of phytochrome-mediated processes in bean roots. In Runeckles VC, TS Tso (eds): "Recent Advances in Phytochemistry, Vol 5: Structural and Functional Aspects of Phytochemistry." New York: Academic Press, pp 81–104.

Jaffe MJ (1976): Phytochrome-controlled acetylcholine synthesis at the endoplasmic reticulum. In Smith H (ed): "Light and Plant Development," London: Butterworths, pp 333–334.

Jaffe MJ, Thoma L (1973): Rapid-phytochrome-mediated changes in the uptake by bean roots of sodium acetate ($1 - {}^{14}C$) and their modification by cholinergic drugs. Planta 113:282–291.

Jones RS, CA Stutte (1984): The acetylcholine-ethylene connection. Proc Plant Growth Regulators Soc Am 11:46–51.

Jones RS, CA Stutte (1986): Acetylcholine and red-light influence of ethylene evolution from soybean leaf tissues. Ann Bot 57:887–900.

Jones RW, Sheard RW (1975): Phytochrome, nitrate movement, and induction of nitrate reductase in etiolated pea terminal buds. Plant Physiol 55:954–959.

Juliona RL (1973): The proteins of the erythrocyte membrane. Biochem Biophys Acta 300:341–378.

Kamisada S (1979): Catecholamine stimulation of the gibberellin action that induces lettuce hypocotyl elongation. Plant Cell Physiol 20:1199–1207.

Kamisada S, K Shibata (1982): Identification in lettuce seedlings of a catecholamine active synergistically enhancing the gibberellin effect on lettuce hypocotyl elongation. Plant Growth Regulation 1:3–10.

Kandeler R (1972): Die Wirkung von Acetylcholin auf die photoperiodische Steuerung der Blütenbildung dei Lemnaceen. Z Pflanzenphysiol 67:86–92.

Kandeler R, H Löppert, T Rottenberg, K Scharfetter (1980): Early effects of phytochrome in *Lemna*. In De Greef J (ed): "Photoreceptors and Plant Development." Antwerpen, Antwerpen University Press, pp 485–492.

Kasemir H, H Mohr (1972): Involvement of acetylcholine in phytochrome-mediated processes. Plant Physiol 49:453–454.

Kasturi R (1978): Alleviation of phytotoxicity of fensulfothion (Dasanit) on pea. J Ind Inst Sci 60:9–31.

Kasturi R (1979): Influence of light, phytohormones and acetylcholine on the de novo synthesis of acetylcholinesterase in roots of *Pisum sativum*. Ind J. Biochem Biophys 16:14–17.

Kasturi R, VN Vasantharajan (1976): Properties of acetylcholinesterase from *Psium sativum*. Phytochemistry 15:1345–1347.

Kauss H (1981): Lectins and their physiological role in slime molds and higher plants. In Tanner W, FA Loewus (eds): "Encyclopedia of Plant Physiology, New Series, Vol 13B, Plant Carbohydrates. II. Extracellular carbohydrates." New York: Springer Verlag, pp 628–656.

Keynes RD (1979): Ion channels in the nerve-cell membrane. Sci Am 240:98–107.

Khurana JP (1982): In vitro control of flowering in duckweeds. Ph.D. Thesis, University of Delhi.

Khurana JP, BK Tamot, N Maheshwari, SC Maheshwari (1987); Role of catecholamines in promotion of flowering in a short-day duckweed, *Lemna paucicostata* 6746. Plant Physiol 85:10–12.

Kieffer I (1986): Untersuchungen zur Aktivität der Acetylcholinesterase bei der Regeneration von Soja-Protoplasten. Master Thesis, University of Mainz.

Kirshner RL, JM White, CS Pike (1975): Control of bean bud ATP levels by regulatory molúcules and phytochrome. Physiol Plant 34:373–377.

Kopcewicz J, M Cymerski, Porazinski (1977): Influence of red and far-red irradiation on acetylcholine and gibberellin content in Scots pine seedlings. Bull Acad Pol Sci Sevr Sci Biol 25:111–117.

Kostir J, J Klenha, V Jiracek (1965): The influence of choline and acetylcholine on the germination of seeds of economically important plants. I Rostlinna Vyroba (Praha) 11:1239–1280.

Kumaravel M, R Sundaran, DS Rao (1979): Studies on active chemical principles responsible for nyctinastic behavior of *Samanea saman*. Ind J Biochem Biophys 16 (1) Suppl:89–90.

Laing AC, HR Miller, KS Bricknell (1967): Purification and properties of the inducible cholinesterase of *Pheudomonas fluorescens* (Goldstein). Can J Biochem 45:1711–1724.

Lawson VR, RM Brady, Campbell A, BG Knox, GD Knox, RL Walls (1978): Interaction of acetylcholine chloride with IAA GA₃, and red light in the growth of excised apical coleoptile segments. Bull Torr Bot Club 105:187–191.

Lees GL, JE Thompson (1975): The effects of germination on the subcellular distribution of cholinesterase in cotyledons of *Phaseolus vulgaris*. Physiol Plant 34:230–237.

Lester HA (1977): The response to acetylcholine. Sci Am 236:107-118.

Lin RCY (1955): Presence of acetylcholine in the Malayan jack-fruit plant, *Artocarpus integra*. Br J Pharmacol 10:247–253.

Lin RCY (1957): Distribution of acetylcholine in the Malayan jack-fruit plant, *Artocarpus integra*. Br J Pharmacol 12:265–269.

Llinas RR (1982): Calcium in synaptic transmission. Sci Am 247:38–47.

Löpert H, W Kronberger, R Kandeler (1978): Phytochrome-mediated changes in the membrane potential of subepidermal cells of *Lemna paucicostata* 6746. Planta 138:133–136.

Lüttge U, N Higinbotham ((1979):" Transport in Plants." New York: Springer Verlag.

Maheshwari SC, R Gupta, PK Gharyal (1982): Cholinesterases in plants. In Sen SP (ed): "Recent Developments in Plant Sciences: SM Sircar Memorial Volume," New Delhi: Today and Tomorrows Printers and Publishers, pp 145–160.

Mansfield DH, G Webb, DC Clark, IEP Taylor (1978): Partial purification and some properties of a cholinesterase from bush bean (*Phaseolus vulgaris* L.) roots. Biochem J 175:769–777.

Marmé D (1977): Phytochrome: Membranes as possible site of primary action. Annu Rev Plant Physiol 28:173–198.

Martin FW (1972): In vitro measurement of pollen tube growth inhibition. Plant Physiol 49:924–925.

Michell RH (1975): Inositol phospholipids and cell surface receptor function. Biochem Biophys Acta 415:81–147.

Morré DJ, B Gripshover, A Monroe, JT Morré (1984): Phosphatidylinositol turnover in isolated soybean membranes stimulated by synthetic growth hormone 2,4-dichlorophenoxyacetic acid. J Biol Chem 259:15364-15366.

Mukherjee I (1980): The effect of acetylcholine on hypocotyl elongation in soybean. Plant Cell Physiol 21:1657–1660.

Nachmansohn D (1966): Chemical control of the permeability cycle in excitable membranes during activity. In Shapiro B, M Prywes (eds): "Impact of Basic Sciences on Medicine." New York; Academic Press, pp 123–141.

Nachmansohn D (1976): Acetylcholine: its role in nerve excitability. Trends Biochem Sci 1:237–238.

Nathanson JA, P Greengard (1977): "Second messengers" in the brain. Sci Am 237:108–119.

Oota Y (1977): Removal by chemicals of photoperiodic light requirements of Lemna gibba G3. Plant Cell Physiol 18:95–105.

Oota Y, T Hoshino (1974): Diurnal change in temperature sensitivity of Lemna gibba G3 induced by acetylcholine in continuous light. Plant Cell Physiol 15:1063–1072.

Oota Y, H Nakashima (1978): Photoperiodic flowering in Lemna gibba G3: Time measurement. Bot Mag Tokyo (special issue) 1:177–198.

Oury A, ZM Bacq (1937): Presence d'un ester instable de la choline chez Lactarius blennis. CR Soc Biol (Paris) 126:1263–1264.

Penel C, E Darimont, H Greppin, T Gaspar (1976): Effect of acetylcholine on growth and isoperoxidases of the lentil (Lens culinaris) root. Biol Plant 18:293–298.

Penel C, H Greppin (1973): Action des lumières rouge et infrarouge sur l'activite peroxydasique des feuilles d'epinards avant et apres l'induction florale. Ber Schweiz Bot Ges 83:253–261.

Penel C, H Greppin (1974): Variation de la photostimulation de elactivite des peroxidases basiques chez lépinard. Plant Sci Lett 3:75–80.

Pickard BG (1973): Action potentials in higher plants. Bot Rev 39:172–201.

Putney JW (1981): Recent hypothesis regarding the phosphatidylinositol effect. Life Sci 29:1183–1194.

Racusen RH, B Etherton (1975): Role of membrane-bound, fixed-charge changes in phytochrome-mediated mung bean root tip adherence phenomenon. Plant Physiol 55:491–495.

Racusen R, K Miller (1972): Phytochrome-induced adhesion of mung bean root tips to platinum electrodes in a direct current field. Plant Physiol 49:654–655.

Ramaswamy O (1979): Role of light on the regulation of nitrate reductase in wheat (Triticum aestivum). M. Phil. Thesis. New Delhi: Jawaharlal Nehru University.

Rau W (1980): Blue light-induced carotenoid biosynthesis in microorganisms. In Senger H (ed): "The Blue Light Syndrome," Berlin: Springer Verlag, pp 283–298.

Reed WA, BA Bonner (1971): Investigations on the rapid phytochrome induced inhibition of Tropaelum stem elongation. Plant Physiol 47(suppl):2.

Riov J, JM Jaffe (1973a): A cholinesterase from bean roots and its inhibition by plant growth retardants. Experientia 29:264–265.

Riov J, MJ Jaffe (1973b): Cholinesterases from plant tissues. I. Purification and characterization of cholinesterase from mung bean roots. Plant Physiol 51:520–528.

Roshchina VV, EN Mukhin (1985): Acetylcholine action on the photochemical reactions of pea chloroplasts. Plant Sci 42:95–98.

Roshchina VV (1986): Cholinesterase from chloroplasts. Dokl USSR Akad Sci 290:486–489.

Roux SJ, RO Wayne, N Datta (1986): Role of calcium ions in phytochrome responses: an update. Physiol Plant 66:344–348.

Sakmann B, J Patlak, E Neher (1980): Single acetylcholine-activated channels show burst-kinetics in presence of desensitizing concentration of agonist. Nature 286:71–73.

Sastry BVR, C Sadavongvivad (1979): Cholinergic system in nonnervous tissues. Pharmacol Rev 30:65–132.

Satter RL, PB Applewhite, AW Galston (1972): Phytochrome-controlled nyctinasty in *Albizzia julibrissin*. V. Evidence against acetylcholine participation. Plant Physiol 50:523–525.

Satter R, AW Galston (1981): Mechanisms of control of leaf movements. Annu Rev Plant Physiol 32:83–110.

Saunders JA, JW McClure (1973): Acetylcholine inhibition of phytochrome-mediated increase in a flavonoid and in phenylalanine ammonia-lyase activity of etiolated barley plumules. Plant Physiol 51:407–408.

Sibaoka T (1966): Action potentials in plant organs. Symp Soc Exp Biol 20:49–74.

Silver A (1974): "The Biology of Cholinesterases." Amsterdam: North-Holland.

Simons PJ (1981): The role of electricity of plant movements. New Phytol 87:11–37.

Smallman BN, A Maneckjee (1981): The synthesis of acetylcholine in plants. Biochem J 194:361–364.

Smith TA (1977): Phenethylamine and related compounds in plants. Phytochemistry 16:9–18.

Spanswick RM (1972): Electrical coupling between cells of higher plants: A direct demonstration of intercellular communication. Planta 102:215–227.

Stevens CF: (1979): The neuron. Sci Am 241:48–59.

Tanada T (1968a): A rapid photoreversibile response of barley root tips in the presence of 3-indoleacetic acid. Proc Natl Acad Sci USA 59:376–380.

Tanada T (1968b): Substances essential for a red, far-red light reversible attachment of mung bean root tips to glass. Plant Physiol 43:2070–2071.

Tanada T (1972): On the involvement of acetylcholine in phytochrome action. Plant Physiol 49:860–862.

Tanada T (1973): Indoleacetic acid and abscisic acid antagonism. I. On the phytochrome-mediated attachment of mung bean root tips on glass. Plant Physiol 51:150–153.

Toriyama H (1975): The effects of acetylcholine and noradrenalin upon the immature sieve tube of root tip. In XII International Botanical Congress, Moscow, p 374 (abstract).

Toriyama H (1978): Observational and experimental studies of meristem of luguminous plants. I. Effects of acetylcholine, red light, and far-red light on protoplasts of root tip meristem. Cytologia 43:325–337.

Toriyama H, MJ Jaffe (1972): Migration of calcium and its role in the regulation of seismonasty in the motor cell of *Mimosa pudica* L. Plant Physiol 49:72–81.

Tung HF, V Raghavan (1968): Effects of growth retardants on the growth of excised roots of *Dolichos lablab* L. in culture. Ann Bot 32:509–519.

Umrath K (1959): Der Erregungsvorgang. In Ruhland W (ed): "Handbuch der Pflanzenphysiologie," Vol 17, Part I, Berlin: Springer Verlag, pp 24–110.

Verbeke M, JC Vendrig (1977): Are acetylcholine-like cotyledon factors involved in the growth of cucumber hypocotyl? Pflanzenphysiol 83:335–340.

Weber M (1981): Untersuchungen zur Bindung von Acetylcholin bei Pflanzen. Master Thesis, University of Mainz.

White JM, CS Pike (1974): Rapid phytochrome-mediated changes in adenosine 5'-triphosphate content of etiolated bean buds. Plant Physiol 53:76–79.

Whittaker VP (1963): Identification of acetylcholine and related esters of biological origin. In Koelle GB (ed): "Cholinesterases and Acetylcholinesterases Agents." Berlin: Springer Verlag, pp 1–39.

Yang SF, NE Hoffmann (1984): Ethylene biosynthesis and its regulation in higher plants. Annu Rev Plant Physiol 35:155–189.

Yunghans H, MJ Jaffe (1970): Phytochrome controlled adhesion of mung bean root tips to glass: A detailed characterization of the phenomenon. Physiol Plant 23:1004–1016.

Yunghans H, MJ Jaffe (1972): Rapid respiratory changes due to red light of acetylcholine during the early events of phytochrome-mediated photomorphogenesis. Plant Plysiol 49:1–7.

Zawadzki, T, K Trebacz (1982): Action potentials in *Lupinus angustifolius* L. Shoots. VI. Propagation of action potential in the stem after the application of mechanical block. J Exp Bot 33:100–110.

Zawadzki T, K Trebacz (1985): Extra- and intracellular measurements of action potentials in the liverwort *Coenocephalum conicum*. Physiol Plant 64:477-481.

Second Messengers in Plant Growth
and Development, pages 289–313
© 1989 Alan R. Liss, Inc.

12. The Potential for Second Messengers in Light Signaling

Benjamin A. Horwitz

Department of Biology, Technion–Israel Institute of Technology, Haifa 32000, Israel

ADVANTAGES OF LIGHT AS AN INDUCER

Light is a well-defined physical stimulus, perceived by photoreceptor molecules that undergo photochemical reactions leading to a (relatively) stable photoproduct. In contrast to chemical inducers, light can be instantaneously switched on and off. This unique property makes it possible to study independently the biochemical reactions induced by a pulse of light and the initial photoevents. With the possible exceptions of fluorescence, phosphorescence, and chemiluminescence, light stimuli are extracellular.

I shall refer to the first step in the transduction chain after the photoproduct as the *second* messenger. One absorbed photon can produce, at most, one molecule of photoproduct. Each photoproduct molecule might catalyze the formation or mobilization of many messenger molecules, so the second messenger could act as the first amplification step in the response chain. Our knowledge of this chain is still incomplete, and thus many of the second messengers discussed here could really be third or later in the sequence of reactions initiated by light.

Light, often at levels too low to drive significant photosynthesis, has many regulatory functions in plants and fungi. Phototropism of higher plants and of fungi, photomovement of some microorganisms, sporulation of many fungi, induction of carotenoid synthesis in fungi, morphogenesis of *Acetabularia,* rapid changes in rates of stem growth, and induction of planar growth in fern gametophytes are typical of responses to blue and near-ultraviolet light. The photoreversible pigment phytochrome, whose two forms have absorption maxima in the red and far-red regions of the visible

Address reprint requests to Dr. Benjamin A. Horwitz, Department of Biology, Technion–Israel Institute of Technology, Haifa 32000, Israel.

spectrum, mediates seed germination, induction of flowering, phototropism in mosses, growth rate in etiolated and green plants, elimination of the lag phase in greening of angiosperms, anthocyanin synthesis, movement of chloroplasts within cells, movement of leaves, enzyme activities, and expression of specific genes. In contrast with blue light receptors, both the chromophore and apoprotein of phytochrome are well characterized. Many responses are jointly controlled by red and blue light, acting together or sequentially. The above are only a few examples. The literature on photomorphogenesis has been extensively reviewed (Mohr, 1972; Johnson, 1982; Shropshire and Mohr, 1983; Kendrick and Kronenberg, 1986; for emphasis on blue light, Presti and Delbrück, 1978; Senger and Briggs, 1981; Senger, 1987).

It is reasonable to consider the earliest detectable effects of light as the most primary of the response chain. Following this view, many rapid responses to light were discovered. The rapid actions of phytochrome have been reviewed by Quail (1983), and early effects of blue light have been discussed by Horwitz and Gressel (1987). In many ways, the study of these early effects is synonymous with studies of second messengers in other systems. A substance (or cell property) that changes quickly in response to light and influences a physiologically relevant cell process fits a (broad) definition of a second messenger. Most of the messengers discussed in this volume could, in principle, play some role in light signaling. The following sections cover a sampling of photoresponses and messenger candidates for which there has been recent progress.

CANDIDATES FOR MESSENGERS ACTING IN PHOTORESPONSES
Modulation of Growth Rate and Direction: Phytohormones

Light can modify the direction of growth (phototropism) or its rate (light–growth response). At the output (growth), phytohormones act as extracellular messengers, transmitting a response perceived in one region to another region of the organ where it is expressed. Growth regulators might act as intracellular second messengers when a light signal is perceived in the same cells that respond (Johnson, 1982; Blowers and Trewavas, Chapter 1). Diffusible growth regulators may mediate phototropism and growth rate changes in higher plants and, perhaps, in *Phycomyces* sporangiophores as well. The following examples illustrate photoresponses that occur within minutes or seconds and are sufficiently rapid to establish a time frame within which to seek second messengers.

Higher plant phototropism. The Cholodny-Went hypothesis for phototropism explains asymmetric growth on the basis of asymmetric auxin (indole-3-acetic acid, IAA) redistribution following exposure to unilateral (blue) light. Central to the model, at least in its original version, is the idea that redistribution at the apex is preserved as auxin is transported from the apex to the growing zone. Although much accepted, this model for phototropism has come under critical scrutiny (Firn and Digby, 1980). The main aspects questioned are the role of the apex, the need for a redistribution of auxin (or other growth effectors), and whether the observed auxin redistribution is sufficient to account for bending. A theoretical calculation showed that at least in *Avena* coleoptiles, the Cholodny-Went theory cannot be faulted on the basis that the auxin concentration differences are too low to cause bending (Gressel and Horwitz, 1982).

Recent work with corn coleoptiles supports the Cholodny-Went model (Iino and Briggs, 1984; Baskin et al., 1986). Following unilateral blue illumination, there was a redistribution of growth, increased on the shaded side and decreased on the lighted side of the coleoptile. In addition, asymmetric application of auxin mimicked light or canceled the effect of unilateral light applied to the illuminated side. The most direct measurement possible would be to follow the local, free IAA concentrations in all parts of the coleoptile before and during the bending response. The growth and bending rates predicted from the endogenous IAA levels could then be compared with those actually observed. Such an ideal measurement is, of course, impossible. Even if practical, the conclusions would be limited by the fact that some IAA is conjugated and presumably unavailable as a growth regulator (Bandurski et al., 1986).

The experiments with applied auxin indirectly provide some information on the internal IAA concentration: It is in a range that allows the coleoptile to respond in a sensitive manner to small changes in auxin (Baskin et al., 1986). Coleoptiles are a rather special tissue type, belonging only to germinating grass seedlings. Phototropism in other plant tissues, especially dicot stems, or other light conditions (photosynthetically competent plants) may make use of different mechanisms.

Phycomyces. The constraints on second messengers, however, are well illustrated in the giant unicellular sporangiophores of the Zygomycete fungus *Phycomyces* (Fig. 1A). To explain the short lag times, one model would be direct modulation of a growth-limiting enzyme by light. Growth, chitin synthetase activity, and cell wall extensibility are sequentially altered (Fig. 1A). The results are consistent with a hypothetical chain of events in which light leads to loosening of the *Phycomyces* sporangiophore cell wall, followed by acceleration of growth, activation of chitin synthetase, increased rigidity of the wall, and slower growth (Herrera-Estrella and Ruiz-Herrera, 1983). Activation of chitinase, which would certainly be expected to make

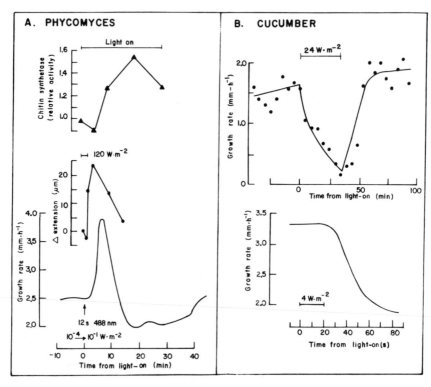

Fig. 1. Rapid effects of blue light on growth rate. **A:** Light-growth response of *Phycomyces* sporangiophores; **bottom curve:** response to pulse, recorded with an automated tracking machine (Lipson and Block, 1983): **middle curve:** mechanical extensibility of the sporangiophore (Ortega and Gamow, 1976); **top curve:** activity of chitin synthetase in sporangiophore extracts (Herrera-Estrella and Ruiz-Herrera, 1983). Even though the three curves are not strictly comparable, they all demonstrate a rapid response to light. **B:** Light inhibition of growth of cucumber hypocotyls. **Top:** partially greened seedlings, data of Gaba and Black (1979). **Bottom:** cucumber seedlings grown in total darkness. Early kinetics were resolved, showing lag of about 20 sec (Cosgrove, 1981). If a dark-grown seedling was continuously irradiated with blue light, its growth rate decreased further, with a time course similar to that in the upper curve.

the wall less rigid, was inferred from an increase of the number of reducing ends in the chitin chains. The various parameters in Figure 1A have never been measured together under the same growth conditions and light stimuli. In fact, the activation of chitinase has been inferred only indirectly, and it appears significant only by 10-min illumination (Herrera-Estrella and Ruiz-Herrera, 1983). The extensibility of the sporangiophore increases well before this (Fig. 1A).

The chitinase change thus may not be the first biochemical result of illumination. A growth regulator or another cell-wall-strengthening or loosening enzyme could act between light and chitinase. The existence of a regulator that can be redistributed by a unilateral light stimulus has been inferred from the different kinetics of light–growth and bending responses in *Phycomyces* (Galland et al., 1985). If no growth regulator were involved, one would expect the phototropic response to be merely a linear superposition of local light–growth responses. The chitin synthetase and chitinase changes (Herrera-Estrella and Ruiz-Herrera, 1983) were lacking in photomutants, evidence that they belong on the response chain. More work on chitinase and chitin synthetase, under well-controlled light conditions, is needed. However, correlations between growth changes and the two enzymes make the *Phycomyces* system a good one for investigating the nature of second messengers between light perception and enzymatic responses.

Higher plants: growth-rate changes. A wide range of light–growth responses have been observed in higher plant seedlings, with time courses varying from hours to seconds (Cosgrove, 1986). Examples of both slow and fast responses are given here. In a slow (hours) response, elongation of the mesocotyl of etiolated corn seedlings is inhibited by red light acting through phytochrome. This inhibition in intact seedlings can be explained in part by a decrease in auxin supply (mostly from the coleoptile), as shown by time-course and fluence–response studies. An additional direct effect of red light on the mesocotyl itself appeared to be independent of auxin levels (Iino, 1982).

Figure 1B illustrates one of the fastest known responses to (blue) light. As with the corn mesocotyl, the growth of cucumber hypocotyls is inhibited by light. The lag of only 20 sec for the lower curve in Figure 1B puts an upper limit on the time within which second messengers, if they exist, must act. Here, the time interval is probably too short for diffusion and action of hormones. It will be worthwhile to study those light-induced biochemical changes in cucumber stems that occupy this narrow 20-sec time window.

Turgor-Mediated Movements: Membrane Events

Rapid reorientation of leaves and stems, and the control of stomatal opening, are regulated by turgor pressure in specialized cells. These turgor responses are directly related to the flow of water in response to solute redistribution. Thus it is not surprising that attempts to understand turgor responses have focused on membrane properties.

Stomata. The opening of the stomates in the epidermis of a leaf depends on a combination of stimuli, one of which is light (Zeiger, 1983). Stomatal opening in response to light has two components. One is restricted to the blue region of the spectrum, and the other matches the absorption spectrum of chlorophyll. Increased osmotic potential in the guard cells leads to water uptake, increased turgor, and widening of the stomatal pore. The increase in osmotic potential results from uptake of K^+. Uptake of Cl^- and malate synthesis inside the cell preserve electroneutrality. Light apparently does not directly provide energy for K^+ uptake, but, rather, activates proton extrusion, which creates an electrochemical gradient for passive uptake of K^+.

Patch-clamp studies provide an elegant method with which to study the plasma membrane and to alter at will the composition of the solution on either side of the membrane. Whole-cell patch-clamp measurements on *Vicia* guard cell protoplasts showed a transient membrane hyperpolarization after a pulse of blue light (Assmann et al., 1985). When the membrane potential was clamped at -50 mV, a blue-light-stimulated outward positive current was still observed, bearing strong evidence for an electrogenic pump. If the outward current is carried by the H^+ ions responsible for acidification, the current should be proportional to the rate of acidification. The agreement between the rate of H^+ release and the time course of outward current is excellent (Fig. 2). Estimates of the peak magnitude of the current from the rate of acidification also fit the hypothesis that outward current is proportional to the acidification rate (Shimazaki et al., 1986).

A proton ATPase may be responsible for the light-mediated proton current. Evidence for this comes from studies with the plasma membrane ATPase inhibitor vanadate. Vanadate blocks light-induced acidification of the medium by epidermal peels (Gepstein et al., 1982). A high vanadate concentration (1 mM) was required, as expected, in view of its limited penetration into the cell. Only guard cells remain viable after peeling the epidermis, so that the acidification was attributed to the guard cells. The vanadate data suggested that the light-induced proton pumping could be attributed to a plasma membrane ATPase. Work with guard cell protoplasts confirmed that the H^+ efflux belongs to the guard cells (Shimazaki et al., 1986).

Vanadate, however, was ineffective in inhibiting the light-induced proton efflux from protoplasts. Perhaps vanadate was not taken up by the protoplasts, the cell surface of which might be different from that of guard cells in epidermal peels. In a patch-clamp study on the red-light portion of the response, an outward proton current was detected, as was found for blue light (Serrano et al., 1988). In this case, vanadate (at micromolar concentrations) blocked the response when applied inside the patch-clamp pipette. Vanadate has not yet been applied in this way to block blue-light-induced proton flow, but it is likely that the patch clamp technique will completely circumvent permeability problems.

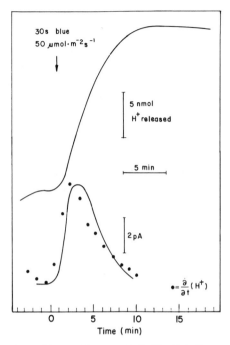

Fig. 2. Ionic current as a possible second messenger for blue light in stomatal guard cells. *Vicia faba* leaf stomatal guard cell protoplasts were studied using two different methods: **Upper curve:** H^+ efflux measured in the medium (Shimazaki et al., 1986); **lower curve:** whole-cell-membrane current at fixed membrane potential, measured with the patch-clamp technique (Assmann et al., 1985). The circles on the membrane current curve are values of the derivative with respect to time of the H^+ curve (tangent estimated graphically) plotted relative to the maximum.

The biochemical coupling between absorption of light and activation of the H^+ pump is still unknown, but it is quite clear that ions do act as second messengers for the blue light signal in stomatal opening. One way in which light might be coupled to the proton pump is through a light-modulated electron transport chain. Potential second messengers are provided by electron and/or proton movements via plasma membrane electron transport chains, for example, Crane, Chapter 5.

Pulvini: inositol phosphates as second messengers. Leaf movements in some plants are mediated by the turgor in cells of the pulvinus, a specialized motor organ at the base of the petiole (or rachis in a compound leaf). *Samanea saman,* a tropical legume that folds its paired leaflets at night, is an example. Ions, principally K^+ and Cl^-, play a central role in the interaction of light (red or blue) with the circadian control of leaf opening (Satter et al.,

1974). Recent work with *Samanea* has focused on what might be one of the steps between light absorption and the control of ion distribution, which, in turn, modulates the volume of the pulvinar cells through the movement of water.

Increased phosphatidylinositol (PI) turnover accompanied by an increase in inositol-1,4,5,-trisphosphate (IP_3) and diacylglycerol (DAG) may play an essential role in signal transduction in plants, as in animals (see Boss, Chapter 2). In *Samanea*, PI turnover may be part of the phototransduction chain. PI, phosphatidylinositol-4-monophosphate (PIP), phosphatidylinositol-4,5-bisphosphate (PIP_2), and the aqueous breakdown products inositol-1-phosphate (IP), inositol-1,4-bisphosphate (IP_2), and IP_3 have been identified in extracts of *Samanea* pulvini prelabeled with [^{3}H]-myo-inositol (Morse et al., 1987a; Coté et al., 1987). A short (15–30 sec) pulse of white light changes the chromatographic profile of both the polyphosphoinositides and the water-soluble inositol phosphates, by decreasing PIP and PIP_2 and increasing IP_2 and IP_3 (Morse et al., 1987b). Thus the PI cycle may function similarly in plants and in animals.

Microinjection of a second messenger into an animal cell can sometimes elicit the response in the absence of the stimulus. Injection of IP_3 into ventral photoreceptors of the horseshoe crab *Limulus* mimicked several electrophysiological effects of light (Brown et al., 1984). Application of exogenous IP_3 (if the experiment is technically feasible) might substitute for light in *Samanea* also. If such an experiment is successful, the hypothesis that IP_3 is a second messenger would be still further strengthened. This approach might be limited if DAG is required as well, as DAG might be difficult to apply to its endogenous membrane site of action.

In the *Limulus* visual system, microinjection of the Ca^{2+} buffer EGTA along with 1,4,5-IP_3 blocked the electrical response normally associated with injection of 1,4,5-IP_3 alone (Rubin and Brown, 1985). Injection of EGTA alone did not block a light-induced electrical response. The authors suggest that a light-induced increase in 1,4,5-IP_3 probably does not belong on a linear chain of responses mediating excitation. Another possibility is that it is calcium that does not belong on the excitation chain. In plants, the phytochrome system provides an additional "handle." One can ask whether far-red light can reverse the response even after an IP_3 change has occurred. In *Samanea* pulvini, as in the *Limulus* photoreceptor, light causes membrane potential changes, and thus might be a plant system amenable to such experiments. Far-red reversibility and the contribution of blue-light receptors have not yet been studied in *Samanea*. Further experiments on the PI cycle could extend the findings of Morse et al. (1987b) to other systems, such as stomatal guard cells, the blue-light-sensitive pulvini of the sun-tracking leaves of *Lavatera cretica* (Schwartz and Koller, 1978), or to responses that do not involve turgor at all.

Light-Induced Absorbance Changes (LIAC): Coupling Between Irradiation and Electron Transport

The photochemical phenomena reviewed in this section are examples of a second messenger in search of a response. Intense blue irradiation can mediate redox reactions, which, with some exceptions, can be attributed to flavin-mediated cytochrome b reduction. Such photoresponses are detected by light-minus-dark difference spectroscopy in vivo or in vitro and are thus referred to as "LIACs" (LIACs have been recorded in vivo in the blue-light-sensitive fungi *Neurospora, Phycomyces,* and *Trichoderma,* in wheat and corn coleoptiles in vivo, and in membrane fractions from *Neurospora* and corn (review: Senger and Briggs, 1981; Gressel and Rau, 1983). In corn coleoptiles the herbicide acifluorfen enhanced phototropism and LIAC, but not gravitropism (Leong and Briggs, 1982). In *Trichoderma,* acifluorfen also enhanced a blue light response, induction of sporulation (Gaba and Gressel, 1987). The herbicide even increased sporulation when applied 30 min after the inductive light pulse. Acifluorfen thus interacted with the dark reactions following photoinduction. *Trichoderma* LIAC measured at 440 nm decayed rapidly (within 1 min, Horwitz et al., 1986), so that acifluorfen could not have interacted with the LIAC 30 min after the light. The effect of acifluorfen on the *Trichoderma* LIAC has not yet been studied.

Blue-light-induced electron flow could be coupled to proton flow across the membrane, leading to charge separation, depending how the electron carriers are arranged in the membrane. In an in vitro model, this is photochemically feasible (Schmidt, 1984). The model consisted of artificial membrane vesicles to which flavins were incorporated by attaching them covalently to hydrocarbon chains. The vesicles were loaded with an electron acceptor (cytochrome c), and an electron donor was supplied outside. On illumination with blue light, redox equivalents were carried across the membrane.

In short, no blue-light response has been shown conclusively to depend on a LIAC, but the biochemical possibility is there. LIACs seem quite universal in tissue extracts, which always contain flavins and cytochromes. Thus, one approach to demonstrate physiological relevance of LIACs may be to show their absence or alteration in photoresponse mutants. Altered flavin/cytochrome LIAC was found for a *Neurospora* mutant with a b-type cytochrome defect (Brain et al., 1977) and *Trichoderma* photosporulation mutants (Horwitz et al., 1986) but not *Phycomyces* phototropism mutants (Lipson and

Presti, 1977). A possible course to follow would be to attempt to isolate specific flavocytochromes, which are altered in the mutants. The plasma membrane is a likely subcellular fraction in which to look, but it is certainly not the only possibility.

Electrical Potentials and Currents as Messengers

Electrical currents and voltages might, in themselves, be a causal step in photoresponses. There is no lack of reports of rapid electrical changes following a light stimulus. A few examples of such studies are given below (for review of other work, see Racusen and Galston, 1983; Horwitz and Gressel, 1987; Blatt, 1987). Most reports fall into two major classes: studies of currents in the medium surrounding submerged organs or cells and studies of membrane potentials. Measurement of the current has the advantage that spatial mapping of the current indicates heterogeneity in the membrane, presumably in the distribution of pumps and ion channels. Sites of outward or inward current can predict subsequent development. This led to the hypothesis that the electrical pattern causes the developmental pattern (Jaffe and Nuccitelli, 1977). A difficulty has always been to find a biochemical link between rapid electrical changes and morphogenesis, which is generally much slower. Clear evidence that an exogenous electric current can replace light in a plant is still lacking.

Recent work linking light modulation of electrical parameters with morphogenesis made use of two lower plants. In the siphonaceous alga *Vaucheria sessilis,* a blue-light-dependent outward current induced by local irradiation is associated with chloroplast aggregation (Blatt et al., 1981). Such sites of outward current correspond well with the sites of subsequent branching, also induced by local blue irradiation (Weisenseel and Kicherer, 1981). The correlation in time and location fits the idea that the current may actually cause the branching. In gametophytes of many ferns, red light allows filamentous growth, while blue light stimulates bidimensional growth. Red light depolarized while blue light hyperpolarized membrane potential (Racusen and Cooke, 1982; Cooke et al., 1983). As in *Vaucheria,* the effects of light on electrical parameters and on morphogenesis were consistent. If the currents or voltages act as messengers between light and response, it should be possible to block development by canceling the endogenous electric current with an externally applied one.

There is also evidence for electrical controls in the photoperiodic induction of flowering. The sensitivity of spinach leaf mesophyll cell membrane potentials to white light increased following transfer of the plants from short-day to inductive (continuous light) conditions (Montavon and Greppin,

1985). The light-fluence rates (20 Wm^{-2}) were certainly high enough to allow photosynthesis, so it will be interesting to see whether the wavelength dependence will point to phytochrome or photosynthetic pigments and whether inductive night-break light pulses also can produce the same electrical modifications.

One of the clearest examples of electrical potentials as second messengers is found in the cyanophyte *Phormidium*. This organism responds to a step-down in irradiance by stopping its movement, then gliding off in the reverse direction. The light signal, perceived via photosynthetic pigments, induces a potential change, and the photoinduced reversal of movement can be simulated by an appropriate external electric field (Häder, 1981). Such experiments, analogous in some ways to microinjecting a putative second messenger like IP_3 into a cell, are more difficult to investigate in the context of the more complex photomorphogenetic responses outlined in the previous paragraph. Yet work on "simple" unicellular models (see Blatt, 1987) may suggest mechanisms acting in the complex multicellular ones.

Light Modulation of Calcium Levels, and the Role of Calmodulin

There is considerable evidence that Ca^{2+}, together with the regulatory protein calmodulin, is an exceedingly important second messenger in plant cells (Blowers and Trewavas, Chapter 1; Marmé, Chapter 3; Hepler and Wayne, 1985). The role of Ca^{2+} in photomorphogenesis and photomovement has recently been reviewed (Hepler and Wayne, 1985; Blatt, 1987). Evidence for the calcium concentration as a messenger in light signaling will be discussed below for three systems. 1) Light-mediated Ca^{2+} fluxes across chloroplast membranes show that plant membranes can transduce light-to-calcium changes. 2) The filamentous green alga *Mougeotia* is a convenient model for calcium studies. 3) Studies on germination of spores of the fern *Onoclea* allowed temporal separation between photoinduction and calcium transport.

Ca^{2+} as a regulator of photosynthesis. Intact, isolated chloroplasts of wheat and spinach take up Ca^{2+} when illuminated with bright broad-band light; the uptake is reversible in the dark (Muto et al., 1982). In these studies, Ca^{2+} fluxes were followed with either $^{45}Ca^{2+}$ or a Ca^{2+}-specific electrode. Clearly, light can be coupled to Ca^{2+} fluxes across plant membranes. The role of these changes in intact plants is not yet known. One possibility is photocontrol of the supply of $NADP^+$ for photosynthesis via calmodulin-activated NAD^+ kinase (Black and Brand, 1986, Cormier et al., 1985). Another important role could be control of protein kinases. Although action spectroscopy has not been done, the photoreceptors for these effects are probably chlorophylls and associated pigments.

Mougeotia. Chloroplast rotation in the filamentous green alga *Mougeotia* (Haupt, 1982; Wagner and Grolig, 1985; Blatt, 1987) is a convenient system for the study of the role of Ca^{2+}. The organism is aquatic, so the flux between inside and outside is relatively easy to measure, and inhibitors can be applied in the bathing solution. Also, in *Mougeotia,* there is an obvious role for Ca^{2+}: activation of the actin–myosin contractile system, leading to movement of the chloroplast (see Wagner and Grolig, 1985). In other systems, for example, spore germination, the calcium concentration may change, but the identities of the target enzymes for Ca^{2+}/calmodulin are still only speculative.

Red, far-red light reversible $^{45}Ca^{2+}$ influx into the filaments was demonstrated by autoradiography (Dreyer and Weisenseel, 1979). These observations indicated that the source for calcium is probably outside the cell. Fluorescent staining showed that the density of calcium-containing vesicles near the edge of the chloroplast increased within 1–5 min after induction of reorientation by red light (Wagner and Grolig, 1985). Far-red reversal of this effect has not yet been reported. The role of intracellular sequestering of Ca^{2+} in the rotation response now needs to be considered as well. It is not clear whether light controls the amount of Ca^{2+} sequestered in the vesicles.

A protein with all the characteristics expected for calmodulin was isolated by affinity chromatography on a column to which a calmodulin-binding phenothiazine drug had been attached (Wagner et al., 1984). Another calmodulin-binding phenothiazine interfered with the chloroplast rotational response, as did a hydrophilic calmodulin antagonist. The use of antagonists from two different chemical classes, and their low effective concentrations, support the conclusion that the inhibition was specific. Because the response can only be studied in intact cells, it is difficult to check the more stringent specificity requirement for enzyme work, namely, that the inhibitor did not affect the basal (calmodulin-independent) activity (Gietzen and Bader, 1985). Local application of a calcium-specific ionophore to *Mougeotia* filaments promoted chloroplast rotation, replacing red light (Serlin and Roux, 1984). This result is strong evidence that Ca^{2+} is a second messenger in *Mougeotia* chloroplast rotation. Perhaps a calcium-specific microelectrode could be used to show that Ca^{2+} influx actually precedes the onset of chloroplast rotation.

Onoclea spore germination. Light triggers the germination of spores of the fern *Onoclea sensibilis*. In contrast to *Mougeotia* chloroplast rotation, this phytochrome-mediated response is not rapid; germination is assayed 48 h after the photoinductive red-light treatment. It would thus seem more difficult to test the involvement of Ca^{2+} as a second messenger, yet several lines of evidence implicated Ca^{2+} in the early steps after light (Wayne and

Hepler, 1984). In fact, the long response time allowed uncoupling of Ca^{2+} transport and phytochrome phototransformation. In the absence of Ca^{2+}, the spores "remembered" the induction by red for 4 hr. The *Onoclea* example will be discussed in detail below, because it illustrates rather well how the properties of the phytochrome system can be used to obtain information about the signal transduction chain.

External Ca^{2+} was required for germination of spores prewashed with the chelator EGTA to remove cell-wall-bound Ca^{2+}. Germination followed a saturation curve, with a threshold at 1 μM calcium and saturation above 10 μM. These concentrations are in the range expected for activation of calmodulin (assuming, of course, that the Ca^{2+} concentrations reached inside the spore are similar to the applied ones). The inorganic Ca^{2+} antagonists La^{3+} and Co^{2+} inhibited germination in the presence of 1 mM Ca^{2+} (50% inhibition at about 10 μM La^{3+} or 0.1–1 mM Co^{2+}). The inhibition could be reversed by increasing the external Ca^{2+} concentration. La^{3+} needed to be present just before or during the 5-min inductive red irradiation (Fig. 3).

This result suggests that a transient influx of Ca^{2+} is part of the signal chain. Furthermore, if Ca^{2+} was present before red light was given, no far-red light reversal could be detected. The effect of red was reversible to 50% in the absence of Ca^{2+}. After 30 min in the dark, the action of red could no longer be reversed by far-red light (Fig. 3). The simplest interpretation of this "escape" from reversibility is that the action of the far-red absorbing form of phytochrome (Pfr) is complete within 30 min. According to this view, after Pfr has completed its action, it is no longer required, and its removal by far-red light cannot reverse the response. In the presence of Ca^{2+}, the action of Pfr is apparently completed within minutes. Under this interpretation for the "escape," Ca^{2+} added 4 hr after illumination must be taken up by a mechanism set in place by Pfr but which does not require the presence of Pfr. The process that escapes from far-red reversal over 30 min does not seem strictly Ca^{2+} dependent, because expression of the response is normal with addition of Ca^{2+} 4 hr after illumination.

Two phenothiazine drugs inhibited germination, implicating calmodulin (to the extent that the inhibitors are specific). As mentioned above for *Mougeotia,* the strongest case for specificity can be made if the calmodulin antagonists do not inhibit the basal (non-calmodulin-dependent) activity. For *Onoclea* spore germination, this important control experiment was done, with the result that dark germination was not inhibited. Unfortunately, the dark germination levels were very low (about 3%), even without the inhibitors. An ideal system would be one in which there is significant basal (calcium and light-independent) activity.

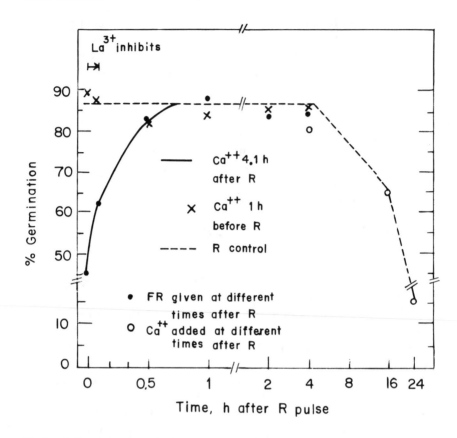

Fig. 3. Ca^{2+} and phototransduction in light-induced germination of spores of the fern *Onoclea sensibilis*. Germination was induced by a saturating pulse of red light (R). Ca^{2+} need not be present at the time of irradiation in order to reach the maximum level, but it must be added within 4 hr. After 4 hr, the "memory" is lost (o). The effect of saturating R can be partially reversed by far-red light (FR), provided Ca^{2+} is added 4 hr after light. The solid line shows the escape from reversibility (●). The escape is apparently very rapid if Ca^{2+} is already present before R (x). La^{3+}, a Ca^{2+} antagonist, inhibits germination only during the first 5 min from the start of R irradiation, suggesting a rapid Ca^{2+}-requiring step. Note linear time scale below, log scale above 1 hr. Combined data from Figures 5 and 6, and Table 4 of Wayne and Hepler (1984); see text for further details.

A major unaddressed problem in most studies of Ca^{2+} and light is the measurement of intracellular free Ca^{2+} concentrations, as quickly as possible after light stimulation. For example, in detailed studies on the effects of external Ca^{2+} on *Chlamydomonas* phototaxis (Morel-Laurens, 1987), intracellular Ca^{2+} concentrations are still the missing parameter. Internal free

Ca^{2+} concentrations would be difficult to measure in fern spores. Large algal cells should be more amenable to such measurements. Selective micro-electrodes and a fluorescent indicator were used to detect a gradient of cytoplasmic free Ca^{2+} within rhizoid cells of the marine brown alga *Fucus* (Brownlee and Wood, 1986). Developing zygotes of *Fucus* and *Pelvetia* were among the first organisms for which extracellular electrical currents were measured (Jaffe and Nuccitelli, 1977). Light is used as the stimulus to polarize the cells, but little photobiological work has been done to correlate the ionic changes with illumination or in search of the photoreceptors.

Protein Phosphorylation

Activation of specific protein kinases is an important signal in animal and plant cells (Blowers and Trewavas, Chapter 1). There is at least one clear example to show that phosphorylation can mediate an important effect of light in plants. This is summarized in the next section, after which I will discuss data for phytochrome.

Thylakoid protein phosphorylation. Phosphorylation of several thylakoid proteins was detected when isolated intact chloroplasts were incubated in the light (Bennett, 1983; Millner et al., 1986). The 24 kDa and 26 kDa light-harvesting chlorophyll a/b binding proteins (LHCP) of photosystem II (PSII) are among the major phosphoproteins (Fig. 4). The protein kinase that phosphorylates the LHCP is sensitive to the redox state of the plastoquinone pool. The redox state of the plastoquinone pool is, in turn, a measure of light quality, because the pigments transferring energy to photosystem I (PSI) absorb, on the average, at longer wavelengths than those of PSII. PSII fluorescence at room temperature was decreased by LHCP phosphorylation, that is, when there is more flow of excitation energy to PSI. At 77°K, fluorescence from PSI can be directly seen as a long-wavelength emission peak. Again, increased energy transfer to PSI was found with phosphorylated membranes. Bennett (1983) showed that the red (R, λ_{max} 645 nm), far-red (FR, λ_{max} 714 nm) reversible phosphorylation of the LHCPs also occurs in vivo, in detached barley leaves incubated with ^{32}P-ortho-phosphate (Fig. 4). Taken together, these results are good evidence that protein phosphorylation transduces a light signal (ratio between far-red and red light) to a biochemically meaningful change (redistribution of energy between the photosystems). As with phytochrome, red and far-red light have opposing effects. The analogy is superficial, though, because experiments could be designed to distinguish between phytochrome and photosynthetic effects, with the help of action spectroscopy, bleaching with herbicides, or mutants. Plants appear to have evolved more than one way to measure the red/far-red light ratio. There is also an intriguing, untested

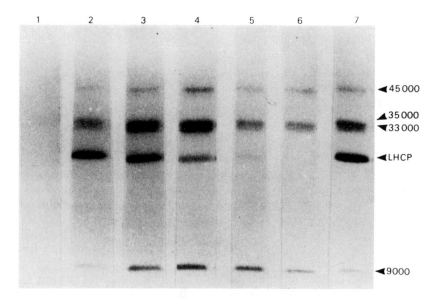

Fig. 4. Light-controlled phosphorylation of chloroplast polypeptides in barley leaves. Detached leaves were allowed to take up ^{32}P-orthophosphate under different light treatments prior to electrophoresis and autoradiography. **Lane 1:** 4 hr under 714 nm light (Far-red light [FR], preferentially absorbed by photosystem I [PSI]). **Lane 2:** 4 hr FR followed by 10 min 645 nm light (red light [R], preferentially absorbed by photosystem II [PSII]). **Lane 3:** 4 hr R. **Lanes 4–6:** the leaves received 4 hr R, followed by 8 min, 16 min, or 32 min FR. **Lane 7:** corresponds to 4 hr R, followed by 32 min FR, then 10 min R. Note specific photoreversible phosphorylation; the light-harvesting chlorophyll a/b binding protein (LHCP) is more rapidly dephosphorylated under 714 nm light than are the 33- and 35-kDa phosphoproteins, while phosphorylation of the 9 kDa polypeptide shows yet a different pattern. (Reproduced from Bennett, (1983), with permission of the publisher.)

possibility that the chloroplast protein kinase might be Ca^{2+}, calmodulin regulated.

Phytochrome-mediated protein phosphorylation. An attractive hypothesis would be that light acting via phytochrome is also transduced by phosphorylation of specific proteins. The phosphorylation status of several unidentified proteins in etiolated oats is rapidly modulated by phytochrome (Otto and Schäfer, 1986).

Several proteins in nuclei isolated from plumules of etiolated peas were reversibly phosphorylated by red and far-red light (Datta et al., 1985). In contrast with the situation with chloroplasts (Fig. 4), the intensity of all the labeled bands was about equally modulated by the light treatments. This makes it difficult to tell whether light triggered phosphorylation of specific proteins or merely altered the total photophosphorylation per unit protein loaded on the gel. Copurification of a protein kinase with phytochrome certainly complicates the picture. Ideally, one would like to measure transcription of specific light-regulated genes as a function of the phosphorylation state of the proposed regulatory proteins. A more feasible experiment would be to show that nuclear proteins are reversibly phosphorylated in vivo, in a plant where uptake of radioactive phosphate is not a problem. The results could then be compared with nuclear runoff transcription for known light-regulated genes.

Phytochrome can be phosphorylated, and there is now some evidence that it is itself a protein kinase. Three mammalian protein kinases, as well as an endogenous polycation-dependent protein kinase, catalyzed phosphorylation of *Avena* phytochrome in vitro (Wong et al., 1986). Mammalian kinases C and G preferentially phosphorylated the Pr form of phytochrome. The endogenous kinase copurified with phytochrome itself. These results provide a useful probe for light-dependent structural rearrangements, but, of course, cannot yet show whether light-dependent phosphorylation of phytochrome is important in the responses of the intact plant.

Cyclic Nucleotides and G Proteins

The search for plant messengers has often been strongly guided by the data from animal systems. The presence in plants of cAMP, the molecule that led to the term *second messenger* in animal cells, has been reported many times, but its role in signaling is not clear. All attempts have failed to demonstrate cAMP-dependent protein kinase in plants (Blowers and Trewavas, Chapter 1; Hepler and Wayne, 1985). The history of reports of light-induced changes in cyclic nucleotide levels in plants and fungi is one of controversy (for the latest, compare Shaw and Harding, 1987; Hasunuma et al., 1987).

Perhaps the most promising plant system in which to look for homologues of animal cyclic nucleotide signaling pathways is the unicellular flagellate *Chlamydomonas*. The photoreceptor for phototaxis in this alga is a rhodopsin (Foster et al., 1984). In vertebrate vision, cGMP opens sodium channels in the retinal rods, and the protein (transducin) linking excited vertebrate rhodopsin to cGMP phosphodiesterase activity (and thus to sodium channel closure) has also been identified (Stryer, 1986). Transducin, a GTP-binding protein (G protein), is homologous to signal-transducing proteins found in other animal systems. A recent report of G proteins in extracts of the duckweed *Lemna* (Hasunuma and Funadera, 1987) is limited by the lack of gel separations of the putative labeled G proteins. Cholera toxin-catalyzed

ADP-ribosylation (Cassel and Pfeuffer, 1978) would be a preferable way to label G-proteins.

At present, there is too much conflict in the literature to strongly implicate either cAMP or cGMP as a second messenger in plant or fungal photoresponses. The search for a G protein in plants is just now beginning.

Phytochrome as its Own Second Messenger

It is possible that phytochrome acts in the nucleus (Mohr, 1972). One way in which this hypothesis can now be phrased in molecular terms is as follows: does Pfr bind to specific DNA sequences and directly modulate transcription? A less direct mechanism would be for Pfr to interact with other factors (proteins?) that affect transcription or stability of specific mRNAs. [^{3}H]-UTP incorporation into the mRNA fraction of nuclei isolated from etiolated plants was increased by the addition of purified 124 kDa rye phytochrome as Pfr (Ernst and Oesterhelt, 1984). The red-absorbing form of phytochrome Pr had only a very weak effect. Bovine serum albumin and cytochrome c slightly decreased incorporation, and partially degraded (114/118 kDa) Pfr had no effect. Furthermore, Pfr only stimulated incorporation by nuclei from dark-grown seedlings. The stimulation seems specific to Pfr, although bovine serum albumin, cytochrome c, and Pr seem a rather small set of representatives for the multitude of physical properties exhibited by proteins.

Working with barley nuclei and either oat or rye phytochrome, Mösinger et al. (1987) carried this analysis further. A pulse of red light in vivo caused transient changes in the transcription rates for specific genes by the subsequently isolated nuclei. A second pulse, given after the effect of the first had begun to decay, increased the duration of the transient changes in transcription. Purified phytochrome (Pfr) was added in an attempt to simulate the second pulse. Exogenously added oat phytochrome (Pfr) simulated the second in vivo irradiation: transcription of LHCP mRNA sequences increased, and transcription of NADPH-photochlorophyllide oxidoreductase sequences decreased. Positive and negative modulation of these two different genes in the same assay is the strongest evidence that the in vitro assay may be physiologically relevant. Still, one may question the high concentrations of Pfr required (250 μg/ml), much higher than that found in nuclei. It would be important to try to follow the Pfr concentration in nuclei (by immunoassay) following red pulses at different fluences. The rapid relocalization of phytochrome in the cells of dark-grown seedlings after a pulse of red light (Speth et al., 1986) gives no clue to action in the nucleus. The second red-light pulse (in vivo) was effective, even when given less than a minute prior to extraction of the nuclei, and could be reversed by far-red light (Mösinger et al., 1987), leaving little time for a second messenger to act (Schäfer and Briggs, 1986). These observations do not fit easily into the general view outlined in this chapter, according to which gene expression should be at the end of a cascade of second (and later) messengers. It is worthwhile to remember that the link between rapid photoresponses, on which most second messenger evidence is based, and slow responses requiring gene expression, is still not established. In vitro reactions like this one may be the first opportunity to resolve the biochemical mechanism of phytochrome action, provided their relevance in the intact plant can be established.

OUTLOOK: CRITERIA FOR IMPLICATION OF SECOND MESSENGERS

It is clear that there are numerous candidates for second messengers in light signaling (Table I). Some of these are physiologically relevant in their own right, for example, ion movements in stomatal opening. Ca^{2+} concentration changes, protein phosphorylation, electric current patterns, and the like have been studied in different organisms. The time scales of the responses are different in each example, so one cannot even postulate a temporal order for the activity of the various "second" messengers. The trivial criterion, that the messenger must precede the start of the response, cannot always be evaluated from available data, and more time courses are needed. The different messengers may act in various combinations: for example, PI turnover could activate a protein kinase through an increase in DAG and at the same time mobilize intracellular Ca^{2+} via IP_3. Ca^{2+} would not necessarily need to depend on IP_3: a voltage-gated Ca^{2+} channel could transduce membrane potential changes directly into intracellular Ca^{2+} concentration changes. Observations summarized in Table I will suggest other combinations.

Demonstration of messenger involvement in a response is facilitated by manipulation of the pathway. This can be done biochemically, photobiologically, through variation of the light fluence, or by the use of mutants with defects in the response or the messenger. Photoresponse mutants provide a powerful tool to implicate or exclude specific second messenger candidates. One way to show that a putative second messenger or rapid effect does not belong to the response pathway is to eliminate the effect and show that the pathway still functions. Conversely, it should be possible to take a mutant that lacks the response and carry out biochemical comparisons in order to identify the defective step. The genetic approach has been most

TABLE I. Second Messenger Candidates in Photoresponses

System response	Short-term	Wavelength band[a]	Implicated messengers	Reference
Samanea pulvini	Pulvini movement	Red or blue	IP/IP$_2$/IP$_3$	Morse et al. (1987b)
Stomata	Opening	blue	Ion pump	Assmann et al. (1985)
Phycomyces	Phototropism	Blue growth regulator	Unknown	Galland et al. (1985)
Corn coleoptile	Phototropism	Blue	Auxin	Baskin et al. (1986)
Mougeotia	Chloroplast rotation	Red (blue)	Ca^{2+} calmodulin	Wagner et al. (1984) Serlin and Roux (1984)
Onoclea	Spore germination	Red	Ca^{2+} calmodulin	Wayne and Hepler, (1984)
Chlamydomonas	Phototaxis	Blue/green	Ca^{2+}	Morel-Laurens (1987)
Barley	Gene transcription	Red	Pfr	Mösinger et al. (1987)
Vaucheria	Chloroplast aggregation	Blue	Ionic current	Blatt et al. (1981)
Barley	PSII/PSI energy distribution	Red/blue	LHCP phosphorylation	Bennett (1983)

[a]"Red" effects have all been attributed to phytochrome, except for LHCP phosphorylation, which is mediated by photosynthetic pigments. "Blue" effects belong to cryptochrome(s) or other unidentified receptors (Senger, 1987). Phototaxis in *Chlamydonomas* is controlled by a rhodopsin (Foster et al., 1984).
IP, inositol-1-phosphate; IP$_2$, inositol-1,4-bisphosphate; IP$_3$, inositol-1,4,5-trisphosphate; PSII, photosystem II; PSI, photosystem I; Pfr, far-red form of phytochrome; LHCP, light-harvesting chlorophyll a/b binding proteins.

extensively used for blue-light responses, because blue-sensitive fungi were most amenable to genetic analysis (Gressel and Rau, 1983). Work with higher plant photomutants has recently gained momentum (Adamse et al., 1987; Parks et al., 1987). Advances in molecular genetic technology have made it possible in other systems to complement a mutant with the wild-type gene, allowing identification of the gene, as well as opening the possibility for site-directed mutagenesis. A purely biochemical strategy is to replace light by application of the putative second messenger, often by microinjection into a cell. Large cells are most convenient. Many plant systems (particularly some fungi and lower plants) should be amenable to this approach. An ideal organism in which to combine the genetic and biochemical approaches has not yet been found.

In conclusion, there is ample potential for second messengers in light signaling. There is time, even in the fastest responses, for rapid reactions between photoproduct and effect. A recent, and promising, addition to the long list of candidates (Table I) is PI turnover, which could be a trigger for two other messengers, calcium and the phosphorylation state of key proteins. The challenge for future work will be to use biochemical and molecular genetic methods to order the various messengers along a (not necessarily linear) transduction chain.

ACKNOWLEDGMENTS

I am grateful to Dr. J. Gressel for comments on the manuscript and to Drs. M.J. Morse, R. Satter, and E. Zeiger for suggestions and access to reports prior to publication.

REFERENCES

Adamse P, PAPM Jaspers, RE Kendrick M Koornneef (1987): Photomorphogenetic responses of a long-hypocotyl mutant of *Cucumis sativus* L. J Plant Physiol 127:481–491.

Assmann S, L Simoncini, JI Schroeder (1985): Blue light activates electrogenic ion pumping in guard cell protoplasts of *Vicia faba*. Nature 318:285–287.

Bandurski RS, A Schulze, DM Reinecke (1986): Biosynthetic and metabolic aspects of auxins. In Bopp M (ed): "Plant Growth Substances 1985." Berlin: Springer-Verlag, pp 83–91.

Baskin TI, WR Briggs, M Iino (1986): Can lateral redistribution of auxin account for phototropism of maize coleoptiles? Plant Physiol 18:306–309.

Bennett J (1983): Regulation of photosynthesis by reversible phosphorylation of the light-harvesting chlorophyll a/b protein. Biochem J 212:1–13.

Black CC, JJ Brand (1986): Roles of calcium in photosynthesis. In Cheung WY (ed): "Calcium and Cell Function," Vol VI. Orlando, FL: Academic Press, pp 327–355.

Blatt MR, MH Weisenseel, W Haupt (1981): A light-dependent current associated with chloroplast aggregation in the alga *Vaucheria sessilis*. Planta 152:513–526.

Blatt MR (1987): Yearly review: Toward the link between membranes, transport and photoperception in plants. Photochem Photobiol 45:933–938.

Brain RD, DO Woodward, WR Briggs (1977): Correlative studies of light sensitivity and cytochrome content in *Neurospora crassa*. Carnegie Inst Wash Year Book 76:295–299.

Brown JE, LJ Rubin, AJ Ghalayini, AP Tarver, RF Irvine, MJ Berridge, RE Anderson (1984): *myo*-Inositol polyphosphate may be a messenger for visual excitation in *Limulus* photoreceptors. Nature 311:160–162.

Brownlee C, JW Wood (1986): A gradient of cytoplasmic free calcium in growing rhizoid cells of *Fucus serratus*. Nature 320:624–626.

Cassel D, T Pfeuffer (1978): Mechanism of cholera toxin action: Covalent modification of the guanyl nucleotide-binding protein of the adenylate cyclase system. Proc Natl Acad Sci USA 75:2669–2673.

Cooke TJ, RH Racusen, WR Briggs (1983): Initial events in the tip-swelling response of the filamentous gametophyte of *Onoclea sensibilis* L. to blue light. Planta 159:300–307.

Cormier MJ, A Harmon, C Putnam-Evans (1985): Ca^{++}-dependent regulation of NAD kinase and protein phosphorylation in plants. In Hidaka H, DJ Hartshorne (eds): "Calmodulin Antagonists and Cellular Physiology." Orlando, FL: Academic Press, pp 445–456.

Cosgrove DJ (1981): Rapid suppression of growth by blue light: Occurrence, time course and general characteristics. Plant Physiol 67:584–590.

Cosgrove DJ (1986): Photomodulation of growth. In Kendrick RE, GHM Kronenberg (eds): "Photomorphogenesis in plants." Dordrecht, The Netherlands: Martinus Nijhoff, pp 341–365.

Coté GG, MJ Morse, RC Crain, RL Satter (1987): Isolation of soluble metabolites of the phosphatidylinositol cycle from *Samanea saman*. Plant Cell Rep (in press).

Datta N, Y-R Chen, SJ Roux (1985): Phytochrome and calcium stimulation of protein phosphorylation in isolated pea nuclei. Biochem Biophys Res Commun 128:1403–1408.

Dreyer EM, MH Weisenseel (1979): Phytochrome-mediated uptake of calcium in *Mougeotia* cells. Planta 146:31–39.

Ernst D, D Oesterhelt (1985): Purified phytochrome influences in vitro transcription in rye nuclei. EMBO J 3:3075–3078.

Firn RD, J Digby (1980): The establishment of tropic curvatures in plants. Annu Rev Plant Physiol 31:131–148.

Foster K, J Saranak, N Patel, G Zarilli, M Okabe, T Kline, K Nakanishi (1984): A rhodopsin is the functional photoreceptor for phototaxis in the unicellular eukaryote *Chlamydomonas*. Nature 311:756–759.

Gaba V, M Black (1979): Two separate photoreceptors control hypocotyl growth in green seedlings. Nature 278:51–54.

Gaba V, J Gressel (1987): Acifluorfen enhancement of cryptochrome-modulated sporulation following an inductive light pulse. Plant Physiol 83:225–227.

Galland P, A Palit, ED Lipson (1985): *Phycomyces:* Phototropism and light-growth response to pulse stimuli. Planta 165:538–547.

Gepstein S, M Jacobs, L Taiz (1982): Inhibition of stomatal opening in *Vicia faba* epidermal tissue by vanadate and abscisic acid. Plant Sci Lett 28:63–72.

Gietzen K, H Bader (1985): Effects of calmodulin antagonists on Ca^{++}-transport ATPase. In Hidaka H, DJ Hartshorne (eds): Calmodulin Antagonists and Cellular Physiology, Orlando, FL: Academic Press, pp 347–362.

Gressel J, Horwitz BA (1982): Gravitropism and phototropism. In Smith H, D Grierson (eds): "The Molecular Biology of Plant Development: Botanical Monographs," Vol 18. Oxford: Blackwell, pp 405–433.

Gressel J, W Rau (1983): Photocontrol of fungal development. In Shropshire W, H Mohr (eds): "Photomorphogenesis: Encyclopedia of Plant Physiology, New Series," Vol 16B. Berlin: Springer-Verlag, pp 603–639.

Häder D-P (1981): Electrical and proton gradients in the sensory transduction of photophobic responses in the blue-green alga, *Phormidum uncinatum*. Arch Microbiol 130:83–86.

Hasunuma K, K Funadera (1987): GTP-binding protein(s) in green plant, *Lemna paucicostata*. Biochem Biophys Res Commun 143:908–912.

Hasunuma K, K Funadera, Y Shinohara, K Furukawa, M Watanabe (1987): Circadian oscillation and light-induced changes in the concentration of cyclic nucleotides in *Neurospora*. Curr Genet 12:127–133.

Haupt W. (1982): Light-mediated movement of chloroplasts. Annu Rev Plant Physiol 33:205–233.

Hepler PK, RO Wayne (1985): Calcium and plant development. Annu Rev Plant Physiol 36:397–439.

Herrera-Estrella L, J Ruiz-Herrera (1983): Light response in *Phycomyces blakesleeanus:* Evidence for roles of chitin biosynthesis and breakdown. Exp Mycol 7:362–369.

Horwitz BA, J Gressel (1987): First measurable effects following photoinduction of morphogenesis. In Senger H (ed): "Blue Light Responses: Phenomena and Occurrence in Plants and Microorganisms," Boca Raton, FL: CRC Press, pp 53–70.

Horwitz BA, CH Trad, ED Lipson (1986): Modified light-induced absorbance changes in *dimY* photoresponse mutants of *Trichoderma*. Plant Physiol 81:726–730.

Iino M (1982): Inhibitory action of red light on the growth of the maize mesocotyl: Evaluation of the auxin hypothesis. Planta 156:388–395.

Iino M, WR Briggs (1984): Growth redistribution during first positive phototropic curvature of maize coleoptiles. Plant Cell Environ 7:97–104.

Jaffe LF, R Nuccitelli (1977): Electrical controls of development. Annu Rev Biophys Bioeng 6:445–476.

Johnson CB (1982): Photomorphogenesis. In Smith H, D Grierson (eds): "The Molecular Biology of Plant Development, Botanical Monographs," Vol 18. Oxford: Blackwell, pp 365–404.

Kendrick RE, GHM Kronenberg (eds) (1986): Photomorphogensis in Plants. Dordrecht, The Netherlands: Martinus Nijhoff.

Leong TY, WR Briggs (1982): Evidence from studies with acifluorfen for participation of a flavin–cytochrome complex in blue light photoreception for phototropism of oat coleoptiles. Plant Physiol 70:875–881.

Lipson ED, SM Block (1983): Light and dark adaptation in *Phycomyces* light-growth response. J Gen Physiol 81:845–859.

Lipson ED, D Presti (1977): Light-induced absorbance changes in *Phycomyces* photomutants. Photochem Photobiol 25:203–208.

Millner PA, JB Marder, K Gounaris, J Barber (1986): Localization and identification of phosphoproteins within the photosystem II core of higher plant thylakoid membranes. Biochim Biophys Acta 852:30–37.

Mohr H (1972): "Lectures on Photomorphogenesis." Berlin: Springer-Verlag.

Montavon M, H Greppin (1985): Potentiel intracellulaire du mesophyll d'épinard (*Spinacia oleracea* L. cv. Nobel) en relation avec la lumière et l'induction florale. J Plant Physiol 118:471–475.

Morel-Laurens N (1987): Calcium control of phototactic orientation in *Chylamydomonas reinhardtii:* Sign and strength of response. Photochem Photobiol 45:119–128.

Morse MJ, RC Crain, RL Satter (1987a): Phosphatidylinositol cycle metabolites in *Samanea saman* pulvini. Plant Physiol 83:640–644.

Morse MJ, RC Crain, RL Satter (1987b): Light stimulated phosphatidylinositol turnover in *Samanea saman* leaf pulvini. Proc Natl Acad Sci USA 84:7075–7078.

Mösinger E, A Batschauer, R Vierstra, K Apel, E Schäfer (1987): Comparison of the effects of exogenous native phytochrome and in vivo irradiation on in vitro transcription in isolated nuclei from barley (*Hordeum vulgare*) Planta 170:505–514.

Muto S, S Izawa, S Miyachi (1982): Light-induced Ca^{++} uptake by intact chloroplasts. Fed Eur Biochem Soc Lett 139:250–254.

Ortega JKE, RI Gamow (1976): An increase in mechanical extensibility during the period of light-stimulated growth. Plant Physiol 57:456–457.

Otto V, E Schäfer (1986): Rapid, phytochrome controlled protein phosphorylation and dephosphorylation in *Avena sativa* L. In Furuya M (ed): Phytochrome and Plant Photomorphogenesis: Proceedings of the XVI Yamada Conference." Okazaki, Japan: Yamada Science Foundation, p 106.

Parks BM, AM Jones, P Adamse, M Koornneef, RE Kendrick, PH Quail (1987): The *aurea* mutant of tomato is deficient in spectrophotometrically and immunocytochemically detectable phytochrome. Plant Mol Biol 9:97–107.

Presti D, M Delbrück (1978): Photoreceptors for biosynthesis, energy storage and vision. Plant Cell Environ 1:81–100.

Quail PH (1983): Rapid action of phytochrome in photomorphogensis. In Shropshire W, H Mohr (eds): "Photomorphogensis, Encyclopedia of Plant Physiology, New Series," Vol 16B. Berlin: Springer-Verlag, pp 178–212.

Racusen RH, TJ Cooke (1982): Electrical changes in the apical cell of the fern gametophyte during irradiation with photomorphogenetically-active light. Plant Physiol 70:331–334.

Racusen RH, AW Galston (1983): Developmental significance of light-mediated electrical responses in plant tissue. In Shropshire W, H Mohr (eds): "Photomorphogensis, Encyclopedia of Plant Physiology, New Series" Vol 16B. Berlin: Springer-Verlag, pp 687–703.

Rubin LJ, JE Brown (1985): Intracellular injection of calcium buffers blocks IP_3-induced but not light-induced electrical responses of *Limulus* ventral photoreceptors. Biophys J 47:38a.

Satter RL, GT Geballe, AW Galston (1974): Potassium flux and leaf movement in *Samanea saman*. II. Phytochrome-controlled movement. J Gen Physiol 64:431–442.

Schäfer E, WR Briggs (1986): Photomorphogenesis from signal perception to gene expression. Photochem Photobiophys 12:305–320.

Schmidt W (1984): Blue light-induced, flavin-mediated transport of redox equivalents across artificial bilayer membranes. J Membr Biol 82:113–122.

Schwartz A, D Koller (1978): Phototropic response to vectorial light in leaves of *Lavatera cretica* L. Plant Physiol 61:924–928.

Senger H (ed) (1987): "Blue Light Responses: Phenomena and Occurrence in Plants and Microorganisms." Boca Raton, FL: CRC Press.

Senger H, WR Briggs (1981): The blue light receptor(s): Primary reactions and subsequent metabolic changes. In Smith KC (ed): Photochemical and Photobiological Reviews, Vol 6, New York: Plenum, pp 1–37.

Serlin BS, SJ Roux (1984): Modulation of chloroplast movement in the green alga *Mougeotia* by the Ca^{++} ionophore A23187 and by calmodulin antagonists. Proc Natl Acad Sci USA 81:6368–6372.

Serrano E, E Zeiger, S Hagiwara (1988): Red light stimulates an electrogenic proton pump in *Vicia* guard cell protoplasts. Proc Natl Acad Sci USA 85:436–440.

Shaw NM, RW Harding (1987): Intracellular and extracellular cyclic nucleotides in wild-type and white collar mutant strains of *Neuorspora crassa*. Plant Physiol 83:377–383.

Shimazaki K, M Iino, E Zeiger (1986): Blue light-dependent proton extrusion by guard cell protoplasts of *Vicia faba* L. Nature 319:324–326.

Shropshire W, H Mohr (eds) (1983): "Photomorphogenesis: Encyclopedia of Plant Physiol, New Series," Vol. 16. Berlin: Springer-Verlag.

Speth V, V Otto, E Schäfer (1986): Intracellular localization of phytochrome in oat coleoptiles by electron microscopy. Planta 168:299–304.

Stryer L. (1986): Cyclic GMP cascade of vision. Annu Rev Neurosci 9:87–119.

Wagner G, F Grolig (1985): Molecular mechanisms of photoinduced chloroplast movements. In Colombetti G, F Lenci, P-S Song (eds): "Sensory Perception and Transduction in Aneural Organisms." New York: Plenum, pp 281–298.

Wagner G, P Valentin, P Dieter, D Marmé (1984): Identification of calmodulin in the green alga *Mougeotia* and its possible function in chloroplast reorientational movement. Planta 162:62–67.

Wayne R, PK Hepler (1984): The role of calcium ions in phytochrome-mediated germination of spores of *Onoclea sensibilis* L. Planta 160:12–20.

Weisenseel MH, RM Kicherer (1981): Ionic currents as control mechanism in cytomorphogenesis. In Kiermayer O (ed): "Cell Biology Monographs, Vol 8, Cyto-morphogensis in Plants." Vienna: Springer-Verlag, pp 379–399.

Wong Y-S, H-C Cheng, DA Walsh, JC Lagarias (1986): Phosphorylation of *Avena* phytochrome in vitro as a probe of light-induced conformational changes. J Biol Chem 261:12089–12097.

Zeiger E (1983): The biology of stomatal guard cells. Annu Rev Plant Physiol 34:441–475.

Second Messengers in Plant Growth
and Development, pages 315–326
© 1989 Alan R. Liss, Inc.

13. Second Messengers and Gene Expression

Tom J. Guilfoyle

*Department of Biochemistry, University of Missouri, Columbia, Missouri
65211*

INTRODUCTION

The role of second messengers, particularly Ca^{2+}, in regulating plant growth, developmental processes, and metabolism is the subject of a number of chapters in this volume and other recent reviews (Hepler and Wayne, 1985; Roux and Slocum, 1982; Poovaiah and Reddy, 1987) and will not be discussed in detail in this chapter. Although alterations in gene expression must be a prerequisite for many, if not most, growth and developmental transitions in plants, our current understanding of such changes in gene expression is sketchy, at best. The mechanisms underlying altered plant gene expression are unknown. What is known is that a variety of environmental signals and plant hormones, including synthetic plant growth regulators, can trigger changes in plant growth and development commensurate with changes in gene expression (Guilfoyle, 1986; Kuhlemeier et al., 1987).

Circumstantial evidence (albeit far from definitive) suggests that certain environmental or hormonal signals might require second messengers at some point in the transduction pathway between signal perception and gene activation or repression. Changes in intracellular Ca^{2+} concentrations, which accompany hormone application (Poovaiah and Reddy, 1987; Brummel and Hall, 1987) or environmental signals such as light (Hepler and Wayne, 1985; Roux and Slocum, 1982; Poovaiah and Reddy, 1987), suggest that Ca^{2+} may act as such a second messenger in plant cells. To date, however, the role of any second messenger in the regulation of plant gene expression is purely hypothetical. This differs from animal gene expression,

Address reprint requests to Tom J. Guilfoyle, Department of Biochemistry, 117 Schweitzer Hall, University of Missouri, Columbia, MO 62511.

where there is considerable evidence for direct regulation of specific genes by second messengers, including Ca^{2+}.

In this chapter, I will describe a few of the animal genes that are responsive to second messengers. I will then discuss some plant genes that might be responsive to second messengers. Finally, I will consider some approaches that might provide insight into gene regulation by both second messengers and primary signals in plant systems.

I will not assume that plant hormones act as second messengers, as suggested by Blowers and Trewavas in Chapter 1. Instead, I will limit my discussion of second messengers primarily to the most likely candidate in plants, Ca^{2+}, and cAMP as a second messenger in the regulation of animal gene expression. I will also describe gene expression as regulated by phorbol ester tumor promoters. My discussion on the regulation of gene expression will be restricted to the transcriptional regulation of specific genes. I will not discuss global effects (e.g., effects on overall nuclear transcription rates or RNA polymerase activities or translational efficiencies) of second messengers on gene expression, since these are unlikely mechanisms to explain subtle transitions in growth and development. I will only briefly discuss the role of protein kinases in the transduction process and the role of protein phosphorylation in the regulation of gene expression. It is worthwhile pointing out, however, that phosphorylation/dephosphorylation of specific transcription factors (e.g., enhancer binding proteins) may play an important role in the regulation of gene expression in both animals and plants.

EXAMPLES OF SECOND MESSENGERS AS REGULATORS OF GENE EXPRESSION IN ANIMALS

Ca^{2+} as a Second Messenger

There is recent evidence that changes in intracellular calcium concentration, whether induced by serum, growth factors, mitogens, stress, or Ca^{2+} ionophores, can result in activation of a number of animal genes. The mechanism for gene activation may vary with different systems. Several examples of gene regulation by Ca^{2+} in animal cells are described below.

In quiescent T cells or thymocytes stimulated by mitogens such as concanavalin A, there is a rapid increase in intracellular Ca^{2+} concentration (Tsien et al., 1982). In this system, concanavalin A is thought to mimic the action of antigens and to cause an increase in cytosolic Ca^{2+} within 60 sec after mitogen stimulation.

Within 24 to 48 hr after application of the mitogen, the cells enter the S phase of the cell cycle. Prior to the entry into S phase, the abundance of a number of messenger RNAs (including those for actin, tubulin, ornithine decarboxylase, and the cellular oncogenes c-*myc* and c-*fos*) increase in T cells. For the c-*myc* gene, the transcription rate, measured by nuclear run-on in vitro transcription assays, increases severalfold following mitogen stimulation, while transcription rates for tubulin, actin, and ornithine decarboxylase genes remain constant or decline slightly after mitogen addition (White et al., 1987).

The increase in abundance of c-*myc* and ornithine decarboxylase mRNAs elicited by concanavalin A is inhibited by 2–10 mM EGTA. EGTA addition blocks the Ca^{2+} increase normally induced by the mitogen. The transition from G_0 to S phase in thymocytes also can be induced by the calcium ionophore, A23187 (Moore et al., 1986). This treatment results in a rapid increase in c-*fos* and c-*myc* mRNAs. In the thymocyte system, where the increases in abundance of c-*myc* and orthinine decarboxylase mRNAs have been studied in detail, there appear to be two different mechanisms regulating the expression of the c-*myc* and ornithine decarboxylase genes. The increase in c-*myc* mRNA abundance is regulated primarily at the level of transcription, while the ornithine decarboxylase mRNA appears to be regulated posttranscriptionally, at the level of mRNA stability. The increase in abundance of both c-*myc* and ornithine decarboxylase mRNAs requires increased cytosolic Ca^{2+}, but the genes are likely responding to two different signal-transduction circuits. With the c-*myc* gene, expression is blocked by Cd^{2+} (White et al., 1987), which is believed to be a calmodulin antagonist by competing with Ca^{2+} for binding to the protein (Andersson et al., 1982). On the other hand, Cd^{2+} does not block the accumulation of ornithine decarboxylase following mitogen stimulation. Mitogen stimulation also causes a breakdown of phosphatidylinositol bisphosphate to diacylglycerol and inositol trisphosphate in thymocytes (Imboden and Stobo, 1985). This breakdown in concert with an increase in intracellular Ca^{2+} is thought to result in the activation of protein kinase C, which, in turn, modulates the abundance of ornithine decarboxylase mRNA at some posttranscriptional step.

In a number of different cell types, transcription of the cellular oncogene, c-*fos* is increased rapidly following the addition of growth factors to tissue cultures (Fisch et al., 1987). The c-*fos* gene transcription is also activated by addition of A23187 to these cells when cultured. Recent studies have been directed at defining the *cis*-acting DNA elements and the *trans*-acting nuclear protein factors that confer growth factor and Ca^{2+} inducibility to this cellular oncogene. A 20-base-pair transcriptional enhancer element with dyad symmetry has been located between −317 and −298 base pairs upstream of the c-*fos* mRNA cap site. This enhancer element is required for serum or growth factor inducibility (Fisch et al., 1987). Nuclear extracts derived from a number of different cell types contain one or more proteins that specifically bind to this enhancer element, as determined by gel mobility shift experiments (Prywes and Roeder, 1986). With some cell types, the binding activity of nuclear proteins that interact with the enhancer element can be increased by prior administration of growth factor to the cells. In contrast with growth factors, A23187 does not induce enhancer binding in nuclear extracts, suggesting that Ca^{2+} activation of the c-*fos* gene is regulated by a mechanism different from that of growth factor or serum activation. In transfection experiments using a c-*fos* chloramphenicol acetyltransferase (c-*fos*-CAT) fusion gene, CAT activity was inducible with epidermal growth factor, but not with A23187 (Fisch et al., 1987). In these experiments the c-*fos* promoter containing up to $-2,000$ base pairs upstream and $+48$ pairs downstream of the c-*fos* cap site was fused to the reporter gene, CAT. This result contrasted with the endogenous c-*fos* gene present in the cells used for transfection that was induced by both epidermal growth factor and A23187 in transfected and untransfected cells. These results, along with other recent results, suggest that the Ca^{2+}-activating DNA element differs from the growth-factor-enhancer element and, further, suggests that the Ca^{2+} element (if one exists) resides beyond $-2,000$ base pairs upstream or $+48$ downstream of the mRNA cap site in the human c-*fos* gene. Since the endogenous c-*fos* gene, but not the transfected gene, was activated by the Ca^{2+} ionophore, A23187, it is also possible that the chromatin structure of the c-*fos* gene confers Ca^{2+} inducibility, as opposed to a specific DNA sequence.

Several different tissue culture types, including chicken, mouse, hamster, rat, and human, have been shown to synthesize a set of proteins in response to glucose starvation or calcium ionophore treatment (Lee, 1987). Two of these proteins have been characterized and are referred to as GRP78 and GRP94 (i.e., glucose-regulated proteins (GRPs) of 78 kDa and 94 kDa, respectively). In the uninduced condition, GRP78 is localized within the endoplasmic reticulum and GRP94 is located within the Golgi apparatus. After induction, however, both proteins can be found in the nucleus (Lee, 1987). A number of different stresses will induce the synthesis of the GRPs. While GRP78 shares some sequence similarity and immunological cross-reactivity with HSP70 (i.e., the heat-shock protein of 70 kDa), it is a protein distinct from heat-shock proteins. In addition to stresses, the calcium ionophores, ionomycin and A23187, will induce the sythesis of the GRPs. Shifting cells from low Ca^{2+} (e.g., 0.15 mM) to high Ca^{2+} (e.g., 1.8 mM) also induces the synthesis of GRPs. Other agents that block protein glycosylation (i.e., tunicamycin, glucosamine, and 2-deoxyglucose) will also induce synthesis of GRPs.

Two cDNA clones that encode the GRP78 and GRP94 proteins have been isolated (Resendez et al., 1985). A23187-treated cells show higher transcription rates of the genes that encode GRP78 and GRP95, and both genes appear to be activated coordinately (Lee, 1987; Resendez et al., 1985). The mRNAs corresponding to these GRPs increase 20–30-fold by 3–5 hr after administration of A23187. The transcriptional induction is specific to calcium ionophores, since other proton and cation ionophores such as gramicidin, valinomycin, and nigericin do not induce expression of the GRP genes.

A Ca^{2+}/calmodulin pathway appears to be involved in expression of the GRP genes, since the calmodulin antagonist W-7 affects the abundance of the GRP mRNAs under both uninduced and induced conditions. Deletion analysis of the GRP78 gene indicates that a region between -480 and -85 upstream of the mRNA cap site is sufficient to confer inducibility of the gene to the variety of stresses dicscussed above (Lin et al., 1986). It is unlikely in this system that Ca^{2+} itself is the inducer because of the long time required for induction by A23187 (i.e., 2–4 hrs) and because new protien sysnthesis is required for ionophore induction. This contrasts with the c-*fos* gene discussed above, which is induced within minutes after ionophore application and superinduced in the presence of protein synthesis inhibitors. Therefore, it is not likely that the GRP genes contain Ca^{2+}-inducible *cis*-acting elements. It is more likely that these genes contain elements responsive to nuclear proteins that accumulate or are activated by a number of stress agents that modulate Ca^{2+} stores within the cell and disrupt the protein glycosylation processes occurring within the endoplasmic reticulum and the Golgi apparatus.

Like the cell systems described in the preceding paragraph, rat exocrine pancreas cells show induction (and repression) of specific genes in response to glucose starvation and calcium ionophore treatment (Stratowa and Rutter, 1986). With pancreas cells, however, trypsin mRNA increases as opposed to the stress-related GRPs. Cells treated with A23187 show a severalfold increase in the abundance of trypsin mRNA while chymotrypsin, amylase, and carboxypeptidase A1 mRNAs show a decline in abundance. Gene fusion assays, using a 1,600-base-pair fragment upstream of the trypsin mRNA cap site fused to the CAT reporter gene showed that this 5′-flanking fragment of the rat trypsin gene was sufficient to confer A23187 inducibility to the CAT gene. This contrasted with results using the chymotrypsin gene 5′-flanking sequences fused to the CAT gene where CAT expression was decreased a fewfold in response to A23187. The DNA sequences that respond to A23187 have yet to be defined.

cAMP as a Second Messenger

Although it is unlikely that cyclic nucleotides play a significant regulatory role in plant growth and development or metabolism, it may still be beneficial to examine the mechanism of action of this second messenger with specific animal genes. Recent results suggest that certain nuclear transcription factors may be at the receiving end of the cAMP signal-transduction pathway.

A number of animal genes appear to be activated by the second messenger, cAMP. These genes include human proenkephalin, human glycoprotein hormone α-subunit, rat somatostatin, human prolactin, and rat phosphoenol pyruvate carboxykinase, to mention a few. Detailed analysis with a number of these genes has led to the identification of putative *cis*-acting DNA sequences that confer inducibility by cAMP (Comb et al.,1986; Delegeane et al., 1987; Montminy et al., 1986; Silver et al., 1987). These studies have been carried out by using CAT fusion genes where the 5'-flanking or putative promoter regions of different cAMP-responsive genes have been fused to the CAT reporter gene. The CAT fusion genes are then transfected into appropriate cell lines, and the transfected cells are assayed for both basal and cAMP-inducible CAT activity. If a cAMP-responsive element is present in the 5'-flanking region of the CAT fusion genes, then CAT activity is generally increased severalfold following applications of cAMP or active cAMP analogues to the transfected cells. In some cases, it has been demonstrated that cholera toxin or forskolin (i.e., agents that stimulate adenyl cyclase) also activate CAT activity in cells transfected with CAT fusion genes containing the cAMP-responsive element.

cis-Acting DNA sequences in the 5'-flanking region of cAMP-responsive genes, which are required for cAMP regulation, have been localized by deletion analysis of the 5'-flanking regions. cAMP *cis*-acting elements are located within the first 300 base pairs upstream of the mRNA cap sites for enkephalin, somatostatin, and glycoprotein hormone α-subunit genes. These elements have enhancer-like properties and generally retain activity if moved to another position in the 5'-flanking region of the gene (i.e., the cAMP-responsive element is moved either closer to or farther from the mRNA cap site), if placed in opposite orientation within the 5'-flanking region of the gene, or if moved to the 3'-flanking region of the CAT fusion gene. If the cAMP-responsive element is inserted within, adjacent to, or close to a noninducible heterologous promoter (e.g., a promoter not normally inducible by cAMP), such chimeric promoter CAT fusions become inducible by cAMP. Synthetic oligodeoxyribonucleotides of 20–30 base pairs, which are copies of the natural cAMP enhancer found in cAMP-responsive genes, confer cAMP regulation to noninducible heterologous promoters when fused as one or more copies to these promoters. By analyzing a number of 5'-flanking regions of cAMP-regulated genes, a consensus cAMP enhancer element has been identified. In its simplest form, this cAMP enhancer element has the sequence 5'-CTGACGTCAG-3'; a number of cAMP-regulated genes contain a perfect or nearly perfect copy of this cAMP enhancer element within 300 base pairs of the mRNA cap sites.

With the somatostatin gene, the cAMP-regulated element has been shown to function in wild-type cultured cells but not in mutant cells that lack cAMP-dependent protein kinase activity (Montminy et al., 1986). This observation suggests that activation of the somatostatin gene requires phosphorylation of some protein(s), possibly a nuclear transcription factor(s) that binds to the cAMP enhancer element. DNAase I protection assays on the 5'-flanking region of the somatostatin gene, have identified a nuclear protein that binds to the cAMP enhancer DNA sequence (Montminy and Bilezikjian, 1987). This protein has been purified by sequence-specific DNA affinity chromatography and shown to be a polypeptide of 43 kDa on sodium dodecylsulfate (SDS)-polyacrylamide gels. When the purified 43-kDa polypeptide is incubated with the catalytic subunit of cAMP-dependent protein kinase, the nuclear DNA binding protein becomes phosphorylated. Other experiments have demonstrated that this cAMP enhancer binding protein is more highly phosphorylated when purified from cultured cells that have been stimulated by the adenyl cyclase activator, forskolin, compared with the protein purified from unstimulated cells. In total, the results observed with the somatostatin gene suggest that the cAMP-dependent signal-transduction pathway may regulate specific gene transcription in response to hormonal administration by stimulating the phosphorylation of a cAMP enhancer binding protein.

Montminy and Bilezikjian (1987) have proposed that the previously described nuclear transcription factor, AP-1 (Lee et al., 1987), which recognizes the consensus sequence 5'-TGACTCA-3' in a number of promoters (e.g., SV40, human metallothionein), may be related to the cAMP enhancer protein that recognizes a similar sequence element 5'-TGACGTCA-3'. More recent evidence indicates that the 47-kDa AP-1 nuclear transcription factor is not the cAMP enhancer binding protein (Imagawa et al., 1987). Instead, AP-1 is identical to a nuclear protein that binds to an enhancer element found in some phorbol-ester-responsive genes. On the other hand, another independent and previously characterized nuclear transcription factor, AP-2, which is a 50-kDa protein and binds to the consensus sequence 5'-TCCCCAN(G/C)(C/G)(G/C)-3' (Imagawa et al., 1987; Mitchell et al., 1987), appears to mediate induction by both protein kinase C and cAMP signal-transduction pathways. The binding of AP-2 to some cAMP-regulated genes suggests that more than one type of cAMP enhancer element exists, since the AP-2 sequence element differs from the cAMP enhancer element described above.

Phorbol Esters and Protein-Kinase-C-Regulated Gene Expression

Protein kinase C is a Ca^{2+} and phospholipid-dependent protein kinase. This kinase is a component of a second messenger system, the phosphatidylinositol pathway. In addition, this kinase appears to be activated directly by the tumor promoters, the phorbol esters (e.g., 12-O-tetradecanylphorbol-13-acetate [TPA]), which are structurally similar to diacylglycerol. In this section, I will describe recent studies that suggest a role for protein kinase C in the activation of TPA-responsive genes by phosphorylation of specific nuclear transcription factors.

Several TPA-responsive genes have been studied. These include collagenase, stromelysin, metallothionein IIA, and c-*fos*. By employing gene fusion transfection technology similar to that described above for cAMP responsive genes, an 8-base-pair consensus sequence has been identified in the 5'-flanking regions of a number of TPA-regulated genes (Angel et al., 1987; Chiu et al., 1987; Imbra and Karin, 1987). The TPA regulatory element consists of the sequence 5'-TGAGTCAG-3'. Synthetic oligodeoxyribonucleotide copies of this motif, when fused to promoters not normally responsive to TPA, confer TPA inducibility to these promoter elements. DNA footprint analysis of TPA responsive genes indicates that a nuclear protein specifically binds to the TPA regulatory element. This nuclear protein has been purified to homogeneity and shown to be identical to the AP-1 transcription factor (Angel et al., 1987; Lee et al., 1987b) that was previously characterized by Lee et al. (1987a).

The TPA induction response at the gene expression level is not inhibited by translational inhibitors such as cycloheximide (Imbra and Karin, 1987). In addition, the rapid increase in AP-1 binding to the TPA enhancer element is resistant to cycloheximide treatment. The resistance of the induction and binding responses to protein synthesis inhibitors suggests that a posttranslational mechanism operates in the activation of the TPA-responsive genes. A probable mechanism regulating TPA gene induction is the phosphorylation of AP-1 by protein kinase C (Angel et al., 1987; Lee et al., 1987b). It is important to note that AP-1 is not the only transcription factor involved in phorbol ester-induced gene expression. Other transcription factors identified to date include AP-2 (Imagawa et al., 1987; Mitchell et al., 1987) and NF-κB (Lenardo et al., 1987; Sen and Baltimore, 1986).

The above results suggest that the transcription factor, AP-1, is at the receiving end of the transmission pathway elicited by phorbol esters. This pathway starts at the plasma membrane and terminates with the posttranslational modification of this specific transcription factor. Phosphorylation of other transcription factors might by catalyzed also by protein kinase C and some of these, along with AP-1, could involve the phosphatidylinositol pathway.

PROBLEMS WITH AND PERSPECTIVES ON SECOND MESSENGERS IN THE REGULATION OF PLANT GENE EXPRESSION

In the Introduction, I commented briefly on our general ignorance in the area of the regulation of plant gene expression. This is true despite the fact that a large number of plant genes have been sequenced, and growing numbers of plant genes are being analyzed for regulatory elements. In vitro mutagenesis, along with either transient expression assays in isolated protoplasts or *Agrobacterium*-mediated stable transformation of suitable hosts such as tobacco or petunia, have been used for this purpose. Furthermore, a number of putative regulatory *cis*-acting DNA elements have been shown to bind nuclear proteins in a developmentally specific manner or after plants are induced with a particular stimulus (e.g., light, heat) (Jofuku et al., 1987; Kuhlemeier et al., 1987). Two major underdeveloped research areas are likely to continue to stifle progress on the regulation of plant gene expression until critical new advances are made.

One of these research areas is that involving signal-transduction pathways. This applies whether the primary signal is a particular wavelength of light (e.g., phytochrome reception, blue light reception, and so on), water stress, heat stress, pathogen infection, or hormonal stimulation. Of the primary signals listed above, we have little or no information as to how the primary signal is transduced to the level of the genome and/or to the transcriptional apparatus. Although it has been suggested on occasion that phytochrome-mediated (Poovaiah and Reddy, 1987) and auxin-induced (Brummel and Hall, 1987; Poovaiah and Reddy, 1987) gene expression might involve Ca^{2+} as a second messenger, this remains speculative at this time. A number of phytochrome-responsive genes have now been analyzed for possible regulatory or enhancer-like elements (Kuhlemeir et al., 1987), and it seems plausible to test the possible role of Ca^{2+} in the regulation of these genes. It should be noted, however, that testing may not be straightforward, since such elements have not been described in animal genes as yet, and they may not exist, or they may respond to a cascade of events initiated by Ca^{2+}. In the case of auxin, a number of genes have recently been isolated and characterized, but little or nothing is known about the function of these genes (or the gene products encoded) (Guilfoyle, 1986). While Ca^{2+} has been suggested to play a role in auxin-induced cell extension in elongating regions of hypocotyls, epicotyls, or coleoptiles (Poovaiah and Reddy, 1987), none of the currently available auxin-regulated genes has been shown to play any role in cell extension (although it should be noted that a role has not been ruled out either). Therefore, with the auxin genes, it is unclear whether they are good candidates in the search for Ca^{2+} regulatory elements. In any case, it is all too obvious that ignorance in the area of signal-transduction in plants is likely to be a severe limitation to future research on the regulation of environmentally or hormonally induced genes.

The second underdeveloped research area is soluble in vitro transcription systems. Currently, there exist no soluble in vitro transcription systems derived from plant cells that could be generally applied to determine whether nuclear protein binding to a particular DNA element provides a meaningful assay for gene induction. In other words, we have no functional assay to determine whether a particular nuclear protein functions to activate or repress transcription of a gene once it binds to a regulatory element. Such an assay is important, since it is becoming apparent from animal studies that protein binding to a regulatory element does not necessarily indicate that the binding is functional (Angel et al., 1987; Lee et al., 1987b). For example, unphosphorylated and inactive transcription factors may have the same affinity for a DNA-enhancer element as do phosphorylated and active factors, but it may be the phosphorylation event that makes the factor able to induce transcription of the gene. This type of regulation can be tested in animals because the appropriate soluble in vitro transcription systems either are available or are being developed (Gorski et al., 1986; Lee et al., 1987b). Without equivalent systems of plant origin, the functionality of nuclear binding proteins will be difficult, if not impossible, to assess.

REFERENCES

Andersson A, T Drakenberg, S Forsen, E Thulin (1982): Characterization of the Ca^{2+} binding sites of calmodulin from bovine testis using ^{43}Ca and ^{113}Cd NMR. Eur J Biochem 126:510–505.

Angel P, M Imagawa, R Chiu, B Stein, RJ Imbra, HJ Rahmsdorf, C Jonat, P Herrlich, M Karin (1987): Phorbol ester-inducible genes contain a common *cis* element recognized by a TPA modulated trans-acting factor. Cell 49:729–739.

Brummell DA, JL Hall (1987): Rapid cellular responses to auxin and the regulation of growth. Plant Cell Environ 10:523–543.

Chiu R, M Imagawa, RJ Imbra, JR Bockoven, M Karin (1987): Multiple *cis*- and *trans*-acting elements mediate the transcriptional response to phorbol esters. Nature 329:648–651.

Comb M, NC Birnberg, A Seasholtz, E Herbert, HM Goodman (1986): A cyclic AMP- and phorbol ester-inducible DNA element. Nature 323:353–356.

Delegeane AM, LH Ferland, PL Mellon (1987): Tissue-specific enhancer of the human glycoprotein hormone α-subunit gene: Dependence on cyclic AMP-inducible elements. Mol Cell Biol 7:3994–4002.

Fisch TM, R Prywes, RG Roeder (1987): c-*fos* Sequences necessary for basal expression and induction by epidermal growth factor, 12-O-tetradecanoyl phorbol-13-acetate and calcium ionophore. Mol Cell Biol 7:3490–3502.

Gorski K, M Carneiro, U Schibler (1986): Tissue-specific in vitro transcription from the mouse albumin promoter. Cell 47:767–776.

Guilfoyle TJ (1986): Auxin-regulated gene expression in higher plants. CRC Crit Rev Plant Sci 4:247–276.

Hepler PK, RO Wayne (1985): Calcium and plant development. Annu Rev Plant Physiol 36:397–439.

Imagawa M, R Chiu, M Karin (1987): Transcription factor AP-2 mediates induction by two different signal-transduction pathways. Protein kinase C and cAMP. Cell 51:251–260.

Imboden JB, JD Stobo (1985): Transmembrane signaling by the T cell antigen receptor: Perturbation of the T3-antigen receptor complex generates inositol phosphates and releases calcium ions from intracellular stores. J Exp Med 161:446–456.

Imbra RJ, M Karin (1987): Metallothionein gene expression is regulated by serum factors and activators of protein kinase C. Mol Cell Biol 7:1358–1363.

Jofuku KD, JK Okamuro, RB Goldberg (1987): Interaction of an embryo DNA binding protein with a soybean lectin gene upstream region. Nature 328:734–737.

Kuhlmeier C, PJ Green, N-M Chua (1987): Regulation of gene expression in higher plants. Annu Rev Plant Physiol 38:221–257.

Lee AS (1987): Coordinate regulation of a set of genes by glucose and calcium ionophore in mammalian cells. Trends Biochem Sci 12:20–23.

Lee W, A Haslinger, M Karin, R Tjian (1987a): Two factors that bind and activate the human metallothionein IIA gene in vitro also interact with the SV40 promoter and enhancer regions. Nature 325:368–372.

Lee W, P Mitchell, R Tjian (1987b): Purified transcription factor AP-1 interacts with TPA-inducible enhancer elements. Cell 49:741–752.

Lenardo M, JW Pierce, D Baltimore (1987): Protein-binding sites in the Ig gene enhancer determine transcriptional activity and inducibility. Science 236:1573–1577.

Lin Ay, SC Chang, AS Lee (1986): A calcium ionophore-inducible cellular promoter is highly active and has enhancer-like properties. Mol Cell Biol 6:1235–1243.

Mitchell PJ, C Wang, R Tjian (1987): Positive and negative regulation of transcription in vitro: Enhancer-binding protein AP-2 is inhibited by SV-40 T antigen. Cell 50:847–861.

Montminy MR, LM Bilezikjian (1987): Binding of a nuclear protein to the cyclic AMP response element of the somatostatin gene. Nature 328:175–178.

Montminy MR, KA Sevarino, JA Wagner, G Mandel, RH Goodman (1986): Identification of a cyclic AMP reponsive element within the rat somatostatin gene. Proc Natl Acad Sci USA 83:6682–6686.

Moore JP, JA Todd, TR Hesketh, JC Metcalf (1986): c-*fos* and c-*myc* Gene activation, ionic signals, and DNA synthesis in thymocytes. J Biol Chem 261:8158–8162.

Poovaiah BW, ASN Reddy (1987): Calcium messenger system in plants. CRC Crit Rev Plant Sci 6:47–103.

Prywes R, RG Roeder (1986): Inducible binding of a factor to the c-*fos* enhancer. Cell 47:777-784.

Resendez E Jr, JW Attenello, A Grafsky, CS Chang, AS Lee (1985): Calcium ionophore A 23187 induces expression of glucose-regulated genes and their heterologous fusion genes. Mol Cell Biol 5:1212–1219.

Roux SJ, RD Slocum (1982): Role of calcium in mediating cellular functions important for growth and development in higher plants. In W.Y. Cheung (ed): "Calcium and Cell Function," Vol. III. New York: Academic Press, pp 409–453.

Sen R, D Baltimore (1986): Inducibility of the κ immunoglobulin enhancer-binding protein NF-κB by a posttranslational mechanism. Cell 47:921–928.

Silver BJ, JA Bokar, JB Virgin, EA Vallen, A Milsted, JH Nilson (1987): Cyclic regulation of the human glycoprotein hormone α-subunit gene is mediated by an 18-base-pair element. Proc Natl Acad Sci USA 84:2198–2202.

Stratowa C, WJ Rutter (1986): Selective regulation of trypsin gene expression by calcium and by glucose starvation in a rat exocrine pancresa cell line. Proc Natl Acad Sci USA 83:4292–4296.

Tsien RY, T Pozzan, TJ Rink (1982): T-cell mitogens cause early changes in cytoplasmic free Ca^{2+} and membrane potential in lymphocytes. Nature 295:68–71.

White MW, AK Oberhauser, A Kuepfer, DR Morris (1987): Different early-signaling pathways coupled to transcriptional and posttranscriptional regulation of gene expression during mitogenic activation of T lymphocytes. Mol Cell Biol 7:3004–3007.

Index